Hydroprocessing of Heavy Oils and Residua

CHEMICAL INDUSTRIES

A Series of Reference Books and Textbooks

Founding Editor

HEINZ HEINEMANN
Berkeley, California

Series Editor

JAMES G. SPEIGHT
Laramie, Wyoming

1. *Fluid Catalytic Cracking with Zeolite Catalysts,* Paul B. Venuto and E. Thomas Habib, Jr.
2. *Ethylene: Keystone to the Petrochemical Industry,* Ludwig Kniel, Olaf Winter, and Karl Stork
3. *The Chemistry and Technology of Petroleum,* James G. Speight
4. *The Desulfurization of Heavy Oils and Residua,* James G. Speight
5. *Catalysis of Organic Reactions,* edited by William R. Moser
6. *Acetylene-Based Chemicals from Coal and Other Natural Resources,* Robert J. Tedeschi
7. *Chemically Resistant Masonry,* Walter Lee Sheppard, Jr.
8. *Compressors and Expanders: Selection and Application for the Process Industry,* Heinz P. Bloch, Joseph A. Cameron, Frank M. Danowski, Jr., Ralph James, Jr., Judson S. Swearingen, and Marilyn E. Weightman
9. *Metering Pumps: Selection and Application,* James P. Poynton
10. *Hydrocarbons from Methanol,* Clarence D. Chang
11. *Form Flotation: Theory and Applications,* Ann N. Clarke and David J. Wilson
12. *The Chemistry and Technology of Coal,* James G. Speight
13. *Pneumatic and Hydraulic Conveying of Solids,* O. A. Williams

14. *Catalyst Manufacture: Laboratory and Commercial Preparations,* Alvin B. Stiles
15. *Characterization of Heterogeneous Catalysts,* edited by Francis Delannay
16. *BASIC Programs for Chemical Engineering Design,* James H. Weber
17. *Catalyst Poisoning,* L. Louis Hegedus and Robert W. McCabe
18. *Catalysis of Organic Reactions,* edited by John R. Kosak
19. *Adsorption Technology: A Step-by-Step Approach to Process Evaluation and Application,* edited by Frank L. Slejko
20. *Deactivation and Poisoning of Catalysts,* edited by Jacques Oudar and Henry Wise
21. *Catalysis and Surface Science: Developments in Chemicals from Methanol, Hydrotreating of Hydrocarbons, Catalyst Preparation, Monomers and Polymers, Photocatalysis and Photovoltaics,* edited by Heinz Heinemann and Gabor A. Somorjai
22. *Catalysis of Organic Reactions,* edited by Robert L. Augustine
23. *Modern Control Techniques for the Processing Industries,* T. H. Tsai, J. W. Lane, and C. S. Lin
24. *Temperature-Programmed Reduction for Solid Materials Characterization,* Alan Jones and Brian McNichol
25. *Catalytic Cracking: Catalysts, Chemistry, and Kinetics,* Bohdan W. Wojciechowski and Avelino Corma
26. *Chemical Reaction and Reactor Engineering,* edited by J. J. Carberry and A. Varma
27. *Filtration: Principles and Practices: Second Edition,* edited by Michael J. Matteson and Clyde Orr
28. *Corrosion Mechanisms,* edited by Florian Mansfeld
29. *Catalysis and Surface Properties of Liquid Metals and Alloys,* Yoshisada Ogino
30. *Catalyst Deactivation,* edited by Eugene E. Petersen and Alexis T. Bell
31. *Hydrogen Effects in Catalysis: Fundamentals and Practical Applications,* edited by Zoltán Paál and P. G. Menon
32. *Flow Management for Engineers and Scientists,* Nicholas P. Cheremisinoff and Paul N. Cheremisinoff
33. *Catalysis of Organic Reactions,* edited by Paul N. Rylander, Harold Greenfield, and Robert L. Augustine
34. *Powder and Bulk Solids Handling Processes: Instrumentation and Control,* Koichi Iinoya, Hiroaki Masuda, and Kinnosuke Watanabe
35. *Reverse Osmosis Technology: Applications for High-Purity-Water Production,* edited by Bipin S. Parekh
36. *Shape Selective Catalysis in Industrial Applications,* N. Y. Chen, William E. Garwood, and Frank G. Dwyer

37. *Alpha Olefins Applications Handbook,* edited by George R. Lappin and Joseph L. Sauer
38. *Process Modeling and Control in Chemical Industries,* edited by Kaddour Najim
39. *Clathrate Hydrates of Natural Gases,* E. Dendy Sloan, Jr.
40. *Catalysis of Organic Reactions,* edited by Dale W. Blackburn
41. *Fuel Science and Technology Handbook,* edited by James G. Speight
42. *Octane-Enhancing Zeolitic FCC Catalysts,* Julius Scherzer
43. Oxygen in Catalysis, Adam Bielanski and Jerzy Haber
44. *The Chemistry and Technology of Petroleum: Second Edition, Revised and Expanded,* James G. Speight
45. *Industrial Drying Equipment: Selection and Application,* C. M. van't Land
46. *Novel Production Methods for Ethylene, Light Hydrocarbons, and Aromatics,* edited by Lyle F. Albright, Billy L. Crynes, and Siegfried Nowak
47. *Catalysis of Organic Reactions,* edited by William E. Pascoe
48. *Synthetic Lubricants and High-Performance Functional Fluids,* edited by Ronald L. Shubkin
49. *Acetic Acid and Its Derivatives,* edited by Victor H. Agreda and Joseph R. Zoeller
50. *Properties and Applications of Perovskite-Type Oxides,* edited by L. G. Tejuca and J. L. G. Fierro
51. *Computer-Aided Design of Catalysts,* edited by E. Robert Becker and Carmo J. Pereira
52. *Models for Thermodynamic and Phase Equilibria Calculations,* edited by Stanley I. Sandler
53. *Catalysis of Organic Reactions,* edited by John R. Kosak and Thomas A. Johnson
54. *Composition and Analysis of Heavy Petroleum Fractions,* Klaus H. Altgelt and Mieczyslaw M. Boduszynski
55. *NMR Techniques in Catalysis,* edited by Alexis T. Bell and Alexander Pines
56. *Upgrading Petroleum Residues and Heavy Oils,* Murray R. Gray
57. *Methanol Production and Use,* edited by Wu-Hsun Cheng and Harold H. Kung
58. *Catalytic Hydroprocessing of Petroleum and Distillates,* edited by Michael C. Oballah and Stuart S. Shih
59. *The Chemistry and Technology of Coal: Second Edition, Revised and Expanded,* James G. Speight
60. *Lubricant Base Oil and Wax Processing,* Avilino Sequeira, Jr.
61. *Catalytic Naphtha Reforming: Science and Technology,* edited by George J. Antos, Abdullah M. Aitani, and José M. Parera

62. *Catalysis of Organic Reactions*, edited by Mike G. Scaros and Michael L. Prunier
63. *Catalyst Manufacture,* Alvin B. Stiles and Theodore A. Koch
64. *Handbook of Grignard Reagents,* edited by Gary S. Silverman and Philip E. Rakita
65. *Shape Selective Catalysis in Industrial Applications: Second Edition, Revised and Expanded,* N. Y. Chen, William E. Garwood, and Francis G. Dwyer
66. *Hydrocracking Science and Technology,* Julius Scherzer and A. J. Gruia
67. *Hydrotreating Technology for Pollution Control: Catalysts, Catalysis, and Processes,* edited by Mario L. Occelli and Russell Chianelli
68. *Catalysis of Organic Reactions,* edited by Russell E. Malz, Jr.
69. *Synthesis of Porous Materials: Zeolites, Clays, and Nanostructures,* edited by Mario L. Occelli and Henri Kessler
70. *Methane and Its Derivatives,* Sunggyu Lee
71. *Structured Catalysts and Reactors,* edited by Andrzej Cybulski and Jacob A. Moulijn
72. *Industrial Gases in Petrochemical Processing,* Harold Gunardson
73. *Clathrate Hydrates of Natural Gases: Second Edition, Revised and Expanded,* E. Dendy Sloan, Jr.
74. *Fluid Cracking Catalysts,* edited by Mario L. Occelli and Paul O'Connor
75. *Catalysis of Organic Reactions,* edited by Frank E. Herkes
76. *The Chemistry and Technology of Petroleum: Third Edition, Revised and Expanded,* James G. Speight
77. *Synthetic Lubricants and High-Performance Functional Fluids: Second Edition, Revised and Expanded,* Leslie R. Rudnick and Ronald L. Shubkin
78. *The Desulfurization of Heavy Oils and Residua, Second Edition, Revised and Expanded,* James G. Speight
79. *Reaction Kinetics and Reactor Design: Second Edition, Revised and Expanded,* John B. Butt
80. *Regulatory Chemicals Handbook,* Jennifer M. Spero, Bella Devito, and Louis Theodore
81. *Applied Parameter Estimation for Chemical Engineers,* Peter Englezos and Nicolas Kalogerakis
82. *Catalysis of Organic Reactions,* edited by Michael E. Ford
83. *The Chemical Process Industries Infrastructure: Function and Economics,* James R. Couper, O. Thomas Beasley, and W. Roy Penney
84. *Transport Phenomena Fundamentals,* Joel L. Plawsky

85. *Petroleum Refining Processes*, James G. Speight and Baki Özüm
86. *Health, Safety, and Accident Management in the Chemical Process Industries*, Ann Marie Flynn and Louis Theodore
87. *Plantwide Dynamic Simulators in Chemical Processing and Control*, William L. Luyben
88. *Chemical Reactor Design*, Peter Harriott
89. *Catalysis of Organic Reactions*, edited by Dennis G. Morrell
90. *Lubricant Additives: Chemistry and Applications*, edited by Leslie R. Rudnick
91. *Handbook of Fluidization and Fluid-Particle Systems*, edited by Wen-Ching Yang
92. *Conservation Equations and Modeling of Chemical and Biochemical Processes*, Said S. E. H. Elnashaie and Parag Garhyan
93. *Batch Fermentation: Modeling, Monitoring, and Control*, Ali Çinar, Gülnur Birol, Satish J. Parulekar, and Cenk Ündey
94. *Industrial Solvents Handbook, Second Edition*, Nicholas P. Cheremisinoff
95. *Petroleum and Gas Field Processing*, H. K. Abdel-Aal, Mohamed Aggour, and M. Fahim
96. *Chemical Process Engineering: Design and Economics*, Harry Silla
97. *Process Engineering Economics*, James R. Couper
98. *Re-Engineering the Chemical Processing Plant: Process Intensification*, edited by Andrzej Stankiewicz and Jacob A. Moulijn
99. *Thermodynamic Cycles: Computer-Aided Design and Optimization*, Chih Wu
100. *Catalytic Naphtha Reforming: Second Edition, Revised and Expanded*, edited by George T. Antos and Abdullah M. Aitani
101. *Handbook of MTBE and Other Gasoline Oxygenates*, edited by S. Halim Hamid and Mohammad Ashraf Ali
102. *Industrial Chemical Cresols and Downstream Derivatives*, Asim Kumar Mukhopadhyay
103. *Polymer Processing Instabilities: Control and Understanding*, edited by Savvas Hatzikiriakos and Kalman B . Migler
104. *Catalysis of Organic Reactions*, John Sowa
105. *Gasification Technologies: A Primer for Engineers and Scientists*, edited by John Rezaiyan and Nicholas P. Cheremisinoff
106. *Batch Processes*, edited by Ekaterini Korovessi and Andreas A. Linninger
107. *Introduction to Process Control*, Jose A. Romagnoli and Ahmet Palazoglu

108. *Metal Oxides: Chemistry and Applications*, edited by J. L. G. Fierro
109. *Molecular Modeling in Heavy Hydrocarbon Conversions*, Michael T. Klein, Ralph J. Bertolacini, Linda J. Broadbelt, Ankush Kumar and Gang Hou
110. *Structured Catalysts and Reactors, Second Edition*, edited by Andrzej Cybulski and Jacob A. Moulijn
111. *Synthetics, Mineral Oils, and Bio-Based Lubricants: Chemistry and Technology*, edited by Leslie R. Rudnick
112. *Alcoholic Fuels*, edited by Shelley Minteer
113. *Bubbles, Drops, and Particles in Non-Newtonian Fluids, Second Edition*, R. P. Chhabra
114. *The Chemistry and Technology of Petroleum, Fourth Edition*, James G. Speight
115. *Catalysis of Organic Reactions*, edited by Stephen R. Schmidt
116. *Process Chemistry of Lubricant Base Stocks*, Thomas R. Lynch
117. *Hydroprocessing of Heavy Oils and Residua*, edited by Jorge Ancheyta and James G. Speight
118. *Chemical Process Performance Evaluation*, Ali Cinar, Ahmet Palazoglu, and Ferhan Kayihan

Hydroprocessing of Heavy Oils and Residua

Jorge Ancheyta
Instituto Mexicano del Petroleo
Instituto Politécnico Nacional
Mexico City, Mexico

James G. Speight
CD & W Inc.
Laramie, Wyoming, U.S.A.

CRC Press
Taylor & Francis Group
Boca Raton London New York

CRC Press is an imprint of the
Taylor & Francis Group, an **Informa** business

CRC Press
Taylor & Francis Group
6000 Broken Sound Parkway NW, Suite 300
Boca Raton, FL 33487-2742

Printed in the United States of America on acid-free paper
10 9 8 7 6 5 4 3 2 1

International Standard Book Number-10: 0-8493-7419-7 (Hardcover)
International Standard Book Number-13: 978-0-8493-7419-7 (Hardcover)

Library of Congress Cataloging-in-Publication Data

Hydroprocessing of heavy oils and residua / editors, Jorge Ancheyta, James G. Speight.
p. cm. -- (Chemical industries ; v. 117)
Includes bibliographical references and index.
ISBN-13: 978-0-8493-7419-7 (alk. paper)
1. Petroleum--Refining. 2. Hydrogenation. I. Ancheyta, Jorge. II. Speight, J. G. III. Title. IV. Series.

TP690.45.H93 2007
665.5'3--dc22 2006039245

Visit the Taylor & Francis Web site at
http://www.taylorandfrancis.com

and the CRC Press Web site at
http://www.crcpress.com

Preface

With the inception of hydrogenation as a process by which both coal and petroleum could be converted into lighter products, it was also recognized that hydrogenation would be effective for the simultaneous removal of nitrogen, oxygen, and sulfur compounds from the feedstock. However, with respect to the prevailing context of fuel industries, hydrogenation did not seem to be economical for application to petroleum fractions. At least two factors dampened interest: (1) the high cost of hydrogen and (2) the adequacy of current practices for meeting the demand for high-value products by refining conventional crude oil.

On the other hand, heavy oils and residua are generally considered to be low-value feedstocks by the refining industry. Hence, they are the focus of many conversion scenarios. Indeed, there has been a tendency for the quality of crude oil feedstocks to deteriorate insofar as the *average* refinery feedstock is of lower American Petroleum Institute (API) gravity and higher sulfur content than the average refinery feedstock of two decades ago. This means higher quantities of residua and more heavy oils to process.

There are several valid reasons for using hydrogen to process heavy oils: (1) reduction, or elimination, of sulfur from the products; (2) production of products having acceptable specifications; (3) increasing the performance (and stability) of gasoline; (4) decreasing smoke formation in kerosene; and (5) improvement in the burning characteristics of fuel oil to a level that is environmentally acceptable.

Generally, over the past three decades we have seen a growing dependence on high-sulfur heavier oils and residua emerge as a result of continuing increases in the prices of the conventional crude oils coupled with the decreasing availability of these crude oils through the depletion of reserves in the various parts of the world. Furthermore, the ever-growing tendency to convert as much as possible of lower-grade feedstocks to liquid products is causing an increase in the total sulfur content in refined products. Refiners must therefore continue to remove substantial portions of sulfur from the lighter products, but residua and the heavier crude oils pose a particularly difficult problem. Indeed, it is now clear that there are other problems involved in the processing of the heavier feedstocks and that these heavier feedstocks, which are gradually emerging as the liquid fuel supply of the future, need special attention.

Hydroprocessing petroleum fractions has long been an integral part of refining operations, and in one form or another, hydroprocesses are used in every modern refinery. The process is accomplished by the catalytic reaction of hydrogen with the feedstock to produce higher-value hydrocarbon products. The technology of hydroprocessing is well established for gas oil and lower-boiling products, but there is no comprehensive source of information for hydroprocessing heavy crude oils and residua. Indeed, processing heavy oils and residua presents several problems that are not found with distillate processing and that require process modifications to meet the special requirements necessary for heavy feedstock desulfurization.

In addition, petroleum refining has entered a significant transition period as the industry moves into the 21st century. Refinery operations have evolved to include a range of next-generation processes, as the demand for transportation fuels and fuel oil has shown a steady growth. These processes are different from one another in terms of the method and product slates and will find employment in refineries according to their respective features. The primary goal of these processes is to convert the heavy feedstocks to lower-boiling products, and during the conversion there is a reduction in the sulfur content. Thus, these processes need to be given some consideration and are included in separate chapters (9 and 10), in which these more modern processes for upgrading heavy feedstocks are described.

It is therefore the objective of the present text to indicate how the heavier feedstocks may, in the light of current technology, be treated in hydroprocesses to afford maximum yields of high-value products.

This text is designed for those scientists and engineers who wish to be introduced to hydroprocessing concepts and technology as well as those who wish to make more detailed studies of how hydroprocesses might be accomplished. Chapters relating to the composition and evaluation of heavy oils and residua are regarded as necessary for a basic understanding of the types of feedstock that will necessarily need desulfurization treatment. For those readers requiring an in-depth theoretical treatment, a discussion of the chemistry and physics of the hydroprocesses has been included. The effects of reactor type, process variables and feedstock type, catalysts, and feedstock composition on hydroprocesses provide a significant cluster of topics through which to convey the many complexities of the process. In the concluding chapters, examples and brief descriptions of commercial processes are presented, and of necessity, some indications of methods of hydrogen production are also included.

We are indebted to our numerous colleagues who have contributed to this book and to our friends, students, and colleagues who willingly provided support.

Dr. Jorge Ancheyta
Instituto Mexicano del Petroleo
Escuela Superior de Ingeniería Quimica
e Industrias Extractivas (ESIQIE-IPN)
Mexico City, Mexico

Dr. James G. Speight
University of Trinidad and Tobago
Arima, Trinidad and Tobago

The Editors

Jorge Ancheyta, Ph.D., graduated with a bachelor's degree in petrochemical engineering (1989), a master's degree in chemical engineering (1993), and a master's degree in administration, planning, and economics of hydrocarbons (1997) from the National Polytechnic Institute (IPN) of Mexico. He obtained a split Ph.D. from the Metropolitan Autonomous University (UAM) of Mexico and the Imperial College of Science, Technology and Medicine, London (1998), as well as a postdoctoral fellowship in the Laboratory of Catalytic Process Engineering of the CPE-CNRS in Lyon, France (1999).

He has worked for the Mexican Institute of Petroleum (IMP) since 1989, and his present position is research and development project leader. He has also worked as professor at the undergraduate and postgraduate levels for the School of Chemical Engineering and Extractive Industries at the National Polytechnic Institute of Mexico (ESIQIE-IPN) since 1992 and for the IMP postgraduate since 2003. He has been the supervisor of more than seventy B.Sc., M.Sc., and Ph.D. theses.

Ancheyta has been working in the development and application of petroleum refining catalysts, kinetic and reactor models, and process technologies, mainly in catalytic cracking, catalytic reforming, middle distillate hydrotreating, and heavy oils upgrading. He is the author and coauthor of a number of patents, books, and scientific papers. He has been awarded the National Researcher Distinction by the Mexican government and is a member of the Mexican Academy of Science. He has also been guest editor of various international journals, such as *Catalysis Today*, *Petroleum Science and Technology*, and *Fuel*, and chairman of international conferences.

James G. Speight, Ph.D., has more than 35 years experience in areas associated with the properties and processing of conventional and synthetic fuels. He has participated in, as well as led, significant research in defining the use of chemistry of heavy oil and coal. He has well over 400 publications, reports, and presentations detailing these research activities and has taught more than 70 related courses.

Dr. Speight is currently Editor of the journal *Petroleum Science and Technology* (formerly *Fuel Science and Technology International*) and Editor of the journal *Energy Sources* (Part A and Part B). He is recognized as a world leader in the areas of fuels characterization and development. Dr. Speight is also Adjunct Professor of Chemistry and Adjunct Professor of Chemical Engineering at the University of Wyoming as well as Adjunct Professor of Chemical and Fuels Engineering at the University of Utah.

Dr. Speight is the author, coauthor, and editor of more than 30 books and bibliographies related to fossil fuel processing and environmental issues.

As a result of his work, Dr. Speight has been honored to receive the following awards: Diploma of Honor, National Petroleum Engineering Society for Outstanding Contributions to the Petroleum Industry, 1995; Gold Medal of the Russian Academy of Sciences (Natural) for Outstanding Work in the Area of Petroleum Science, 1996; Specialist Invitation Program Speakers Award from NEDO (New Energy Development Organization, Government of Japan) for Excellence in Coal Research, 1987; Specialist Invitation Program Speakers Award from NEDO for Excellence in Coal Research, 1996; Doctor of Sciences from the Scientific Research Geological Exploration Institute (VNIGRI), St. Petersburg, Russia for Exceptional Work in Petroleum Science, 1997; Einstein Medal of the Russian Academy of Sciences (Natural) in recognition of Outstanding Contributions and Service in the Field of Geologic Sciences, 2001; Gold Medal — Scientists without Frontiers, Russian Academy of Sciences in Recognition of Continuous Encouragement of Scientists to Work Together across International Borders, 2005; and Gold Medal — Giants of Science and Engineering, Russian Academy of Sciences, in Recognition of Continued Excellence in Science and Engineering, 2006.

Contributors

Syed Ahmed Ali
Center for Refining and Petrochemicals
The Research Institute
King Fahd University of Petroleum and Minerals
Dhahran, Saudi Arabia

Geoffrey E. Dolbear
G.E. Dolbear & Associates, Inc.
Diamond Bar, California

Esteban López-Salinas
Instituto Mexicano del Petroleo
Mexico D.F., Mexico

S.K. Maity
Instituto Mexicano del Petroleo
Mexico D.F., Mexico

Jorge Ramírez
Departamento de Ingenieria Quimica
UNICAT
Mexico D.F., Mexico

Mohan S. Rana
Instituto Mexicano del Petroleo
Mexico D.F., Mexico

Paul R. Robinson
Albemarle Catalysts LLC
Houston, Texas

Jaime S. Valente
Instituto Mexicano del Petroleo
Mexico D.F., Mexico

Miguel A. Valenzuela
Escuela Superior de Ingenieria Quimica e Industrias Extractivas (ESIQIE-IPN)
Mexico D.F., Mexico

Beatriz Zapata
Instituto Mexicano del Petroleo
Mexico D.F., Mexico

Table of Contents

1 Heavy Oils and Residua

Jorge Ancheyta and James G. Speight

CONTENTS

1.1 INTRODUCTION

Petroleum is found in various countries (Table 1.1) and scattered throughout the earth's crust, which is divided into natural groups or strata, categorized in order of their antiquity (Speight, 2006). These divisions are recognized by the distinctive systems of organic debris (as well as fossils, minerals, and other characteristics) that form a chronological time chart that indicates the relative ages of the earth's strata. It is generally acknowledged that carbonaceous materials such as petroleum occur in all these geological strata from the Precambrian to the recent, and the origin of petroleum within these formations is a question that remains open to conjecture and the basis for much research.

Petroleum is by far the most commonly used source of energy, especially as the source of liquid fuels (Table 1.2). Indeed, because of the wide use of petroleum, the past 100 years could very easily be dubbed the *oil century* (Ryan, 1998), the *petroleum era* (compare the *Pleistocene era*), or the *new rock oil age* (compare the *new stone age*). As a result, fossil fuels are projected to be the major sources of energy for the next 50 years. In this respect, petroleum and its associates (heavy oils and residua) are extremely important in any energy scenario, especially those scenarios that relate to the production of liquid fuels.

For example, the United States imported approximately 6 million barrels per day of petroleum and petroleum products in 1975 and now imports approximately double this amount. Approximately 90% of the products of a refinery are fuels (Pellegrino, 1998), and it is evident that this reliance on petroleum-based fuels and products will continue for several decades (Table 1.2).

TABLE 1.1
Oil Reserves by Country

Country	Reserves (bbl × 10^9)
Canada	179
Iran	126
Iraq	115
Kuwait	102
Russia	60
Saudi Arabia	262
United Arab Emirates	98
United States	21
Venezuela	79
Other	238

Source: U.S. Energy Information Administration, 2006. (With permission.)

It is a fact that in recent years, the average quality of crude oil has deteriorated (Swain, 1991, 1993, 1998). This has caused the nature of crude oil refining to change considerably. This, of course, has led to the need to manage crude quality more effectively through evaluation and product slates (Waguespack and Healey, 1998; Speight, 2006 and references cited therein). Indeed, the declining reserves of lighter crude oil have resulted in an increasing need to develop options to desulfurize and upgrade the heavy feedstocks, specifically heavy oil and bitumen. This has resulted in a variety of process options that specialize in sulfur removal during refining. It is worthy of note at this point that microbial desulfurization

TABLE 1.2
Current and Projected Energy Consumption Scenarios

	Energy Consumption (quads)					
	1995	1996	2000	2005	2010	2015
	Actual		Projected			
Petroleum	34.7	36.0	37.7	40.4	42.6	44.1
Gas	22.3	22.6	23.9	26.3	28.8	31.9
Coal	19.7	20.8	22.3	24.1	26.2	29.0
Nuclear	7.2	7.2	7.6	7.4	6.9	4.7
Hydro	3.4	4.0	3.1	3.2	3.2	3.2
Other	3.2	3.3	3.7	4.0	4.8	5.2

Source: GRI, Baseline, Report GRI-98/0001, Gas Research Institute, Chicago, IL, 1998, p. 1. (With permission.)

is becoming a recognized technology for desulfurization (Monticello, 1995; Armstrong et al., 1997), but it will not be covered in this text.

With the necessity of processing heavy oil, bitumen, and residua to obtain more gasoline and other liquid fuels, there has been the recognition that knowledge of the constituents of these higher-boiling feedstocks is also of some importance. Indeed, the problems encountered in processing the heavier feedstocks can be equated to the *chemical character* and the *amount* of complex, higher-boiling constituents in the feedstock. Refining these materials is not just a matter of applying know-how derived from refining *conventional* crude oils, but requires knowledge of the *chemical structure* and *chemical behavior* of these more complex constituents.

The elemental analysis of oil sand bitumen (*extra-heavy oil*) has also been widely reported (Speight, 1990), but the data suffer from the disadvantage that identification of the source is too general (that is, Athabasca bitumen, which covers several deposits) and is often not site specific. In addition, the analysis is quoted for separated bitumen, which may have been obtained by any one of several procedures and may therefore not be representative of the total bitumen on the sand. However, recent efforts have focused on a program to produce accurate and reproducible data from samples for which the origin is carefully identified (Wallace et al., 1988). It is hoped that this program continues, as it will provide a valuable database for tar sand and bitumen characterization.

With all of the scenarios in place, there is no doubt that *petroleum* and its relatives — residua and heavy oils as well as tar sand bitumen — will be required to produce a considerable proportion of liquid fuels into the foreseeable future. Desulfurization processes will be necessary to remove sulfur in an environmentally acceptable manner to produce environmentally acceptable products. Refining strategies will focus on upgrading the heavy oils and residua and will emphasize the differences between the properties of the heavy crude feedstocks. This will dictate the choice of methods or combinations thereof for conversion of these materials to products (Schuetze and Hofmann, 1984).

Refinery processes can be categorized as hydrogen addition processes (for example, hydroprocesses such as hydrotreating and hydrocracking, hydrovisbreaking, donor-solvent processes) and carbon rejection processes (for example, catalytic cracking, coking, visbreaking, and other processes, such as solvent deasphalting). All have serious disadvantages when applied singly to the upgrading of heavy oils or residua. Removal of heteroatoms and metals by exhaustive hydrodenitrogenation (HDN), hydrodesulfurization (HDS), and hydrodemetallization (HDM) is very expensive. The catalytic processes suffer from the disadvantage of excessive catalyst use due to metals and carbon deposition. Development of more durable and less expensive catalysts and additives to prevent coke laydown has not yet solved the problem. The noncatalytic processes alone yield uneconomically large amounts of coke.

Both *hydrogen addition* and *carbon rejection* processes will be necessary in any realistic scheme of heavy oil upgrading (Suchanek and Moore, 1986). Most coker products require hydrogenation, and most hydrotreated products require some degree of fractionation. For example, to maximize yields of transport fuels from Maya crude, efficient carbon rejection followed by hydrogenation may be necessary. There are various other approaches to the processing of other heavy oil residua (Bakshi and Lutz, 1987; Johnson et al., 1985). As of now, it is not known which combination of processes best converts a heavy feedstock into salable products.

Hydrodesulfurization and *hydrodemetallization* activities can not be predicted by such conventional measurements as total sulfur, metals, or asphaltene contents, or Conradson carbon values (Dolbear et al., 1987). To choose effective processing strategies, it is necessary to determine properties from which critical reactivity indices can be developed. Indeed, properties of heavy oil vacuum residua determined by conventional methods are not good predictors of behavior of substrates in upgrading processes (Dawson et al., 1989). The properties of residua vary widely, and the existence of relatively large numbers of polyfunctional molecules results in molecular associations that can affect reactivity. Therefore, it is evident that more knowledge is needed about the components of residua that cause specific problems in processing, and how important properties change during processing (Gray, 1990).

In summary, upgrading heavy oils and residua must at some stage of the refinery operations utilize hydrodesulfurization. Indeed, HDS processes are used at several places in virtually every refinery to protect catalysts and to meet product specifications related to environmental regulations (Speight, 1996, 2000, 2006). Thus, several types of chemistry might be anticipated as occurring during hydrodesulfurization. Similarly, hydrodenitrogenation is commonly used only in conjunction with hydrocracking, to protect catalysts. Other hydrotreating processes are used to saturate olefins and aromatics to meet product specifications or to remove metals from residual oils.

1.2 PETROLEUM

Petroleum and the equivalent term *crude oil* cover a wide assortment of materials consisting of mixtures of hydrocarbons and other compounds containing variable amounts of sulfur, nitrogen, and oxygen, which may vary widely in volatility, specific gravity, and viscosity. Metal-containing constituents, notably those compounds that contain vanadium and nickel, usually occur in the more viscous crude oils in amounts up to several thousand parts per million and can have serious consequences during processing of these feedstocks (Speight, 2006 and references cited therein). Because petroleum is a mixture of widely varying constituents and proportions, its physical properties also vary widely and the color ranges from colorless to black.

There are other types of petroleum that are different from conventional petroleum insofar as they are much more difficult to recover from the subsurface reservoir. These materials have a much higher viscosity (and lower American Petroleum Institute [API] gravity) than conventional petroleum, and primary recovery of these petroleum types usually requires thermal stimulation of the reservoir (Speight, 2006 and references cited therein).

Most of the petroleum oil currently recovered is produced from underground reservoirs. However, surface seepage of crude oil and natural gas is common in many regions. In fact, it is the surface seepage of oil that led to the first use of the high-boiling material (bitumen) in the Fertile Crescent. It may also be stated that the presence of active seeps in an area is evidence that oil and gas are still migrating.

The majority of crude oil reserves identified to date are located in a relatively small number of very large fields known as *giants*. In fact, approximately 300 of the largest oil fields contain almost 75% of the available crude oil. Although most of the world's nations produce at least minor amounts of oil, the primary concentrations are in the Persian Gulf, North and West Africa, the North Sea, and the Gulf of Mexico. In addition, of the ninety oil-producing nations, five Middle Eastern countries contain almost 70% of the current known oil reserves.

When petroleum occurs in a reservoir that allows the crude material to be recovered by pumping operations as a free-flowing dark- to light-colored liquid, it is often referred to as *conventional petroleum*. Heavy oils are the other types of petroleum; they are different from conventional petroleum insofar as they are much more difficult to recover from the subsurface reservoir. The definition of heavy oils is usually based on the API gravity or viscosity, and the definition is quite arbitrary, although there have been attempts to rationalize it based upon viscosity, API gravity, and density.

In addition to attempts to define petroleum, heavy oil, bitumen, and residua, there have been several attempts to classify these materials by the use of properties such as API gravity, sulfur content, or viscosity (Speight, 2006). However, any attempt to classify petroleum, heavy oil, and bitumen on the basis of a single property is no longer sufficient to define the nature and properties of petroleum and petroleum-related materials, perhaps even being an exercise in futility.

For many years, petroleum and heavy oils were very generally defined in terms of physical properties. For example, heavy oils were considered to be those crude oils that had gravity somewhat less than 20° API, with the heavy oils falling into the API gravity range of 10 to 15°. For example, Cold Lake heavy crude oil has an API gravity equal to 12°, and extra-heavy oils, such as tar sand bitumen, usually have an API gravity in the range of 5 to 10° (Athabasca bitumen = 8° API). Residua vary depending upon the temperature at which distillation was terminated, but usually vacuum residua are in the range of 2 to 8° API.

1.3 HEAVY OIL

When petroleum occurs in a reservoir that allows the crude material to be recovered by pumping operations as a free-flowing dark- to light-colored liquid, it is often referred to as conventional petroleum.

Heavy oil is another type of petroleum that is different from conventional petroleum insofar as it is much more difficult to recover from the subsurface reservoir. Heavy oil has a much higher viscosity (and lower API gravity) than conventional petroleum, and primary recovery of heavy oil usually requires thermal stimulation of the reservoir.

Thus, the term *heavy oil* has also been arbitrarily used to describe both the heavy oils that require thermal stimulation of recovery from the reservoir and the bitumen in bituminous sand (tar sand, q.v.) formations from which the heavy bituminous material is recovered by a mining operation. However, the term *extra-heavy oil* is used to define the subcategory of petroleum that occurs in the near-solid state and is incapable of free flow under ambient conditions. *Bitumen* from tar sand deposits is often called extra-heavy oil.

Venezuela has 47 to 76 billion barrels of proven reserves, according to oil industry/Department of Energy (DOE) estimates. The U.S. Geological Survey (USGS) puts Venezuelan reserves around the same level, at 48 identified and 110 billion ultimately recoverable. Venezuela claims 1.2 trillion (1.2×10^{12}) barrels of unconventional oil reserves in the supergiant heavy oil field stretching from the mouth of the Orinoco River near Trinidad down the east side of the Andes mountains (Arcaya, 2001). The oil is located in a geosynclinal trough that is theorized to be continuous through the Falkland Islands off the coast of Argentina. Only parts of the heavy oil field have been fully explored, but those parts have been estimated at some 3 to 4 trillion barrels of heavy oil in place, with perhaps one third recoverable using current technology.

Venezuela's oil reserves may almost triple in 2 years as oil companies develop the country's Faja, or heavy oil belt, according to Venezuelan President Hugo Chavez (http://www.caribbeannetnews.com/cgi-script/csArticles/articles/000027/002726.htm).

Proven reserves may rise to 171 billion barrels by November 2007 from the current 77 to 81 billion barrels. Further drilling as part of a certification process may increase reserves to 235.6 billion barrels by 2008. It is possible that Venezuela might supplant Saudi Arabia by 2008 as the country with the world's largest oil reserves. The BP PLC (June 2006) Review of Energy Statistics pegged Saudi Arabia's reserves at 264.2 billion barrels. Venezuela's heavy oil deposits may rival Canada's tar sand deposits as an alternative source of oil to meet growing world fuel demand. Venezuela's heavy oil is high in sulfur, coke, and metals and needs more refining than conventional oil.

1.4 TAR SAND BITUMEN

The term *bitumen* (also referred to as *native asphalt*) includes a wide variety of naturally occurring reddish-brown to black materials of semisolid, viscous to brittle character that can exist in nature with no mineral impurity or with mineral matter contents that exceed 50% by weight. Bitumen is frequently found filling pores and crevices of sandstone, limestone, or argillaceous sediments, in which case the organic and associated mineral matrix is known as *rock asphalt* (Abraham, 1945; Hoiberg, 1964). Tar sand bitumen is a high-boiling material with little, if any, material boiling below 350°C (660°F), and the boiling range is close to that of an atmospheric residuum.

It is incorrect to refer to bitumen as tar or pitch. Although the word *tar* is somewhat descriptive of the black bituminous material, its use is best avoided with respect to natural materials. More correctly, the name *tar* is usually applied to the heavy product remaining after the destructive distillation of coal or other organic matter. *Pitch* is the distillation residue of the various types of tar.

Thus, alternative names, such as *bituminous sand* or *oil sand*, are gradually finding usage, with the former name more technically correct. The term *oil sand* is also used in the same way as the term *tar sand*, and these terms are used interchangeably throughout this text.

The definitions of tar sand and heavy oil are diverse because, in many cases, a single property (such as API gravity) is used to define the material when it is often forgotten that the limits of experimental difference in the determination of this property are critical and negate the usefulness of the definition.

Therefore, for the purposes of this report, the definition of tar sand bitumen is derived from the definition of tar sand defined by the U.S. government (FE-76-4):

> The several rock types that contain an extremely viscous hydrocarbon which is not recoverable in its natural state by conventional oil well production methods including currently used enhanced recovery techniques. The hydrocarbon-bearing rocks are variously known as bitumen-rocks oil, impregnated rocks, oil sands, and rock asphalt.

By inference, heavy oil (Section 1.3) is that resource which can be recovered in its natural state by "conventional oil well production methods including currently used enhanced recovery techniques." The term *natural state* means without conversion of the heavy oil or bitumen as might occur during thermal recovery processes.

Currently world oil reserves are spread throughout different countries (Table 1.1), and current estimates place the exhaustion of the remaining known reserves within the next 50 years. Middle Eastern countries have approximately 50% of the known remaining world oil reserve.

These estimates, however, do not include the heavy oil in various deposits and bitumen in tar sand deposits that occur in countries throughout the world, by which the largest are Venezuelan heavy oil that occurs in the Orinoco oil belt (or Faja del Orinoco) and the tar sand deposits that occur in Ft. McMurray (northeast Alberta), Canada.

The bitumen in tar sand deposits is estimated to be at least 1.7 trillion barrels (1.7×10^{12} billions of barrels [bbl] or 270×10^9 m^3) in the Canadian Athabasca tar sand deposits and 1.8 trillion barrels (1.8×10^{12} bbl or 280×10^9 m^3) in the Venezuelan Orinoco tar sand deposits, compared to 1.75 trillion barrels (1.75×10^{12} bbl or 278×10^9 m^3) of conventional oil worldwide, most of it in Saudi Arabia and other Middle Eastern countries.

Therefore, bitumen in tar sand deposits represents a potentially large supply of energy. However, many of these reserves are only available with some difficulty, and optional refinery scenarios will be necessary for conversion of these materials to low-sulfur liquid products because of the substantial differences in character between conventional petroleum and tar sand bitumen. Bitumen recovery requires the prior application of reservoir fracturing procedures before the introduction of thermal recovery methods. Currently, commercial operations in Canada use mining techniques for bitumen recovery.

Because of the diversity of available information and the continuing attempts to delineate the various world tar sand deposits, it is virtually impossible to present accurate numbers that reflect the extent of the reserves in terms of the barrel unit. Indeed, investigations into the extent of many of the world's deposits are continuing at such a rate that the numbers vary from one year to the next. Accordingly, the data quoted here must be recognized as approximate with the potential of being quite different at the time of publication.

Throughout this text, frequent reference is made to tar sand bitumen, but because commercial operations have been in place for over 30 years (Spragins, 1978; Speight, 1990), it is not surprising that more is known about the Alberta (Canada) tar sand reserves than any other reserves in the world. Therefore, when tar sand deposits are discussed, reference is made to the relevant deposit, but when the information is not available, the Alberta material is used for the purposes of the discussion.

Tar sand deposits are widely distributed throughout the world (Speight, 1990, 2006 and references cited therein). The potential reserves of bitumen that occur in tar sand deposits have been variously estimated on a world basis as being in excess of 3 trillion (3×10^{12}) barrels of petroleum equivalent. The reserves that have been estimated for the United States have been estimated to be in excess of 52 million (52×10^{6}) barrels. That commercialization has taken place in Canada does not mean that it is imminent for other tar sand deposits. There are considerable differences between the Canadian and U.S. deposits that could preclude across-the-board application of the Canadian principles to the U.S. sands (Speight, 1990). The key is accessibility and recoverability.

Various definitions have been applied to energy reserves (q.v.), but the crux of the matter is the amount of a resource that is recoverable using current technology. And although tar sands are not a principal energy reserve, they certainly are significant with regard to projected energy consumption over the next several generations.

Thus, in spite of the high estimations of the reserves of bitumen, the two conditions of vital concern for the economic development of tar sand deposits are the concentration of the resource, or the percent bitumen saturation, and its accessibility, usually measured by the overburden thickness. Recovery methods are based on either mining combined with some further processing or operation on the oil sands *in situ*. The mining methods are applicable to shallow deposits, characterized by an overburden ratio (that is overburden depth to thickness of tar sand deposit). For example, indications are that for the Athabasca deposit, no more than 10% of the in-place deposit is mineable within current concepts of the economics and technology of open-pit mining; this 10% portion may be considered the *proven reserves* of bitumen in the deposit.

However, in order to define conventional petroleum, heavy oil, and bitumen, use of a single physical parameter such as viscosity is not sufficient. Other properties, such as API gravity, elemental analysis, composition, and, most important of all, the properties of the bulk deposit, must also be included in any definition of these materials. Only then will it be possible to classify petroleum and its derivatives (Speight, 2006).

1.5 RESIDUA

A *residuum* (also shortened to *resid*) is the residue obtained from petroleum after nondestructive distillation has removed all the volatile materials. The temperature of the distillation is usually maintained below 350°C (660°F) since the rate of thermal decomposition of petroleum constituents is minimal below this temperature, but the rate of thermal decomposition of petroleum constituents is substantial above 350°C (660°F).

Residua are black, viscous materials obtained by distillation of a crude oil under atmospheric pressure (atmospheric residuum) or reduced pressure (vacuum residuum). They may be liquid at room temperature (generally atmospheric residua) or almost solid (generally vacuum residua), depending upon the cut point of the distillation (Table 1.3) or the nature of the crude oil.

Of all of the feedstocks described in this chapter, residua are the only manufactured products. The constituents of residua do occur naturally as part of the native material, but residua are specifically produced during petroleum refining, and the properties of the various residua depend upon the cut point or boiling point at which the distillation is terminated.

As refineries have evolved, distillation has remained the prime means by which petroleum is refined (Speight, 2006). Indeed, the distillation section of a modern refinery is the most flexible unit in the refinery since conditions can be adjusted to process a wide range of refinery feedstocks, from the lighter crude oils to the heavier, more viscous crude oils. However, the maximum permissible temperature (in the vaporizing furnace or heater) to which the feedstock can be subjected is 350°C (660°F). The rate of thermal decomposition increases markedly above this temperature: if decomposition occurs within a distillation unit, it can lead to coke deposition in the heater pipes or in the tower itself, with the resulting failure of the unit.

The *atmospheric distillation* tower is divided into a number of horizontal sections by metal trays or plates, and each is the equivalent of a still. The more trays, the more redistillation, and hence the better the fractionation or separation of the mixture fed into the tower. A tower for fractionating crude petroleum may be 13 feet in diameter and 85 feet high, but a tower stripping unwanted volatile material from a gas oil may be only 3 or 4 feet in diameter and 10 feet high. Towers concerned with the distillation of liquefied gases are only a few feet in diameter but may be up to 200 feet in height. A tower used in the fractionation of crude petroleum may have from 16 to 28 trays, but one used in the fractionation of liquefied gases may have 30 to 100 trays. The feed to a

TABLE 1.3
Properties of Tia Juana Crude Oil and Its 650, 950, and 1050°F Residua

	Whole Crude	Residua		
		650°F+	950°F+	1050°F+
Yield, vol. %	100.0	48.9	23.8	17.9
Sulfur, wt. %	1.08	1.78	2.35	2.59
Nitrogen, wt. %		0.33	0.52	0.60
API gravity	31.6	17.3	9.9	7.1
Conradson carbon residue, wt. %		9.3	17.2	21.6
Metals				
Vanadium, ppm		185		450
Nickel, ppm		25		64
Viscosity				
Kinematic				
At 100°F	10.2	890		
At 210°F		35	1010	7959
Furol				
At 122°F		172		
At 210°F			484	3760
Pour point, °F	–5	45	95	120

Source: Speight, J.G., *The Chemistry and Technology of Petroleum*, 4th ed., CRC Press/Taylor & Francis, Boca Raton, FL, 2006. (With permission.)

typical tower enters the vaporizing or flash zone, an area without trays. The majority of the trays are usually located above this area. The feed to a bubble tower, however, may be at any point from top to bottom with trays above and below the entry point, depending on the kind of feedstock and the characteristics desired in the products.

Liquid collects on each tray to a depth of several inches or so, and the depth is controlled by a dam or weir. As the liquid level rises, excess liquid spills over the weir into a channel (downspout), which carries the liquid to the tray below.

The temperature of the trays is progressively cooler from bottom to top. The bottom tray is heated by the incoming heated feedstock, although in some instances a steam coil (reboiler) is used to supply additional heat. As the hot vapors pass upward in the tower, condensation occurs on the trays until refluxing (simultaneous boiling of a liquid and condensing of the vapor) occurs. Vapors continue to pass upward through the tower, whereas the liquid on any particular tray spills onto the tray below, and so on, until the heat at a particular point is too intense for the material to remain liquid. It then becomes vapor and joins the other vapors passing upward through the tower. The whole tower thus simulates a collection of several (or many) stills, with the composition of the liquid at any one point or on any one tray remaining fairly consistent. This allows part of the refluxing liquid to be tapped off at various points as *sidestream* products.

Thus, in the distillation of crude petroleum, light naphtha and gases are removed as vapor from the top of the tower; heavy naphtha, kerosene, and gas oil are removed as sidestream products; and reduced crude (*atmospheric residuum*) is taken from the bottom of the tower.

The *topping* operation differs from normal distillation procedures insofar as the majority of the heat is directed to the feed stream rather than the material in the base of the tower being reboiled. In addition, products of volatility intermediate between that of the overhead fractions and bottoms (residua) are withdrawn as sidestream products. Furthermore, steam is injected into the base of the column and the sidestream strippers to adjust and control the initial boiling range (or point) of the fractions.

Topped crude oil must always be *stripped* with steam to elevate the flash point or to recover the final portions of gas oil. The composition of the topped crude oil is a function of the temperature of the vaporizer (or *flasher*).

The temperature at which the residue starts to decompose or crack limits the boiling range of the highest boiling fraction obtainable at atmospheric pressure. If stock is required for the manufacture of lubricating oils, further fractionation without cracking may be desirable, and this may be achieved by distillation under vacuum.

Vacuum distillation as applied to the petroleum refining industry is truly a technique of the 20th century and has since gained wide use in petroleum refining. Vacuum distillation evolved because of the need to separate the less volatile products, such as lubricating oils, from the petroleum without subjecting these high-boiling products to cracking conditions. The boiling point of the heaviest cut obtainable at atmospheric pressure is limited by the temperature (ca. 350°C or 660°F) at which the residue starts to decompose or crack, unless *cracking distillation* is preferred. When the feedstock is required for the manufacture of lubricating oils, further fractionation without cracking is desirable, and this can be achieved by distillation under vacuum conditions.

The distillation of high-boiling lubricating oil stocks may require pressures as low as 15 to 30 mmHg, but operating conditions are more usually 50 to 100 mmHg. Volumes of vapor at these pressures are large, and pressure drops must be small to maintain control, so vacuum columns are necessarily of large diameter. Differences in vapor pressure of different fractions are relatively larger than for lower-boiling fractions, and relatively few plates are required. Under these conditions, a heavy gas oil may be obtained as an overhead product at temperatures of about 150°C (300°F). Lubricating oil fractions may be obtained as sidestream products at temperatures of 250 to 350°C (480 to 660°F). The feedstock and residue temperatures must be kept below 350°C (660°F), above which (as has already been noted) the rate of thermal decomposition increases substantially and cracking occurs. The partial pressure of the hydrocarbons is effectively reduced even further by the injection of steam. The steam added to the column, principally for the stripping of bitumen in the base of the column, is superheated in the convection section of the heater.

The fractions obtained by vacuum distillation of reduced crude oil depend on whether the run is designed to produce lubricating or vacuum gas oils. In the former case, the fractions include:

1. *Heavy gas oil*, an overhead product used as catalytic cracking stock or, after suitable treatment, a light lubricating oil.
2. *Lubricating oil* (usually three fractions — light, intermediate, and heavy), obtained as a sidestream product.
3. *Residuum*, the nonvolatile product that may be used directly as asphalt or to asphalt.

The residuum may also be used as a feedstock for a coking operation or blended with gas oils to produce a heavy fuel oil.

The continued use of atmospheric and vacuum distillation has been a major part of refinery operations during this century and no doubt will continue to be employed, at least into the beginning decades of the 21st century, as the primary refining operation.

When a residuum is obtained from a crude oil and thermal decomposition has commenced, it is more usual to refer to this product as *pitch*. The differences between parent petroleum and the residua are due to the relative amounts of various constituents present, which are removed or remain by virtue of their relative volatility.

The chemical composition of a residuum from an asphaltic crude oil is complex. Physical methods of fractionation usually indicate high proportions of asphaltenes and resins, even in amounts up to 50% (or higher) of the residuum. In addition, the presence of ash-forming metallic constituents, including such organometallic compounds as vanadium and nickel, is also a distinguishing feature of residua and heavy oils. Furthermore, the deeper the cut into the crude oil, the greater is the concentration of sulfur and metals in the residuum and the greater the deterioration in physical properties.

Historically, there may be the impression that the industry is coming full circle. The early uses of petroleum focused on the heavier derivatives that remained after the more volatile fractions had evaporated under the prevailing conditions. This is illustrated in the next section, but first a series of definitions for petroleum and its relatives are defined.

The definitions used to describe petroleum *reserves* are often misunderstood because they are not adequately defined at the time of use. Therefore, as a means of alleviating this problem, it is pertinent at this point to consider the definitions used to describe the amount of petroleum that remains in subterranean reservoirs (Speight, 2006).

REFERENCES

Abraham, H. 1945. *Asphalts and Allied Substances*. Van Nostrand, New York.

Arcaya, I. 2001. Venezuela, the United States, and Global Energy Security. WTC Featured Speakers. Address at the Windsor Court Hotel, New Orleans, August 20, http://wtcno.org/speakers/2001/arcaya.htm.

Armstrong, S.M., Sankey, B.M., and Voordouw, G. 1997. *Fuel*, 76: 223.

Bakshi, A. and Lutz, I. 1987. *Oil & Gas Journal*, 85: 84–87.

Dawson, W.H., Chornet, E., Tiwari, P., and Heitz, M. 1989. Preprints. *Div. Petrol. Chem. Am. Chem. Soc.*, 34: 384.

Dolbear, G., Tang, A., and Moorhead, E. 1987. In *Metal Complexes in Fossil Fuels*, R. Filby and J. Branthaver (Editors). American Chemical Society, Washington, DC, p. 220.

Gray, M.R. 1990. *AOSTRA Journal of Research*, 6: 185; 13: 124.

GRI. 1998. Baseline, Report GRI-98/0001. Gas Research Institute, Chicago, p. 1.

Hoiberg, A.J. 1964. *Bituminous Materials: Asphalts, Tars, and Pitches*. John Wiley & Sons, New York.

Johnson, T.E., Murphy, J.R., and Tasker, K.G. 1985. *Oil & Gas Journal*, 83: 50–55.

Monticello, D.J. 1995. U.S. Patent 5,387,523. February 7.

Pellegrino, J.L. 1998. *Energy and Environmental Profile of the U.S. Petroleum Refining Industry*. Office of Industrial Technologies, U.S. Department of Energy, Washington, DC.

Ryan, J.F. 1998. *Today's Chemist at Work*, 7: 84.

Schuetze, B. and Hofmann, H. 1984. *Hydrocarbon Processing*, 63: 75.

Speight, J.G. 1990. In *Fuel Science and Technology Handbook*, J.G. Speight (Editor). Marcel Dekker, New York, chaps. 12–16.

Speight, J.G. 1996. Petroleum refinery processes. In *Kirk-Othmer Encyclopedia of Chemical Technology*, 4th edition. Wiley Interscience, New York, 18: 433.

Speight, J.G. 2000. *The Desulfurization of Heavy Oils and Residua*, 2nd edition. Marcel Dekker, New York.

Speight, J.G. 2006. *The Chemistry and Technology of Petroleum*, 4th edition. CRC Press/ Taylor & Francis, Boca Raton, FL.

Spragins, F.K. 1978. *Development in Petroleum Science*. No. 7. *Bitumens, Asphalts and Tar Sands*, T.F. Yen and G.V. Chilingarian (Editors). Elsevier, New York, p. 92.

Suchanek, A. and Moore, A. 1986. *Oil & Gas Journal*, 84: 36–40.

Swain, E.J. 1991. *Oil & Gas Journal*, 89: 59.

Swain, E.J. 1993. *Oil & Gas Journal*, 91: 62.

Swain, E.J. 1998. *Oil & Gas Journal*, 96: 43.

Waguespack, K.G. and Healey, J.F. 1998. *Hydrocarbon Processing*, 77: 133.

Wallace, D., Starr, J., Thomas, K.P., and Dorrence. S.M. 1988. *Characterization of Oil Sands Resources*. Alberta Oil Sands Technology and Research Authority, Edmonton, Alberta, Canada.

2 Feedstock Evaluation and Composition

James G. Speight

CONTENTS

2.1 INTRODUCTION

Strategies for upgrading petroleum emphasize the difference in their properties, which in turn influences the choice of methods or combinations thereof for conversion of petroleum to various products. Naturally, similar principles are applied to heavy oils and residua, and the availability of processes that can be employed to convert heavy feedstocks to usable products has increased significantly in recent years (Cantrell, 1987; Speight, 2007). Thus, to determine the *processability* of petroleum, a series of consistent and standardized characterization procedures are required (ASTM, 2006). These procedures can be used with a wide variety of feedstocks to develop a general approach to predict processability (Speight, 2001).

Upgrading of heavy oils and residua can be designed in an optimal manner by performing selected evaluations of chemical and structural features of these heavy feedstocks (Schabron and Speight, 1996). The evaluation schemes do not need to be complex, but must focus on key parameters that affect processability. For example, the identification of the important features can be made with a saturates–aromatics–resins–asphaltene separation. Subsequent analysis of the fractions also provides further information about the processability of the feedstock. In this follow-on analysis, there is often emphasis on the asphaltene fraction since the solubility of asphaltene constituents and thermal products has a dramatic effect on solids deposition and coke formation during upgrading. Further evaluation of the asphaltenes can lead to correlation of asphaltene properties with process behavior. In addition, useful information can be gained from knowledge of elemental composition and molecular weight, and size exclusion chromatography and high-performance liquid chromatography profiles.

Heavy oils and residua exhibit a wide range of physical properties, and several relationships can be made between various physical properties (Speight, 2006). Whereas properties such as viscosity, density, boiling point, and color of petroleum may vary widely, the ultimate or elemental analysis varies, as already noted, over a narrow range for a large number of samples. The carbon content is relatively constant, while the hydrogen and heteroatom contents are responsible for the major differences between petroleum. The nitrogen, oxygen, and sulfur can be present in only trace amounts in some petroleum, which as a result consists primarily of hydrocarbons.

On the other hand, petroleum containing 9.5% heteroatoms may contain essentially hydrocarbon constituents insofar as the constituents contain *at least one or more* nitrogen, oxygen, or sulfur atoms within the molecular structures. And it is the heteroelements that can have substantial effects on the distribution of refinery products. Coupled with the changes brought about to the feedstock constituents by refinery operations, it is not surprising that refining the heavy feedstocks is a monumental task.

Thus, initial inspection of the feedstock (conventional examination of the physical properties) is necessary. From this, it is possible to make deductions about the most logical means of refining. In fact, evaluation of crude oils from physical property data as to which refining sequences should be employed for any particular crude oil is a predominant part of the initial examination of any material that is destined for use as a refinery feedstock.

The chemical composition of a feedstock is a much truer indicator of refining behavior. Whether the composition is represented in terms of compound types or generic compound classes, it can enable the refiner to determine the nature of the reactions. Hence, chemical composition can play a large part in determining the nature of the products that arise from the refining operations. It can also play a role in determining the means by which a particular feedstock should be processed (Ali et al., 1985; Wallace and Carrigy, 1988; Wallace et al., 1988; Speight, 2006; Speight and Ozum, 2002).

Therefore, the judicious choice of a feedstock to produce any given product is just as important as the selection of the product for any given purpose. Alternatively, and more in keeping with the current context, *the judicious choice of a processing sequence to convert a heavy feedstock to liquid products is even more important.*

To satisfy specific needs with regard to the type of petroleum to be processed, as well as to the nature of the product, most refiners have, through time, developed their own methods of petroleum analysis and evaluation. However, such methods are considered proprietary and are not normally available. Consequently, various standards organizations, such as the American Society for Testing and Materials (ASTM, 2006) in North America and the Institute of Petroleum in Britain (IP, 2006), have devoted considerable time and effort to the correlation and standardization of methods for the inspection and evaluation of petroleum and petroleum products. A complete discussion of the large number of routine tests available for petroleum fills an entire book (ASTM, 2006). However, it seems appropriate that in any discussion of the physical properties of petroleum and petroleum products reference be made to the corresponding test, and accordingly, the various test numbers have been included in the text.

Thus, initial inspection of the nature of the feedstock will provide deductions about the most logical means of refining or correlation of various properties to structural types present, and hence attempted classification of the petroleum (Speight, 2006). Indeed, careful evaluation from physical property data is a major part of the initial study of any refinery feedstock. Proper interpretation of the data resulting from the inspection of crude oil requires an understanding of their significance.

Not all tests that are applied to the evaluation of conventional petroleum are noted here. It is the purpose of this chapter to enumerate those tests that are regularly applied to the evaluation of the heavier feedstocks (Speight, 2001). For example, tests used to determine *volatility* (a characteristic that is not obvious in the heavy, or high-boiling, feedstocks) are omitted. Refer to a more detailed discussion of petroleum evaluation for these tests (Speight, 2007 and references cited therein).

2.2 FEEDSTOCK EVALUATION

2.2.1 Elemental (Ultimate) Analysis

The analysis of petroleum feedstocks for the percentages of carbon, hydrogen, nitrogen, oxygen, and sulfur is perhaps the first method used to examine the general nature, and perform an evaluation, of a feedstock. The atomic ratios of the various elements to carbon (that is, H/C, N/C, O/C, and S/C) are frequently used for indications of the overall character of the feedstock. It is also of value to determine the amounts of trace elements, such as vanadium and nickel, in a feedstock since these materials can have serious deleterious effects on catalyst performance during refining by catalytic processes.

However, it has become apparent, with the introduction of the heavier feedstocks into refinery operations, that these ratios are not the only requirement for predicting feedstock character before refining. The use of more complex feedstocks (in terms of chemical composition) has added a new dimension to refining operations. Thus, although atomic ratios, as determined by elemental analyses, may be used on a comparative basis between feedstocks, there is now no guarantee that a particular feedstock will behave as predicted from these data. Product slates cannot be predicted accurately, if at all, from these ratios.

The ultimate analysis (elemental composition) of petroleum is not reported to the same extent as it is for coal (Speight, 1994). Nevertheless, there are procedures (ASTM, 2006) for the ultimate analysis of petroleum and petroleum products, but many such methods may have been designed for other materials.

For example, *carbon content* can be determined by the method designated for coal and coke (ASTM D-3178) or for municipal solid waste (ASTM E-777). There are also methods designated for:

1. *Hydrogen content* (ASTM D-1018, ASTM D-3178, ASTM D-3343, ASTM D-3701, and ASTM E-777)
2. *Nitrogen content* (ASTM D-3179, ASTM D-3228, ASTM D-3431, ASTM E-148, ASTM E-258, and ASTM E-778)
3. *Oxygen content* (ASTM E-385)
4. *Sulfur content* (ASTM D-124, ASTM D-1266, ASTM D-1552, ASTM D-1757, ASTM D-2662, ASTM D-3177, ASTM D-4045, and ASTM D-4294)

Of the data that are available, the proportions of the elements in petroleum vary only slightly over narrow limits. And yet, there is a wide variation in physical properties from the lighter, more mobile crude oils at one extreme to the heavier, asphaltic crude oils at the other extreme. The majority of the more aromatic species and the heteroatoms occur in the higher-boiling fractions of feedstocks. The heavier feedstocks are relatively rich in these higher-boiling fractions.

Heteroatoms do affect every aspect of refining. Sulfur is usually the most concentrated and is fairly easy to remove; many commercial catalysts are available that routinely remove 90% of the sulfur. Nitrogen is more difficult to remove than sulfur, and there are fewer catalysts that are specific for nitrogen. If the nitrogen and sulfur are not removed, the potential for the production of nitrogen oxides (NO_x) and sulfur oxides (SO_x) during processing and use becomes real.

Perhaps the more pertinent property in the present context is the sulfur content, which along with the API gravity represents the two properties that have the greatest influence on the value of a heavy oil and residuum. The sulfur content varies from about 0.1% to about 3% by weight for the more conventional crude oils to as much as 5 to 6% for heavy oils and residua (Koots and Speight, 1975; Speight, 1990). Residua, depending on the sulfur content of the crude oil feedstock, may be of the same order or even have a substantially higher sulfur content. Indeed, the very nature of the distillation process by which residua are produced, that is, removal of distillate without thermal decomposition, dictates that the majority of the sulfur, which is predominantly in the higher molecular weight fractions, be concentrated in the residuum.

2.2.2 Metals Content

Metals (particularly vanadium and nickel) are found in most crude oils (Reynolds, 1988). Heavy oils and residua contain relatively high proportions of metals, either in the form of salts or as organometallic constituents (such as the metallo-porphyrins), which are extremely difficult to remove from the feedstock. Indeed, the nature of the process by which residua are produced virtually dictates that all the metals in the original crude oil are concentrated in the residuum (Speight, 2006). Those metallic constituents that may actually volatilize under the distillation conditions and appear in the higher-boiling distillates are the exceptions here.

Metals cause particular problems because they poison catalysts used for sulfur and nitrogen removal as well as other processes, such as catalytic cracking. Thus, serious attempts are being made to develop catalysts that can tolerate a high concentration of metals without serious loss of catalyst activity or life.

A variety of tests (ASTM D-1026, ASTM D-1262, ASTM D-1318, ASTM D-1368, ASTM D-1548, ASTM D-1549, ASTM D-2547, ASTM D-2599, ASTM D-2788, ASTM D-3340, ASTM D-3341, and ASTM D-3605) have been designated for the determination of metals in petroleum products. Determination of metals in whole feeds can be accomplished by combustion of the sample so that only inorganic ash remains. The ash can then be digested with an acid and the solution examined for metal species by atomic absorption (AA) spectroscopy or inductively coupled argon plasma (ICP) spectrometry.

2.2.3 Density and Specific Gravity

Density is the mass of a unit volume of material at a specified temperature and has the dimensions of grams per cubic centimeter (a close approximation to grams per milliliter). *Specific gravity* is the ratio of the mass of a volume of the substance to the mass of the same volume of water and is dependent on two temperatures, those at which the masses of the sample and the water are measured. When the water temperature is 4°C (39°F), the specific gravity is equal to the density in the centimeter-gram-second (cgs) system, since the volume of 1 g of water at that temperature is by definition 1 ml. Thus, the density of water, for example, varies with temperature, and its specific gravity at equal temperatures is always unity. The standard temperature for a specific gravity in the petroleum industry in North America is 60/60°F (15.6/15.6°C).

The density and specific gravity of crude oil (ASTM D-287, ASTM D-1298, ASTM D-941, ASTM D-1217, and ASTM D-1555) are two properties that have found wide use in the industry for preliminary assessment of the character of the crude oil. In particular, the density was used to give an estimate of the most desirable product, that is, kerosene, in crude oil. Thus, a conventional crude oil with a high content of paraffins (high kerosene) and excellent mobility at ambient temperature and pressure may have a specific gravity (density) of about 0.8. A heavy oil with a high content of asphaltenes and resins (low kerosene) and poor mobility at ambient temperature and pressure, thereby requiring vastly different processing sequences, may have a specific gravity (density) of about 0.95.

Specific gravity is influenced by the chemical composition of petroleum, but quantitative correlation is difficult to establish. Nevertheless, it is generally recognized that increased amounts of aromatic compounds result in an increase in density, whereas an increase in saturated compounds results in a decrease in density. It is also possible to recognize certain preferred trends between the API gravity of crude oils and residua and one or more of the other physical parameters. For example, a correlation exists between the API gravity and sulfur content, Conradson carbon residue, and viscosity (Speight, 2000 and references cited therein).

However, the derived relationships between the density of petroleum and its fractional composition were valid only if they were applied to a certain type of petroleum and lost some of their significance when applied to different types of petroleum. Nevertheless, density is still used to give a rough estimation of the nature of petroleum and petroleum products.

The values for density (and specific gravity) cover an extremely narrow range considering the differences in the feedstock appearance and behavior. In an attempt to inject a more meaningful relationship between the physical properties and processability of the various crude oils, the American Petroleum Institute devised a measurement of gravity devised upon the Baumé scale for industrial liquids. The Baumé scale for liquids lighter than water was used initially:

$$°\text{Baumé} = 140/\text{sp gr @ } 60/60°\text{F} - 130$$

However, a considerable number of hydrometers calibrated according to the Baumé scale were found at an early period to be in error by a consistent amount, and this led to the adoption of the equation

$$°\text{API} = 141.5/\text{sp gr @ } 60/60°\text{F} - 131.5$$

The specific gravity of petroleum usually ranges from about 0.8 (45.3° API) for the lighter crude oils to over 1.0 (10° API) for heavy crude oils and bitumen. Residua are expected to have API gravity on the order of 5 to 10° API. This is in keeping with the general trend that increased aromaticity leads to a decrease in API gravity (or, more correctly, an increase in specific gravity).

Density or specific gravity or API gravity may be measured by means of a hydrometer (ASTM D-287 and ASTM D-1298) or a pycnometer (ASTM D-941 and ASTM D-1217). The variation of density with temperature, effectively the coefficient of expansion, is a property of great technical importance, since most petroleum products are sold by volume and specific gravity is usually determined at the prevailing temperature (21°C) (70°F) rather than at the standard temperature (60°F) (15.6°C). The tables of gravity corrections (ASTM D-1555) are based on an assumption that the coefficient of expansion of all petroleum products is a function (at fixed temperatures) of density only. Recent work has focused on the calculation and predictability of density using new mathematical relationships (Gomez, 1995).

The specific gravity of bitumen shows a fairly wide range of variation. The largest degree of variation is usually due to local conditions that affect material close to the faces, or exposures, occurring in surface oil sand beds. There are also variations in the specific gravity of the bitumen found in beds that have not been exposed to weathering or other external factors. The range of specific gravity variation is usually of the order of 0.9 to 1.04.

A very important property of the Athabasca bitumen (which also accounts for the success of the hot-water separation process) is the variation in density (specific gravity) of the bitumen with temperature. Over the temperature range of 30 to 130°C (85 to 265°F) there is a density–water inversion in which the bitumen is lighter than water. Flotation of the bitumen (with aeration) on the water is facilitated — hence the logic of the hot-water separation process (Speight, 2007).

2.2.4 Viscosity

Viscosity is the most important single fluid characteristic governing the motion of petroleum and petroleum products and is actually a measure of the internal resistance to motion of a fluid by reason of the forces of cohesion between molecules or molecular groupings.

By definition, viscosity is the force in dynes required to move a plane of 1 cm^2 area at a distance of 1 cm from another plane of 1 cm^2 area through a distance of 1 cm in 1 sec. In the centimeter-gram-second (cgs) system the unit of viscosity is the *poise* (P) or *centipoise* (cP) (1 cP = 0.01 P). Two other terms in common use are *kinematic viscosity* and *fluidity*. The kinematic viscosity is the viscosity in centipoise divided by the specific gravity. The unit of kinematic viscosity is the *stoke* (cm^2/sec), although *centistokes* (0.01 cSt) is in more common usage; fluidity is simply the reciprocal of viscosity.

In the early days of the petroleum industry viscosity was regarded as the body of an oil, a significant number for lubricants or for any liquid pumped or handled in quantity. The changes in viscosity with temperature, pressure, and rate of shear are pertinent not only in lubrication, but also for such engineering concepts as heat transfer. The viscosity and relative viscosity of different phases, such as gas, liquid oil, and water, are determining influences in producing the flow of reservoir fluids through porous oil-bearing formations. The rate and amount of oil production from a reservoir are often governed by these properties.

The viscosity (ASTM D-445, ASTM D-88, ASTM D-2161, ASTM D-341, and ASTM D-2270) of crude oils varies markedly over a very wide range. Values vary from less than 10 cP at room temperature to many thousands of centipoise at the same temperature. In the present context, oil sands bitumen occurs at the higher end of this scale, where a relationship between viscosity and density between various crude oils has been noted.

Many types of instruments have been proposed for the determination of viscosity. The simplest and most widely used are capillary types (ASTM D-445), and the viscosity is derived from the equation

$$\mu = \mathrm{B}r^4P/8nl$$

where r is the tube radius, l the tube length, P the pressure difference between the ends of a capillary, n the coefficient of viscosity, and μ the quantity discharged in unit time. Not only are such capillary instruments the most simple, but when designed in accordance with known principles and used with known necessary correction factors, they are probably the most accurate viscometers available. It is usually more convenient, however, to use relative measurements, and for this purpose the instrument is calibrated with an appropriate standard liquid of known viscosity.

Batch flow times are generally used; in other words, the time required for a fixed amount of sample to flow from a reservoir through a capillary is the datum that is actually observed. Any features of technique that contribute to longer flow times are usually desirable. Some of the principal capillary viscometers in use are those of Cannon-Fenske, Ubbelohde, Fitzsimmons, and Zeitfuchs.

The *Saybolt universal viscosity* (SUS) (ASTM D-88) is the time in seconds required for the flow of 60 ml of petroleum from a container, at constant temperature, through a calibrated orifice. The *Saybolt furol viscosity* (SFS) (ASTM D-88) is determined in a similar manner except that a larger orifice is employed.

As a result of the various methods for viscosity determination, it is not surprising that much effort has been spent on interconversion of the several scales, especially converting Saybolt to kinematic viscosity (ASTM D-2161),

$$\text{Kinematic viscosity} = a \times \text{Saybolt sec} + b/\text{Saybolt sec}$$

where a and b are constants.

The SUS equivalent to a given kinematic viscosity varies slightly with the temperature at which the determination is made because the temperature of the calibrated receiving flask used in the Saybolt method is not the same as that of the oil. Conversion factors are used to convert kinematic viscosity from 2 to 70 cSt at 38°C (100°F) and 99°C (210°F) to equivalent Saybolt universal viscosity in seconds (ASTM, 2006). Appropriate multipliers are listed to convert kinematic viscosity over 70 cSt. For a kinematic viscosity determined at any other temperature the equivalent Saybolt universal value is calculated by use of the Saybolt equivalent at 38°C (100°F) and a multiplier that varies with the temperature:

$$\text{Saybolt sec at } 100°\text{F } (38°\text{C}) = \text{cSt} \times 4.635$$

$$\text{Saybolt sec at } 210°\text{F } (99°\text{C}) = \text{cSt} \times 4.667$$

Various studies have also been made on the effect of temperature on viscosity since the viscosity of petroleum, or a petroleum product, decreases as the temperature increases. The rate of change appears to depend primarily on the nature or composition of the petroleum, but other factors, such as volatility, may also have a minor effect. The effect of temperature on viscosity is generally represented by the equation

$$\log \log (n + c) = A + B \log T$$

where n is absolute viscosity, T is temperature, and A and B are constants. This equation has been sufficient for most purposes and has come into very general use. The constants A and B vary widely with different oils, but c remains fixed at 0.6 for all oils having a viscosity over 1.5 cSt; it increases only slightly at lower viscosity (0.75 at 0.5 cSt). The viscosity–temperature characteristics of any oil, so plotted, thus create a straight line, and the parameters A and B are equivalent to the intercept and slope of the line. To express the viscosity and viscosity–temperature characteristics of an oil, the slope and the viscosity at one temperature must be known; the usual practice is to select 38°C (100°F) and 99°C (210°F) as the observation temperatures.

Suitable conversion tables are available (ASTM D-341), and each table or chart is constructed in such a way that for any given petroleum or petroleum product the viscosity–temperature points result in a straight line over the applicable temperature range. Thus, only two viscosity measurements need be made at temperatures far enough apart to determine a line on the appropriate chart from which the approximate viscosity at any other temperature can be read.

The charts can be applicable only to measurements made in the temperature range in which petroleum is assumed to be a Newtonian liquid. The oil may cease to be a simple liquid near the cloud point because of the formation of wax particles or near the boiling point because of vaporization. Thus, the charts do not give accurate results when either the cloud point or boiling point is approached. However, they are useful over the Newtonian range for estimating the temperature at which an oil attains a desired viscosity. The charts are also convenient for estimating the viscosity of a blend of petroleum liquids at a given temperature when the viscosity of the component liquids at the given temperature is known.

Since the viscosity–temperature coefficient of lubricating oil is an important expression of its suitability, a convenient number to express this property is very useful, and hence a viscosity index (ASTM D-2270) was derived. It is established that naphthenic oils have higher viscosity–temperature coefficients than do paraffinic oils at equal viscosity and temperatures. The Dean and Davis scale was based on the assignment of a zero value to a typical naphthenic crude oil and that of 100 to a typical paraffinic crude oil; intermediate oils were rated by the formula

$$\text{Viscosity index} = L - U/L - H \times 100$$

where L and H are the viscosities of the zero and 100 index reference oils, both having the same viscosity at 99°C (210°F), and U is that of the unknown, all at 38°C (100°F). Originally the viscosity index was calculated from Saybolt viscosity data, but subsequently figures were provided for kinematic viscosity.

Because of the importance of viscosity in determining the transport properties of petroleum, recent work has focused on the development of an empirical equation for predicting the dynamic viscosity of low molecular weight and high molecular weight hydrocarbon vapors at atmospheric pressure (Gomez, 1995). The equation uses molar mass and specific temperature as the input parameters and offers a means of estimation of the viscosity of a wide range of petroleum fractions. Other work has focused on the prediction of the viscosity of blends of lubricating oils as a means of accurately predicting the viscosity of the blend from the viscosities of the base oil components (Al-Besharah et al., 1989).

2.2.5 Carbon Residue

The carbon residue (ASTM D-189 and ASTM D-524) of a crude oil is a property that can be correlated with several other properties of petroleum. The carbon residue presents indications of the *volatility* or *gasoline-forming propensity* of the feedstock and, for the most part in this text, the *coke-forming propensity* of a feedstock. Tests for carbon residue are sometimes used to evaluate the carbonaceous depositing characteristics of fuels used in certain types of oil-burning equipment and internal combustion engines.

There are two older well-used methods for determining the carbon residue: the Conradson method (ASTM D-189) and the Ramsbottom method (ASTM D-524). Both are equally applicable to the high-boiling fractions of crude oils that decompose to volatile material and coke when distilled at a pressure of 1 atm. Heavy oils and residua that contain metallic constituents (distillation of crude oils concentrates these constituents in the residua) will have erroneously high carbon residues. Thus, the metallic constituents must first be removed from the oil, or they can be estimated as ash by complete burning of the coke after carbon residue determination.

Although there is no exact correlation between the two methods, it is possible to interconnect the data (ASTM, 2006), but caution is advised when using that portion of the curve below 0.1% w/w Conradson carbon residue.

Recently, a newer method (Noel, 1984) has been accepted (ASTM D-4530) that requires smaller sample amounts and was originally developed as a *thermogravimetric method*. The carbon residue produced by this method is often referred to as the *microcarbon residue* (MCR). Agreements between the data from the three methods are good, making it possible to interrelate all of the data from carbon residue tests (Long and Speight, 1989).

Even though the three methods have their relative merits, there is a tendency to advocate use of the more expedient *microcarbon method* to the exclusion of the *Conradson method* and *Ramsbottom method* because of the lesser amounts required in the microcarbon method.

The mechanical design and operating conditions of such equipment have such a profound influence on carbon deposition during service that comparison of carbon residues between oils should be considered as giving only a rough approximation of relative deposit-forming tendencies. Recent work has focused on the carbon residue of the different fractions of crude oils, especially the asphaltenes. A more precise relationship between carbon residue and hydrogen content (H/C atomic ratio), nitrogen content, and sulfur content has been shown to exist. These data can provide more precise information about the anticipated behavior of a variety of feedstocks in thermal processes (Roberts, 1989).

Because of the extremely small values of carbon residue obtained by the Conradson and Ramsbottom methods when applied to the lighter distillate fuel oils, it is customary to distill such products to 10% residual oil and determine the carbon residue thereof. Such values may be used directly in comparing fuel oils, as long as it is kept in mind that the values are carbon residues on 10% residual oil and are not to be compared with straight carbon residues.

2.2.6 Specific Heat

Specific heat is defined as the quantity of heat required to raise a unit mass of material through 1° of temperature (ASTM D-2766).

Specific heats are extremely important engineering quantities in refinery practice because they are used in all calculations on heating and cooling petroleum products. Many measurements have been made on various hydrocarbon materials, but the data for most purposes may be summarized by the general equation

$$C = 1/d\ (0.388 + 0.00045t)$$

where C is the specific heat at t°F of an oil whose specific gravity 60/60°F is d; thus, specific heat increases with temperature and decreases with specific gravity.

2.2.7 Heat of Combustion

The gross heat of combustion of crude oil and its products is given with fair accuracy by the equation

$$Q = 12{,}400 - 2100d^2$$

where d is the 60/60°F specific gravity. Deviation is generally less than 1%, although many highly aromatic crude oils show considerably higher values; the range for crude oil is 10,000 to 11,600 cal/g. For gasoline, the heat of combustion is 11,000 to 11,500 cal/g, and for kerosene (and diesel fuel), it falls in the range of 10,500 to 11,200 cal/g. Finally, the heat of combustion for fuel oil is on the order of 9500 to 11,200 cal/g. Heats of combustion of petroleum gases may be calculated from the analysis and data for the pure compounds. Experimental values for gaseous fuels may be obtained by measurement in a water flow calorimeter, and heats of combustion of liquids are usually measured in a bomb calorimeter.

For thermodynamic calculation of equilibria useful in petroleum science, combustion data of extreme accuracy are required because the heats of formation of water and carbon dioxide are large in comparison with those in the hydrocarbons. Great accuracy is also required of the specific heat data for the calculation of free energy or entropy. Much care must be exercised in selecting values from the literature for these purposes, since many of those available were determined before the development of modern calorimetric techniques.

2.3 CHROMATOGRAPHIC METHODS

Feedstock evaluation by separation into various fractions has been used quite successfully for several decades. The knowledge of the components of a feedstock on a *before refining* and *after refining* basis has been a valuable aid to process development (Altgelt and Gouw, 1979; Altgelt and Boduszynski, 1994).

There are several experimental procedures for feedstock/product evaluation:

1. Determination of aromatic content of olefin-free gasoline by silica gel adsorption (ASTM D-936).
2. Separation of aromatic and nonaromatic fractions from high-boiling oils (ASTM D-2549).
3. Determination of hydrocarbon groups in rubber extender oils by clay–gel adsorption (ASTM D-2007).
4. Determination of hydrocarbon types in liquid petroleum products by a fluorescent indicator adsorption test (ASTM D-1319).

Gel permeation chromatography is an attractive technique for the determination of the number average molecular weight (M_n) distribution of petroleum fractions, especially the heavier constituents, and petroleum products (Altgelt, 1968, 1970; Oelert, 1969; Baltus and Anderson, 1984; Hausler and Carlson, 1985; Reynolds and Biggs, 1988).

Ion exchange chromatography is also widely used in the characterization of petroleum constituents and products. For example, cation exchange chromatography can be used primarily to isolate the nitrogen constituents in petroleum (Drushel and Sommers, 1966; McKay et al., 1974), thereby giving an indication of how the feedstock might behave during refining as well as any potential deleterious effects on catalysts.

Liquid chromatography (also called *adsorption chromatography*) has helped to characterize the group composition of crude oils and hydrocarbon products since the beginning of this century. The type and relative amount of certain hydrocarbon classes in the matrix can have a profound effect on the quality and performance of the hydrocarbon product. The *fluorescent indicator adsorption* (FIA) method (ASTM D-1319) has been used to measure the paraffinic, olefinic, and aromatic contents of gasoline, jet fuel, and liquid products in general (Suatoni and Garber, 1975; Miller et al., 1983; Norris and Rawdon, 1984).

High-performance liquid chromatography (HPLC) has found great utility in separating different hydrocarbon group types and identifying specific constituent types (Colin and Vion, 1983; Drushel, 1983; Miller et al., 1983; Chartier et al., 1986). Of particular interest is the application of the HPLC technique to the identification of the molecular types in the heavier feedstocks, especially the molecular types in the asphaltene fraction. This technique is particularly useful for studying such materials on a before processing and after processing basis (Chmielowiec et al., 1980; Alfredson, 1981; Bollet et al., 1981; Colin and Vion, 1983;

George and Beshai, 1983; Felix et al., 1985; Coulombe and Sawatzky, 1986; Speight, 1986, 2007).

Several recent high-performance liquid chromatographic separation schemes are applicable since they also incorporate detectors not usually associated with conventional hydrocarbon group-type analyses (Matsushita et al., 1981; Miller et al., 1983; Rawdon, 1984; Lundanes and Greibokk, 1985; Schwartz and Brownlee, 1986; Hayes and Anderson, 1987).

The general advantages of high-performance liquid chromatography are:

1. Each sample may be analyzed *as received* even though the boiling range may vary over a considerable range.
2. The total time per analysis is usually on the order of minutes.
3. The method can be adapted for on-stream analysis in a refinery.

In recent years, *supercritical fluid chromatography* has found use in the characterization and identification of petroleum constituents and products. A supercritical fluid is defined as a substance above its critical temperature. A primary advantage of chromatography using supercritical mobile phases results from the mass transfer characteristics of the solute. The increased diffusion coefficients of supercritical fluids compared with liquids can lead to greater speed in separations or greater resolution in complex mixture analyses. Another advantage of supercritical fluids compared with gases is that they can solubilize thermally labile and nonvolatile solutes and, upon expansion (decompression) of this solution, introduce the solute into the vapor phase for detection (Lundanes et al., 1986).

Currently, supercritical fluid chromatography is leaving the stages of infancy. The indications are that it will find wide applicability to the problems of characterization and identification of the higher molecular weight species in petroleum thereby adding an extra dimension to our understanding of refining chemistry. It will still retain the option as a means of product characterization, although the use may be somewhat limited because of the ready availability of other characterization techniques.

2.4 MOLECULAR WEIGHT

Even though refining produces, in general, lower molecular weight species than those originally in the feedstock, there is still the need to determine the molecular weight of the original constituents as well as the molecular weights of the products as a means of understanding the process. For those original constituents and products, for example, resins and asphaltenes, that have little or no volatility, *vapor pressure osmometry* (VPO) has proven to be of considerable value.

A particularly appropriate method involves the use of different solvents (at least two), and the data are then extrapolated to infinite dilution (Schwager et al., 1979). There has also been the use of different temperatures for a particular solvent after which the data are extrapolated to room temperature (Speight et al., 1985; Speight, 1987). In this manner, different solvents are employed and the molecular weight of a petroleum fraction (particularly the asphaltenes) can be determined, for which it can be assumed that there is little or no influence from any intermolecular forces. In summary, the molecular weight may be as close to the real value as possible.

In fact, it is strongly recommended that to negate concentration and temperature effects the molecular weight determination be carried out at three different concentrations at three different temperatures. The data for each temperature are then extrapolated to zero concentration, and the zero concentration data at each temperature are then extrapolated to room temperature (Speight, 1987).

A correlation relating molecular weight, asphaltene content, and heteroatom content with the carbon residues of whole residua has been developed and extended to molecular weight and carbon residue (Schabron and Speight, 1997a, 1997b). The linear correlation holds for the whole residua and residue, forming materials such as asphaltenes, saturates, aromatics, and polar constituents. At molecular weights greater than about 3000, the result suggests that the correlation can be used as a tool to quantitatively gauge association for asphaltenes. Subsequent results suggest that the inclusion of heteroatoms is not necessary in the relationship, since heteroatom associative affects are already taken into account by measuring the apparent molecular weights of the asphaltenes.

2.5 PHYSICAL (BULK) COMPOSITION

The term *physical composition* (or *bulk composition*) refers to the composition of crude oil as determined by various physical techniques. For example, the separation of petroleum using solvents and adsorbents (Speight, 2007) into various bulk fractions determines the physical composition of crude oil. However, in many instances, the physical composition may not be equivalent to the chemical composition. These methods of separation are not always related to chemical properties, and the terminology applied to the resulting fractions is often a terminology of convenience.

2.5.1 Asphaltene Separation

The *asphaltene fraction* is that portion of the feedstock that is precipitated when a large excess (40 volumes) of a low-boiling liquid hydrocarbon (for example, *n*-pentane or *n*-heptane) is added to the crude oil (1 volume) (Speight, 1994, 2006). *n*-Heptane is the preferred hydrocarbon, with *n*-pentane still being used, although hexane is used on occasion (Speight, 2007 and references cited therein).

Thus, although the use of both *n*-pentane and *n*-heptane has been widely advocated, and although *n*-heptane is becoming the deasphalting liquid of choice, this is by no means a hard-and-fast rule. And it must be recognized that large volumes of solvent may be required to effect a *qualitative* and *quantitative* reproducible separation. In addition, whether *n*-pentane or *n*-heptane is employed, the method effects a separation of the chemical components with the most complex structures from the mixture, and this fraction should be correctly identified as *n-pentane asphaltenes* or *n-heptane asphaltenes*.

However, it must be recognized that some of these methods were developed for use with feedstocks other than heavy oil, and adjustments are necessary.

Although *n*-pentane and *n*-heptane are the solvents of choice in the laboratory, other solvents can be used (Speight, 1979) and cause the separation of the asphaltene fraction as a brown-to-black powdery solid material. In the refinery, supercritical low molecular weight hydrocarbons (for example, liquid propane, liquid butane, or mixtures of both) are the solvents of choice, and the product is a semisolid (tacky) to solid asphalt. The amount of asphalt that settles out of the paraffin–residuum mixture depends on the size of the paraffin, the temperature, and the paraffin-to-feedstock ratio (Girdler, 1965; Mitchell and Speight, 1973; Corbett and Petrossi, 1978; Speight et al., 1984).

At constant temperature, the quantity of precipitate first increases with an increasing *ratio of solvent to feedstock* and then reaches a maximum. In fact, there are indications that when the proportion of solvent in the mix is <35%, little or no asphaltene constituents are precipitated.

Very few data have been reported that relate to this aspect of asphaltene separation. There is fragmentary evidence to show that the most polar materials (not necessarily the highest molecular weight material) separate first from the feedstock. This is in keeping with the increased paraffin character of the feedstock as the hydrocarbon is added.

When pentane and the lower molecular weight hydrocarbon solvents are used in large excess, the quantity of precipitate and the composition of the precipitate change with *increasing temperature* (Mitchell and Speight, 1973; Andersen, 1994).

Contact time between the hydrocarbon and the feedstock also plays an important role in asphaltene separation. Yields of the asphaltenes reach a maximum after approximately 8 h, which may be ascribed to the time required for the asphaltene particles to agglomerate into particles of a *filterable size* as well as the diffusion-controlled nature of the process. Heavier feedstocks also need time for the hydrocarbon to penetrate their mass.

For example, if the precipitation method (deasphalting) involves the use of a solvent and a residuum and is essentially a leaching of the heavy oil from the insoluble residue, this process may be referred to as *extraction*. However, under the prevailing conditions now in laboratory use, the term *precipitation* is perhaps more correct and descriptive of the method. Variation of solvent type also causes significant changes in asphaltene yield. Thus, the contact time between the feedstock and the hydrocarbon liquid can have an important influence on the yield and character of the asphaltene fraction.

2.5.2 Fractionation

Fractionation of heavy feedstocks into components of interest and study of the components appears to be a better approach than obtaining data on whole residua. By careful selection of a characterization scheme it may be possible to obtain a detailed overview of feedstock composition that can be used for process predictions.

Thus, fractionation methods also play a role, along with the physical testing methods, of evaluating heavy oils and residua as refinery feedstocks. For example, by careful selection of an appropriate technique it is possible to obtain a detailed *map* of feedstock or product composition that can be used for process predictions (Speight, 2006 and references cited therein).

After removal of the asphaltene fraction, further fractionation of petroleum is also possible by variation of the hydrocarbon solvent. For example, liquefied gases, such as propane and butane, precipitate as much as 50% by weight of the residuum or bitumen. The precipitate is a black, tacky, semisolid material, in contrast to the pentane-precipitated asphaltenes, which are usually brown, amorphous solids. Treatment of the propane precipitate with pentane then yields the insoluble brown, amorphous asphaltenes and soluble, near-black, semisolid resins, which are, as near as can be determined, equivalent to the resins isolated by adsorption techniques.

Separation by adsorption chromatography essentially commences with the preparation of a porous bed of finely divided solid, the adsorbent. The adsorbent is usually contained in an open tube (column chromatography); the sample is introduced at one end of the adsorbent bed and induced to flow through the bed by means of a suitable solvent. As the sample moves through the bed, the various components are held (adsorbed) to a greater or lesser extent depending on the chemical nature of the component. Thus, those molecules that are strongly adsorbed spend considerable time on the adsorbent surface rather than in the moving (solvent) phase, but components that are slightly adsorbed move through the bed comparatively rapidly.

There are many procedures that have received attention over the years (Speight, 2006 and references cited therein), but for the purposes of this text, this section will focus on the standard methods of fractionation (ASTM, 2006).

There are three standard test methods that provide for the separation of a feedstock into four or five constituent fractions (Speight, 2001; ASTM, 2006). It is interesting to note that as the methods have evolved there has been a change from the use of pentane (ASTM D-2006 and ASTM D-2007) to heptane (ASTM D-4124) to separate asphaltenes. This is, in fact, in keeping with the production of a more consistent fraction that represents these higher molecular weight, more complex constituents of petroleum (Girdler, 1965; Speight et al., 1984).

Two of the methods (ASTM D-2007 and ASTM D-4124) use adsorbents to fractionate the deasphaltened oil, but the third method (ASTM D-2006) advocates the use of various grades of sulfuric acid to separate the material into compound types. Caution is advised in the application of this method since it does not work well with all feedstocks. For example, when the *sulfuric acid* method (ASTM D-2006) is applied to the separation of heavy feedstocks, complex emulsions can be produced.

Obviously, there are precautions that must be taken when attempting to separate heavy feedstocks or polar feedstocks into constituent fractions. The disadvantages in using ill-defined adsorbents are that adsorbent performance differs with the same feed and, in certain instances, may even cause chemical and physical modification of the feed constituents. The use of a chemical reactant like sulfuric acid should only be advocated with caution since feeds react differently and may even cause irreversible chemical changes or emulsion formation. These advantages may be of little consequence when it is not, for various reasons, the intention to recover the various product fractions *in toto* or in the original state, but in terms of the compositional evaluation of different feedstocks the disadvantages are very real.

In general terms, group-type analysis of petroleum is often identified by the acronyms for the names: PONA (paraffins, olefins, naphthenes, and aromatics), PIONA (paraffins, *iso*-paraffins, olefins, naphthenes, and aromatics), PNA (paraffins, naphthenes, and aromatics), PINA (paraffins, *iso*-paraffins, naphthenes, and aromatics), or SARA (saturates, aromatics, resins, and asphaltenes). However, it must be recognized that the fractions produced by the use of different adsorbents will differ in content and will also be different from fractions produced by solvent separation techniques.

The variety of fractions isolated by these methods and the potential for differences in the composition of the fractions make it even more essential that the method is described accurately and that it is reproducible not only in one laboratory, but also between various laboratories.

REFERENCES

Al-Besharah, J.M., Mumford, C.J., Akashah, S.A., and Salman, O. 1989. *Fuel*, 68: 809.
Alfredson, T.V. 1981. *J. Chromatogr.*, 218: 715.
Ali, M.F., Hasan, M., Bukhari, A., and Saleem, M. 1985. *Oil Gas J.*, 83: 71.
Altgelt, K.H. 1968. Preprints. *Div. Petrol. Chem. Am. Chem. Soc.*, 13: 37.
Altgelt, K.H. 1970. *Bitumen Teere Asphalte Peche*, 21: 475.
Altgelt, K.H. and Boduszynski, M.M. 1994. *Compositional Analysis of Heavy Petroleum Fractions*. Marcel Dekker, New York.
Altgelt, K.H. and Gouw, T.H. 1979. *Chromatography in Petroleum Analysis*. Marcel Dekker, New York.
Andersen, S.I. 1994. *Fuel Sci. Technol. Int.*, 12: 51.
ASTM. 2006. *Annual Book of Standards*. American Society for Testing and Materials, Philadelphia.
Baltus, R.E. and Anderson, J.L. 1984. *Fuel*, 63: 530.
Bollet, C., Escalier, J.C., Souteyrand, C., Caude, M., and Rosset, R. 1981. *J. Chromatogr.*, 206: 289.
Cantrell, A. 1987. *Oil Gas J.*, 85: 60.
Chartier, P., Gareil, P., Caude, M., Rosset, R., Neff, B., Bourgognon, H.F., and Husson, J.F. 1986. *J. Chromatogr.*, 357: 381.
Chmielowiec, J., Beshai, J.E., and George, A.E. 1980. *Fuel*, 59: 838.
Colin, J.M. and Vion, G. 1983. *J. Chromatogr.*, 280: 152.
Corbett, L.W. and Petrossi, U. 1978. *Ind. Eng. Chem. Prod. Res. Dev.*, 17: 342.
Coulombe, S. and Sawatzky, H. 1986. *Fuel*, 65: 552.
Drushel, H.V. 1983. *J. Chromatogr. Sci.*, 21: 375.
Drushel, H.V. and Sommers, A.L. 1966. *Anal. Chem.*, 38: 19.

Felix, G., Bertrand, C., and Van Gastel, F. 1985. *Chromatographia*, 20: 155.
George, A.E. and Beshai, J.E. 1983. *Fuel*, 62: 345.
Girdler, R.B. 1965. *Proc. Assoc. Asphalt Paving Technologists*, 34: 45.
Gomez, J.V. 1995. *Oil Gas J.*, 26: 60.
Hausler, D.W. and Carlson, R.S. 1985. Preprints. *Div. Petrol. Chem. Am. Chem. Soc.*, 30: 28.
Hayes, P.C. and Anderson, S.D. 1987. *J. Chromatogr.*, 387: 333.
IP. 2006. Standard 143. *Standard Methods for Analysis and Testing of Petroleum and Related Products 1997*. Institute of Petroleum, London.
Koots, J.A. and Speight, J.G. 1975. *Fuel*, 54: 179.
Long, R.B. and Speight, J.G. 1989. *Rev. Inst. Francais Petrole*, 44: 205.
Lundanes, E. and Greibokk, T. 1985. *J. Chromatogr.*, 349: 439.
Lundanes, E., Iversen, B., and Greibokk, T. 1986. *J. Chromatogr.*, 366: 391.
Matsushita, S., Tada, Y., and Ikushige., S. 1981. *J. Chromatogr.*, 208: 429.
McKay, J.F., Cogswell, T.E., and Latham, D.R. 1974. Preprints. *Div. Petrol. Chem. Am. Chem. Soc.*, 19: 25.
Miller, R.L., Ettre, L.S., and Johansen, N.G. 1983. *J. Chromatogr.*, 259: 393.
Mitchell, D.L. and Speight, J.G. 1973. *Fuel*, 52: 149.
Noel, F. 1984. *Fuel*, 63: 931.
Norris, T.A. and Rawdon, M.G. 1984. *Anal. Chem.*, 56: 1767.
Oelert, H.H. 1969. *Erdoel Kohle*, 22: 536.
Rawdon, M. 1984. *Anal. Chem.*, 56: 831.
Reynolds, J.G. 1988. In *Petroleum Chemistry and Refining*, J.G. Speight (Editor). Taylor & Francis, Washington, DC, chap. 3.
Reynolds, J.G. and Biggs, W.R. 1988. *Fuel Sci. Technol. Int.*, 6: 329.
Roberts, I. 1989. Preprints. *Div. Petrol. Chem. Am. Chem. Soc.*, 34: 251.
Schabron, J.F. and Speight, J.G. 1996. *Arab. J. Sci. Eng.*, 21: 663.
Schabron, J.F. and Speight, J.G. 1997a. *Rev. Inst. Franc. Petrole*, 52: 73.
Schabron, J.F. and Speight, J.G. 1997b. Preprints. *Div. Fuel Chem. Am. Chem. Soc.*, 42: 386.
Schwager, I., Lee, W.C., and Yen, T.F. 1979. *Anal. Chem.*, 51: 1803.
Schwartz, H.E. and Brownlee, R.G. 1986. *J. Chromatogr.*, 353: 77.
Speight, J.G. 1979. Information Series 84. Alberta Research Council, Edmonton, Alberta, Canada.
Speight, J.G. 1986. Preprints. *Am. Chem. Soc. Div. Petrol. Chem.*, 31: 818.
Speight, J.G. 1987. Preprints. *Am. Chem. Soc. Div. Petrol. Chem.*, 32: 413.
Speight, J.G. 1990. In *Fuel Science and Technology Handbook*, J.G. Speight (Editor). Marcel Dekker, New York.
Speight, J.G. 1994. In *Asphaltenes and Asphalts*, Vol. 1, *Developments in Petroleum Science*, T.F. Yen and G.V. Chilingarian (Editors). Elsevier, Amsterdam, chap. 2.
Speight, J.G. 2000. *The Desulfurization of Heavy Oils and Residua*, 2nd edition. Marcel Dekker, New York.
Speight, J.G. 2001. *Handbook of Petroleum Analysis*. John Wiley & Sons, New York.
Speight, J.G. 2007. *The Chemistry and Technology of Petroleum*, 4th edition. CRC Press/Taylor & Francis, Boca Raton, FL.
Speight, J.G., Long, R.B., and Trowbridge, T.D. 1984. *Fuel*, 63: 616.
Speight, J.G. and Ozum, B. 2002. *Petroleum Refining Processes*. Marcel Dekker, New York.
Speight, J.G., Wernick, D.L., Gould, K.A., Overfield, R.E., Rao, B.M.L., and Savage, D.W. 1985. *Rev. Inst. Franc. Petrole*, 40: 27.

Suatoni, J.C. and Garber, H.R. 1975. *J. Chromatogr. Sci.*, 13: 367.
Wallace, D. and Carrigy, M.A. 1988. In *The Third UNITAR/UNDP International Conference on Heavy Crude and Tar Sands*, R.F. Meyer (Editor). Alberta Oil Sands Technology and Research Authority, Edmonton, Alberta, Canada.
Wallace, D., Starr, J., Thomas, K.P., and Dorrence, S.M. 1988. *Characterization of Oil Sand Resources*. Alberta Oil Sands Technology and Research Authority, Edmonton, Alberta, Canada.

3 Hydroprocessing Chemistry

Jorge Ancheyta and James G. Speight

CONTENTS

3.1 INTRODUCTION

Understanding refining chemistry not only allows an explanation of the means by which these products can be formed from crude oil, but also offers a chance of predictability. This is very necessary when the different types of crude oil accepted by refineries are considered. And the major processes by which these products are produced from crude oil constituents involve thermal decomposition. There have been many simplified efforts to represent refining chemistry that, under certain circumstances, are adequate to the task. However, refining is much more complicated than such representations would indicate (Speight, 2006 and references cited therein).

The chemical reactions that take place during petroleum upgrading can be very simply represented as reactions involving hydrogen transfer. In the case of hydrotreating, much of the hydrogen is supplied from an external source, and hydrogenation and various hydrogenolysis reactions consume the hydrogen with resulting reduction in the molecular weight of the starting material.

Bond energies offer some guidance about the preferential reactions that occur at high temperature and, for the most part, can be an adequate guide to the thermal reactions of the constituents of petroleum. However, it is not that simple. Often the bond energy data fail to include the various steric effects that are a consequence of complex molecules containing three-dimensional structures.

Furthermore, the complexity of the individual reactions occurring in an extremely complex mixture and the *interference* of the products with those from other components of the mixture are unpredictable. Or the interference of secondary and tertiary products with the course of a reaction, and hence with the formation of primary products, may also be cause for concern. Thus, caution is advised when applying the data from model compound studies to the behavior of petroleum, especially the molecularly complex heavy oils. These have few, if any, parallels in organic chemistry.

Thus, the different reactivity of molecules that contain sulfur, nitrogen, and oxygen can be explained by the relative strength of the carbon–sulfur, carbon–nitrogen, and carbon–oxygen bonds, with respect to the bond, in aromatic and saturated systems. This allows some explanation of the differences in reactivity toward hydrodesulfurization (HDS), hydrodenitrogenation (HDN), and hydrodemetallization (HDM), but it does not account for the reactivity differences due to stereochemistry and the interactions between different molecular species in the feedstock (Speight, 2000).

Petroleum refining is based on two premises: (1) hydrocarbons are less stable than the elements (carbon and hydrogen) from which they are formed at temperatures above 25°C (77°F), and (2) if reaction conditions ensure rapid reaction, any system of hydrocarbons tends to decompose into carbon and hydrogen as a consequence of its thermodynamic instability.

Furthermore, hydrodesulfurization chemistry is based on careful (or more efficient) hydrogen management. This involves not only the addition of hydrogen, but also the removal of (molecular) hydrogen sinks and the addition of a pretreatment step to remove or control any constituents that will detract from the reaction. This latter detraction can be partially resolved by the application of deasphalting, coking, or hydrotreating steps as pretreatment options.

It is therefore the purpose of this chapter to serve as an introduction to the chemistry involved in these conversion processes, so hydrocracking and hydrotreating processes are easier to visualize and understand.

3.2 THERMODYNAMIC ASPECTS

Thermodynamic aspects of hydroprocessing reactions are described in detail in Chapter 4. Only general aspects are mentioned in this section.

The thermodynamics of the hydrodesulfurization reaction have been evaluated from the equilibrium constants of typical desulfurization or partial desulfurization reactions such as:

1. Hydrogenation of model compounds to yield saturated hydrocarbons (R–H) and hydrogen sulfide (H_2S).
2. Decomposition of model compounds to yield unsaturated hydrocarbons R–CH=CH–R^1) and hydrogen sulfide (H_2S).
3. Decomposition of alkyl sulfides to yield thiols (R–SH) and olefins (R–CH=CH–R^1).

4. Condensation of thiols (R–SH) to yield alkyl sulfides ($R–S–R^1$) and hydrogen sulfide (H_2S).
5. Hydrogenation of disulfides ($R–S·S–R^1$) to yield thiols (R–SH, R^1–SH).

The logarithms of the equilibrium constants for the reduction of sulfur compounds to saturated hydrocarbons over a wide temperature range are almost all positive, indicating that the reaction can virtually proceed to completion if hydrogen is present in the stoichiometric quantity.

3.3 KINETICS OF HYDROPROCESSING

Kinetic studies using individual sulfur compounds have usually indicated that simple first-order kinetics with respect to sulfur are the predominant mechanism by which sulfur is removed from the organic material as hydrogen sulfide. However, there is still much to be learned about the relative rates of reaction exhibited by the various compounds present in petroleum.

The reactions involving the hydrogenolysis of sulfur compounds encountered in hydroprocessing are exothermic and thermodynamically complete under ordinary operating conditions. The various molecules have very different reactivity, with mercaptan sulfur much easier to eliminate than thiophene sulfur or dibenzothiophene sulfur.

The structural differences between the various sulfur-containing molecules make it impractical to have a single rate expression applicable to all reactions in hydrodesulfurization. Each sulfur-containing molecule has its own hydrogenolysis kinetics, which are usually complex because several successive equilibrium stages are involved and these are often controlled by internal diffusion limitations during refining.

Thiophenic compounds are the most refractory of the sulfur compounds. Consequently, thiophene is frequently chosen as representative of the sulfur compounds in light feedstocks. The hydrogenolysis of thiophene takes place according to two distinct paths. The first path leads through thiophane to butylmercaptan in equilibrium with butene and dibutylthioether, and finally to butene and hydrogen sulfide. It is considered unlikely that the thiophene and dibutylsulfide can undergo direct hydrogenolysis with production of hydrogen sulfide. However, it is possible that the butyl mercaptan can be decomposed according to the two parallel paths: (1) desulfurization of the mercaptan on the active metal sulfides and acid sites of alumina followed by hydrogenation of the intermediate butene, and (2) direct hydrogenolysis of the C–SH bond on the active metal sulfides.

As complex as the desulfurization of thiophene might appear, projection of the kinetic picture to benzothiphene and dibenzothiphene, and to their derivatives, is even more complex. As has already been noted for bond energy data, kinetic data derived from model compounds cannot be expected to include contributions from the various steric effects that are a consequence of complex molecules containing three-dimensional structures. Indeed, such steric effects can lead to the requirement of additional catalyst and process parameters for sulfur removal (Isoda et al., 1996a, 1996b).

Furthermore, the complexity of the individual reactions occurring in an extremely complex mixture and the interference of the products with those from other components of the mixture are unpredictable. The interference of secondary and tertiary products with the course of a reaction, and hence with the formation of primary products, may also be cause for concern. Thus, caution is advised when applying the data from model compound studies to the behavior of petroleum, especially the molecularly complex heavy oils. These have few, if any, parallels in organic chemistry. And all such contributions may be missing from the kinetic data, which must be treated with some degree of caution.

However, there are several generalizations that come from the available thermodynamic data and investigations of pure compounds as well as work carried out on petroleum fractions (Gray, 1994 and references cited therein). Thus, at room temperature hydrogenation of sulfur compounds to hydrogen sulfide is thermodynamically favorable and the reaction will essentially proceed to completion in the presence of a stoichiometric amount of hydrogen. Sulfides, simple thiophenes, and benzothiophenes are generally easier to desulfurize than the dibenzothiophenes and the higher molecular weight condensed thiophenes.

Nevertheless, the development of general kinetic data for the hydrodesulfurization of different feedstocks is complicated by the presence of a large number of sulfur compounds, each of which may react at a different rate because of structural differences as well as differences in molecular weight. This may be reflected in the appearance of a complicated kinetic picture for hydrodesulfurization in which the kinetics are not apparently first order (Scott and Bridge, 1971). The overall desulfurization reaction may be satisfied by a second-order kinetic expression when it can, in fact, also be considered as two competing first-order reactions. These reactions are (1) the removal of nonasphaltene sulfur and (2) the removal of asphaltene sulfur. It is the sum of these reactions that gives the second-order kinetic relationship.

In addition, the sulfur compounds in a feedstock may cause changes in the catalyst upon contact, and therefore, every effort should be made to ensure that the kinetic data from such investigations are derived under standard conditions. In this sense, several attempts have been made to accomplish standardization of the reaction conditions by presulfiding the catalyst passage of the feedstock over the catalyst until the catalyst is stabilized, obtaining the data at various conditions, and then rechecking the initial data by repetition.

Thus, it has become possible to define certain general trends that occur in the hydrodesulfurization of petroleum feedstocks. One of the more noticeable facets of the hydrodesulfurization process is that the rate of reaction declines markedly with the molecular weight of the feedstock (Scott and Bridge, 1971). For example, examination of the thiophene portion of a (narrow-boiling) feedstock and the resulting desulfurized product provides excellent evidence that benzothiophenes are removed in preference to the dibenzothiophenes and other condensed thiophenes. The sulfur compounds in heavy oils and residua are presumed to react (preferentially) in a similar manner.

It is also generally accepted that the simpler sulfur compounds (for example, thiols, R–SH, and thioethers, $R–S–R^1$) are (unless steric influences offer resistance to the hydrodesulfurization) easier to remove from petroleum feedstocks than the more complex cyclic sulfur compounds, such as the benzothiophenes). It should be noted here that, because of the nature of the reaction, steric influences would be anticipated to play a lesser role in the hydrocracking process.

Residua hydrodesulfurization is considerably more complex than the hydrodesulfurization of model organic sulfur compounds or, for that matter, narrow-boiling petroleum fractions. In published studies of the kinetics of residua hydrodesulfurization, one of three approaches has generally been taken:

1. The reactions can be described in terms of simple first-order expressions.
2. The reactions can be described by use of two simultaneous first-order expressions — one for easy-to-remove sulfur and another for difficult-to-remove sulfur.
3. The reactions can be described using a pseudo-second-order treatment.

Each of the three approaches has been used to describe hydrodesulfurization of residua under a variety of conditions with varying degrees of success, but it does appear that pseudo-second-order kinetics is favored. In this particular treatment, the rate of hydrodesulfurization is expressed by a simple second-order equation:

$$C/1 - C = k\ (1/LHSV)$$

where C is the wt% sulfur in product/wt% sulfur in the charge, k is the reaction rate constant, and LHSV is the liquid hourly space velocity (volume of liquid feed per hour per volume of catalyst).

Application of this model to a residuum desulfurization gave a linear relationship (Beuther and Schmid, 1963). However, it is difficult to accept that the desulfurization reaction requires the interaction of two sulfur-containing molecules (as dictated by the second-order kinetics). To accommodate this anomaly, it has been suggested that, as there are many different types of sulfur compounds in residua and each may react at a different rate, the differences in reaction rates offered a reasonable explanation for the apparent second-order behavior. For example, an investigation of the hydrodesulfurization of an Arabian light-atmospheric residuum showed that the overall reaction could not be adequately represented by a first-order relationship (Scott and Bridge, 1971). However, the reaction could be represented as the sum of two competing first-order reactions, and the rates of desulfurization of the two fractions (the oil fraction and the asphaltene fraction) could be well represented as an overall second-order reaction.

If each type of sulfur compound is removed by a reaction that was first order with respect to sulfur concentration, the first-order reaction rate would gradually, and continually, decrease as the more reactive sulfur compounds in the mix became depleted. The more stable sulfur species would remain, and the residuum would contain the more difficult-to-remove sulfur compounds. This sequence of events will presumably lead to an apparent second-order rate equation that is, in fact, a compilation of many consecutive first-order reactions of continually decreasing rate constant. Indeed, the desulfurization of model sulfur-containing compounds exhibits first-order kinetics, and the concept that the residuum consists of a series of first-order reactions of decreasing rate constant leading to an overall second-order effect has been found to be acceptable.

Application of the second-order rate equation to the hydrodesulfurization process has been advocated because of its simplicity and use for extrapolating and interpolating hydrodesulfurization data over a wide variety of conditions. However, while the hydrodesulfurization process may appear to exhibit second-order kinetics at temperatures near to 395°C (745°F), at other temperatures the data (assuming second-order kinetics) do not give a linear relationship (Ozaki et al., 1963).

On this basis, the use of two simultaneous first-order equations may be more appropriate. The complexity of the sulfur compounds tends to increase with an increase in boiling point, and the reactivity tends to decrease with complexity of the sulfur compounds; residua (and, for that matter, the majority of heavy oils) may be expected to show substantial proportions of difficult-to-desulfurize sulfur compounds. It is anticipated that such an approach would be more consistent with the relative reactivity of various sulfur compound types observed for model compounds and for the various petroleum fractions that have been investigated.

Other kinetic work has shown that, for a fixed level of sulfur removal, the order of a reaction at constant temperature can be defined with respect to pressure:

$$k = 1/\mathrm{LHSV}\ (P_h)^n$$

where P_h is the hydrogen partial pressure, LHSV is the liquid volume hourly space velocity, k is a constant, and n is the order of the reaction. It has been concluded, on the basis of this equation, that the hydrodesulfurization of residuum is first order with respect to pressure over the range of 800 to 2300 psi, although it does appear that the response to pressure diminishes markedly (and may even be minimal) above 1000 psi.

One marked effect of a hydrodesulfurization process is the buildup of hydrogen sulfide, and the continued presence of this reaction product in the reactor reduces the rate of hydrodesulfurization. Thus, using the two first-order models, the effect of hydrogen sulfide on the process can be represented as

$$k \backslash k_0 = 1/1 + k_1\ P_{H_2S}$$

where k is the rate constant with hydrogen sulfide present, k_0 is the rate constant in the absence of hydrogen sulfide, and k_1 is a constant.

Data obtained using this equation showed that a change in hydrogen sulfide concentration from 1 to 12% (by volume) could reduce by 50% the rate constants for the easy-to-desulfurize and the difficult-to-desulfurize reactions. On the basis of the data available from kinetic investigations, the kinetics of residuum hydrodesulfurization may be represented by the following general equation:

$$-ds/dt = [P_H{}^n/(1 + k_aA = k_s\ P_{H_2S})^m]k_iS_i$$

where S is the weight fraction of sulfur in the liquid phase, t is the residence time, P_H is the partial pressure of hydrogen, A is the weight fraction of asphaltenes in the liquid phase, P_{H_2S} is the partial pressure of hydrogen sulfide, S_i is the weight fraction of sulfur associated with component i in the range of i to j, k_a is the adsorption constant for asphaltenes, k_s is the adsorption constant for hydrogen sulfide, and k_i is the specific reaction rate constant for component i.

k_i in the above relationship is a function of the chemistry of the component, the catalyst activity, and the reaction temperature. Therefore:

$$k_i = k_0\ A/A_o\ e^{-E/RT}$$

where k_0 is the reaction rate constant at standard catalyst activity, A_o is the standard catalyst activity, A is the catalyst activity, E is the activation energy, R is the gas constant, and T is the absolute temperature.

This relationship gives activation energies for the hydrodesulfurization of various residua in the range of 27 to 35 kcal g-mol^{-1}. In this context, it was interesting to note that the deasphalting of Khafji residuum had no effect on the activation energy of 30 kcal g-mol^{-1}, and it was suggested that the activation energies of the various components in a particular residuum might be approximately the same.

In spite of all of the work, the kinetics and mechanism of alkyl-substituted dibenzothiophene, where the sulfur atom may be sterically hindered, are not well understood, and these compounds are in general very refractory to hydrodesulfurization. Other factors that influence the desulfurization process, such as catalyst inhibition or deactivation by hydrogen sulfide, the effect of nitrogen compounds, and the effect of various solvents, need to be studied in order to obtain a comprehensive model that is independent of the type of model compound or feedstock used.

The role of various elements in the supported catalyst requires additional work to fully understand their function. There is also a dearth of information dealing with the nature of the surface of the newer catalysts subjected to various pretreatments and exposed to model compounds such as alkyl-substituted dibenzothiophene or other feedstocks. There is a need to correlate those data from such characterization with kinetic and mechanistic studies.

Throughout this section, the focus has been on the kinetic behavior of various organic molecules during refinery operations, specifically desulfurization.

However, it must be remembered that the kinetic properties of the catalyst also deteriorate because of deposits on its surface. Such deposits typically consist of coke and metals that are products of the various chemical reactions.

3.4 HYDROGENATION

Sulfur removal, as currently practiced in the petroleum industry, can be achieved using three options. The first option involves the use of thermal methods such as the various cracking techniques that concentrate the majority of the sulfur into the nonvolatile products, that is, coke. Such processes are located in the *conversion* section of a refinery. The remainder of the sulfur may occur in the gases and as low-boiling organic sulfur compounds. The second option involves the use of chemical methods such as alkali treating (sweetening), as may be located in the product *finishing* section of a refinery. The third option is hydrodesulfurization, which occurs in the *conversion* (hydrocracking) *or finishing* (hydrotreating) section of the refinery. It is this latter process that will receive the major emphasis throughout our text as it relates to the processing of heavy oils and residua.

The purpose of hydrogenating petroleum constituents is to (1) improve existing petroleum products or develop new products or even new uses, (2) convert inferior or low-grade materials into valuable products, and (3) transform higher molecular weight constituents into liquid fuels.

The distinguishing feature of the hydrogenating processes is that, although the composition of the feedstock is relatively unknown and a variety of reactions may occur simultaneously, the final product may actually meet all the required specifications for its particular use (Furimsky, 1983; Speight, 2006).

Hydrogenation processes for the conversion of petroleum and petroleum products may be classified as *destructive* and *nondestructive*. The former (*hydrogenolysis* or *hydrocracking*) is characterized by the rupture of carbon–carbon bonds and is accompanied by hydrogen saturation of the fragments to produce lower-boiling products. Such treatment requires rather high temperatures and high hydrogen pressures, the latter to minimize coke formation. Many other reactions, such as isomerization, dehydrogenation, and cyclization, can occur under these conditions (Dolbear et al., 1987).

On the other hand, nondestructive, or simple, hydrogenation is generally used for the purpose of improving product (or even feedstock) quality without appreciable alteration of the boiling range. Treatment under such mild conditions is often referred to as *hydrotreating* or *hydrofining* and is essentially a means of eliminating nitrogen, oxygen, and sulfur as ammonia, water, and hydrogen sulfide, respectively.

The hydrodesulfurization process can fall into either the destructive or nondestructive category. However, for heavy feedstocks some hydrocracking is preferred, if not necessary, to remove the sulfur. Thus, hydrodesulfurization in this context falls into the hydrocracking or destructive hydrogenation category. The basic chemical concept of the process remains the same: to convert the organic sulfur in the feedstock to hydrogen sulfide.

$$\text{Feedstock}_{\text{sulfur}} + H_2 \wedge H_2S + \text{products}_{\text{sulfur-deficient}}$$

Although the definition of the two processes is purely arbitrary, it is generally assumed that *destructive hydrogenation* (which is characterized by the cleavage of carbon-to-carbon linkage and is accompanied by hydrogen saturation of the fragments to produce lower-boiling products) requires temperatures in excess of 350°C (660°F). *Nondestructive hydrogenation* processes are more generally used for the purpose of improving product quality without any appreciable alteration of the boiling range. Mild processing conditions (temperatures below 350°C or 660°F) are employed so that only the more unstable materials are attached. Thus, sulfur, nitrogen, and oxygen compounds eliminate hydrogen sulfide (H_2S), ammonia (NH_3), and water (H_2O), respectively. Unsaturated thermal products such as olefins ($R{\cdot}CH{=}CH{\cdot}R^1$) are hydrogenated to produce the more stable hydrocarbons ($R{\cdot}CH_2{\cdot}CH_2{\cdot}R^1$).

While the definitions of the various hydroprocesses are (as has been noted above) quite arbitrary, it may be difficult, if not impossible, to limit the process to any one particular reaction in a commercial operation. The prevailing conditions may, to a certain extent, minimize, say, cracking reactions during a hydrotreating operation. However, with respect to the heavier feedstocks, the ultimate aim of the operation is to produce as much low-sulfur liquid products as possible from the feedstock. Any hydrodesulfurization process that has been designed for application to the heavier oils and residua may require that hydrocracking and hydrodesulfurization occur simultaneously.

3.4.1 Hydrocracking

Hydrocracking is a thermal process (>350°C, >660°F) in which hydrogenation accompanies cracking. Relatively high pressure (100 to 2000 psi) is employed, and the overall result is usually a change in the character or quality of the products.

The wide range of products possible from hydrocracking is the result of combining catalytic cracking reactions with hydrogenation and the multiplicity of reactions that can occur. Dual-function catalysts in which the cracking function is provided by silica-alumina (or zeolite) catalysts, and platinum, tungsten oxide catalyze the reactions, or nickel provides the hydrogenation function.

Essentially all of the initial reactions of catalytic cracking occur, but some of the secondary reactions are inhibited or stopped by the presence of hydrogen. For example, the yields of olefins and the secondary reactions that result from the presence of these materials are substantially diminished, and branched-chain paraffins undergo demethanation. The methyl groups attached to secondary carbons are more easily removed than those attached to tertiary carbon atoms, whereas methyl groups attached to quaternary carbons are the most resistant to hydrocracking.

The effect of hydrogen on naphthenic hydrocarbons is mainly that of ring scission followed by immediate saturation of each end of the fragment produced. The ring is preferentially broken at favored positions, although generally all the carbon–carbon bond positions are attacked to some extent. For example, methylcyclopentane is converted (over a platinum–carbon catalyst) to 2-methylpentane, 3-methylpentane, and *n*-hexane.

Aromatic hydrocarbons are resistant to hydrogenation under mild conditions, but under more severe conditions the main reactions are conversion of the aromatic to naphthenic rings and scissions within the alkyl side chains. The naphthenes may also be converted to paraffins.

Polynuclear aromatic hydrocarbons are more readily attacked than the single-ring compounds, the reaction proceeding by a stepwise process in which one ring at a time is saturated and then opened. For example, naphthalene is hydrocracked over molybdenum oxide to produce lower molecular weight paraffins.

The presence of hydrogen changes the nature of the products (especially the coke yield). The yield of coke is decreased by preventing the buildup of precursors that are incompatible in the liquid medium and eventually form coke (Magaril and Aksenova, 1968, 1970; Magaril and Ramazaeva, 1969; Magaril et al., 1970, 1971; Speight and Moschopedis, 1979). In fact, the chemistry involved in the reduction of asphaltenes to liquids using models in which the polynuclear aromatic system borders on graphitic is difficult to visualize. However, the *paper chemistry* derived from the use of a molecularly designed model composed of smaller polynuclear aromatic systems is much easier to visualize (Speight, 2006). But precisely how asphaltenes react with the catalysts is open to much more speculation.

In contrast to the visbreaking process, in which the general principle is the production of products for use as fuel oil, hydroprocessing is employed to produce a slate of products for use as liquid fuels. Nevertheless, the decomposition of asphaltenes is, again, an issue, and just as models consisting of large polynuclear aromatic systems are inadequate to explain the chemistry of visbreaking, they are also of little value for explaining the chemistry of hydrocracking.

Deposition of solids or incompatibility is still possible when asphaltenes interact with catalysts, especially acidic support catalysts, through the functional groups, for example, the basic nitrogen species as they interact with adsorbents. And there is a possibility for interaction of the asphaltene with the catalyst through the agency of a single functional group in which the remainder of the asphaltene molecule stays in the liquid phase. There is also a less desirable option in which the asphaltene reacts with the catalyst at several points of contact, causing immediate incompatibility on the catalyst surface.

3.4.2 Hydrotreating

It is generally recognized that the higher the hydrogen content of a petroleum product, especially the fuel products, the better is the quality of the product. This knowledge has stimulated the use of hydrogen-adding processes in the refinery.

Thus, hydrogenation without simultaneous cracking is used for saturating olefins or for converting aromatics to naphthenes. Under atmospheric pressure, olefins can be hydrogenated up to about 500°C (930°F), but beyond this temperature dehydrogenation commences. Application of pressure and the presence of catalysts make it possible to effect complete hydrogenation at room or even cooler temperature; the same influences are helpful in minimizing dehydrogenation at higher temperatures.

A wide variety of metals are active hydrogenation catalysts; those of most interest are nickel, palladium, platinum, cobalt, iron, nickel-promoted copper, and copper chromite. Special preparations of the first three are active at room temperature and atmospheric pressure. The metallic catalysts are easily poisoned by sulfur- and arsenic-containing compounds, and even by other metals. To avoid such poisoning, less effective but more resistant metal oxides or sulfides are frequently employed, generally those of tungsten, cobalt, chromium, or molybdenum. Alternatively, catalyst poisoning can be minimized by mild hydrogenation to remove nitrogen, oxygen, and sulfur from feedstocks in the presence of more resistant catalysts, such as cobalt-molybdenum-alumina (Co-Mo-Al_2O_3).

3.5 MOLECULAR CHEMISTRY

In a mixture as complex as petroleum, the reaction processes can only be generalized because of difficulties in analyzing not only the products but also the feedstock, as well as the intricate and complex nature of the molecules that make up the feedstock. The concentration of the heteroatoms in the higher molecular weight and polar constituents is detrimental to process efficiency and catalyst performance (Speight, 1987; Dolbear, 1998).

There are a variety of sulfur-containing molecules in a residuum or heavy crude oil that produce different products as a result of the hydrodesulfurization reaction. Although the deficiencies of current analytical techniques dictate that the actual mechanism of desulfurization remain largely speculative, some attempt has been made to determine the macromolecular chemical concepts that are involved in the hydrodesulfurization of heavy oils and residua.

Under the usual commercial hydrodesulfurization conditions (elevated temperatures and pressures, high hydrogen-to-feedstock ratios, and the presence of a catalyst), the various reactions result in the removal of sulfur from the feedstock. Thus, thiols as well as open-chain and cyclic sulfides are converted to saturated or aromatic compounds, depending, of course, on the nature of the particular sulfur compound involved. Benzothiophenes are converted to alkyl aromatics, while dibenzothiophenes are usually converted to biphenyl derivatives. In fact, the major reactions that occur as part of the hydrodesulfurization process involve carbon–sulfur bond rupture and saturation of the reactive fragments (as well as saturation of olefins).

During the course of the reaction, aromatic rings are not usually saturated, even though hydrogenation of the aromatic rings may be thermodynamically favored. The saturation of some of the aromatic rings may appear to have occurred because of the tendency of partial ring saturation to occur prior to carbon–sulfur bond rupture, as is believed to happen in the case of dibenzothiophene derivatives. It is generally recognized that the ease of desulfurization is dependent upon the type of compound, and the lower-boiling fractions are desulfurized more easily than the higher-boiling fractions. The difficulty of sulfur removal increases in the following order:

Paraffins < naphthenes < aromatics

The wide ranges of temperature and pressure employed for the hydrodesulfurization process virtually dictate that many other reactions will proceed concurrently with the desulfurization reaction. Thus, the isomerization of paraffins and naphthenes may occur and hydrocracking will increase as the temperature and pressure increase. Furthermore, at the higher temperatures (but low pressures) naphthenes may dehydrogenate to aromatics and paraffins dehydrocyclize to naphthenes, while at lower temperature (high pressures) some of the aromatics may be hydrogenated.

These reactions do not all occur equally, which is due to some extent to the nature of the catalyst. The judicious choice of a catalyst will lead to the elimination of sulfur (and the other heteroatoms, nitrogen and oxygen), and although some hydrogenation and hydrocracking may occur, the extent of such reactions may be relatively minor. In the present context, the hydrodesulfurization of the heavier feedstocks (heavy oils and residua) may require that part or almost all of the feedstock be converted to lower-boiling products. If this be the case (as is now usual), hydrocracking will, of course, compete on an almost equal footing with the hydrodesulfurization reaction for the production of the low-sulfur, low-boiling products.

Thus, the hydrodesulfurization process is a very complex sequence of reactions due, no doubt, to the complexity of the feedstock. Furthermore, the fact that feedstocks usually contain nitrogen and oxygen compounds (in addition to metal compounds) increases the complexity of the reactions that occur as part of the hydrodesulfurization process. The nitrogen compounds that may be present are typified by pyridine derivatives, quinoline derivatives, carbazole derivatives, indole derivatives, and pyrrole derivatives. Oxygen may be present as phenols (Ar-OH; Ar is an aromatic moiety) and carboxylic acids ($-CO_2H$). The most common metals to occur in petroleum are nickel (Ni) and vanadium (V) (Reynolds, 1998).

In conventional crude oil, these other atoms (nitrogen, oxygen, and metals) may be of little consequence. Heavy oil contains substantial amounts of these atoms, and the nature of the distillation process dictates that virtually all of the metals and substantial amounts of the nitrogen and oxygen originally present in the petroleum will be concentrated in the residua.

The simultaneous removal of nitrogen during the processing is a very important aspect of hydrodesulfurization. Compounds containing nitrogen have pronounced deleterious effects on the storage stability of petroleum products, and nitrogen compounds in charge stocks to catalytic processes can severely limit (and even poison) the activity of the catalyst. Oxygen compounds are corrosive (especially the naphthenic acids) and can promote gum formation as part of the deterioration of the hydrocarbons in the product. Metals in feedstocks that are destined for catalytic processes can poison the selectivity of the catalyst and, like the nitrogen and oxygen compounds, should be removed.

Fortunately, the continued developments of the hydrodesulfurization process over the last two decades have resulted in the production of catalysts that can tolerate substantial amounts of nitrogen compounds, oxygen compounds, and metals without serious losses in catalyst activity or catalyst life. Thus, it is possible to use the hydrodesulfurization process as a means of producing not only low-sulfur liquid products, but also low-sulfur, low-nitrogen, low-oxygen, and low-metal streams that can be employed as feedstocks for processes where catalyst sensitivity is one of the process features.

Refining the constituents of heavy oil and bitumen has become a major issue in modern refinery practice (Gray, 1994). The limitations of processing heavy oils and residua depend to a large extent on the amount of higher molecular weight constituents (that is, asphaltenes) present in the feedstock (Speight, 1984; Ternan, 1983; LePage and Davidson, 1986) that are responsible for high yields of thermal and catalytic coke.

In many attempts to determine the macromolecular changes that occur as a result of hydrodesulfurization, the focus has been on the changes that occur in the bulk fractions, that is, the asphaltenes, resins, and oils. Furthermore, the majority of the attention has been focused on the asphaltene fraction because of the undesirable effects (for example, coke deposition on the catalyst) that this particular fraction (which is found in substantial proportions in residua and heavy oils) has on refining processes.

Indeed, the thermal decomposition of asphaltenes is a complex phenomenon insofar as the asphaltenes are not uniform in composition (Speight, 1994, 2006). Each asphaltene subfraction will decompose to give a different yield of thermal coke, and the thermal decomposition of asphaltenes also produces thermal coke with different solubility profiles that are dependent on the conditions (Speight, 1987).

Because of their high molecular weight and complexity, the asphaltenes remain an unknown entity in the hydrodesulfurization process. There are indications that, with respect to some residua and heavy oils, removal of the asphaltenes prior to the hydrodesulfurization step brings out a several-fold increase in the rate of hydrodesulfurization and that, with these particular residua (or heavy oils), the asphaltenes must actually inhibit hydrodesulfurization. As a result of the behavior of the asphaltenes, there have been several attempts to focus attention on the asphaltenes during hydrodesulfurization studies. The other fractions of a residuum or heavy oil (that is, the resins and the oils) have, on the other hand, been considered higher molecular weight extensions of heavy gas oils about which desulfurization is fairly well understood.

Nevertheless, studies of the thermal decomposition of asphaltenes can provide relevant information about the mechanism by which asphaltenes are desulfurized.

The thermal decomposition of petroleum asphaltenes has received some attention (Magaril and Aksenova, 1968, 1970; Magaril and Ramazaeva, 1969; Magaril et al., 1970, 1971; Schucker and Keweshan, 1980; Shiroto et al., 1983). Special attention has been given to the nature of the volatile products of asphaltene decomposition mainly because of the difficulty of characterizing the nonvolatile coke.

It has been generally assumed that the chemistry of coke formation involves immediate condensation reactions to produce higher molecular weight, condensed aromatic species. However, the initial reactions in the coking of petroleum feedstocks that contain asphaltenes involve the thermolysis of asphaltene aromatic–alkyl systems to produce volatile species (paraffins and olefins) and nonvolatile species (aromatics) (Speight, 1987, 1992, 1998, 2006; Roberts, 1989; Wiehe, 1992, 1994, 1996; Storm et al., 1994, 1997; Mushrush and Speight, 1995; Speight and Long, 1996; Schabron and Speight, 1997; Tojima et al., 1997). In addition, the rate parameters for thermal decomposition vary from feedstock to feedstock and differ even for asphaltenes derived from a similar origin (Neurock et al., 1991). Each feedstock must possess a unique set of rate parameters, which suggests that the molecular composition of the feedstock influences the thermal reactions, and general relationships, such as those derived from elemental analytical data, must be treated with caution and only deduced to be general trends.

An additional corollary to this work is that conventional models of petroleum asphaltenes (which, despite evidence to the contrary, invoked the concept of a large polynuclear aromatic system) offer little, if any, explanation of the intimate events involved in the chemistry of coking. Models that invoke the concept of asphaltenes as a complex solubility class with molecular entities composed of smaller polynuclear aromatic systems (Speight, 2006) are more in keeping with the present data. Indeed, such models are more in keeping with the natural product origins of petroleum (Speight, 1994) and with the finding that the hydrocarbon backbone of the polar fraction of petroleum is consistent with the hydrocarbon types in the nonpolar fractions (Wolny et al., 1997).

If the models, used as examples elsewhere (Speight, 1992, 1994, 1997), can be used as a guide to process chemistry, it should be anticipated that there might be some initial fragmentation and aromatization. The isolatable products from the early stages of asphaltene decomposition are more aromatic than the original asphaltenes and have a slightly reduced molecular weight. By definition and character (Speight, 1987), these are carbenes. The next step will be the more complete fragmentation of the asphaltenes to produce the carboids that are, by definition through solubility, the true precursors to coke.

Molecular breakdown commences at the extremities of both molecules to leave a polar core that become the eventually separates from the liquid reaction medium. In the case of the decomposition of the amphoteric molecule, thermal degradation could just as easily commence at the aliphatic carbon–sulfur bonds followed by thermal scission of the alkyl moieties from the aromatic systems. If the removal of the sulfur-containing moiety is not sterically hindered, this could well be the case, based on the relative strengths of aliphatic carbon–carbon and carbon–sulfur bonds. The end result would be the same. In the case of the neutral polar asphaltene molecule, the nonthiophene sulfur occurs between two aromatic rings, and therefore is somewhat stronger than the aliphatic carbon–sulfur bond, and the molecule at this point receives a degree of steric protection against thermal scission.

There is also the distinct possibility that there will be some reactions that occur almost immediately with the onset of heating. Such reactions will most certainly include the elimination of carbon dioxide from carboxylic fragments and, perhaps, even intermolecular coupling through phenolic moieties. There may even be prompt reactions that occur almost immediately. In other words, these are reactions that are an inevitable consequence of the nature of the asphaltene. An example of such reactions is the rapid aromatization of selected hydroaromatic rings to create a more aromatic asphaltene.

Whereas such reactions will obviously play a role in coking, the precise role is difficult to define. Decarboxylation and the reactions of phenol moieties are not believed to be a major force in the coking chemistry. Perhaps it is the immediate aromatization of selected (or all) hydroaromatic systems that renders inter- and intramolecular hydrogen management difficult and coke formation a relatively simple operation.

As these reactions are progressing, sulfur in aliphatic locations is released as, in the presence of hydrogen, hydrogen sulfide. Alternatively, the formation of hydrogen sulfide in the absence of added hydrogen suggests hydrogen abstraction by the sulfur atoms. The occurrence of sulfur or sulfur atoms in the reaction mixture would complicate the situation by participating in (or catalyzing) intermolecular condensation of the polynuclear aromatic systems.

In keeping with the known data, the sulfur in polynuclear aromatic systems is difficult to remove in the absence of hydrogen. The heterocyclic rings tend to remain intact and are incorporated into the coke. The presence of hydrogen facilitates the removal of heterocyclic sulfur. Indeed, there is also the possibility that the hydrogen, prior to cracking or hydrodesulfurization, may render sulfur removal and hydrocarbon production even more favorable.

REFERENCES

Beuther, H. and Schmid, B.K. 1963. *Proc. 6th World Petrol. Congr.*, 3: 297.

Dolbear, G.E. 1998. In *Petroleum Chemistry and Refining*, J.G. Speight (Editor). Taylor & Francis, Washington, DC, chap. 7.

Dolbear, G.E., Tang, A., and Moorehead, E.L. 1987. *Fuel*, 66: 267.

Furimsky, E. 1983. *Erdol Kohle*, 36: 518.

Gray, M.R. 1994. *Upgrading Petroleum Residues and Heavy Oils*. Marcel Dekker, New York.

Isoda, T., Nagao, S., Ma, X., Korai, Y., and Mochida, I. 1996a. *Energy Fuels*, 10: 482.

Isoda, T., Nagao, S., Ma, X., Korai, Y., and Mochida, I. 1996b. *Energy Fuels*, 10: 487.

LePage, J.F. and Davidson, M. 1986. *Rev. Inst. Franc. Petrole*, 41: 131.

Magaril, R.A. and Aksenova, E.I. 1968. *Int. Chem. Eng.*, 8: 727.

Magaril, R.A. and Aksenova, E.I. 1970. *Khim. Tekhnol. Topl. Masel.*, 7: 22.

Magaril, R.A. and Ramazaeva, L.F. 1969. *Izv. Vyssh. Ucheb. Zaved. Neft Gaz.*, 12: 61.

Magaril, R.L., Ramazaeva, L.F., and Askenova, E.I. 1970. *Khim. Tekhnol. Topl. Masel.*, 15: 15.

Magaril, R.Z., Ramazeava, L.F., and Aksenova, E.I. 1971. *Int. Chem. Eng.*, 11: 250.

Mushrush, G.W. and Speight, J.G. 1995. *Petroleum Products: Instability and Incompatibility.* Taylor & Franicis, Philadelphia.

Neurock, M., Nigam, A., Trauth, D., and Klein, M.T. 1991. In *Tar Sand Upgrading Technology*, AIChE Symposium Series 282, Vol. 87, S.S. Shih and M.C. Oballa (Editors). AIChE, p. 72.

Ozaki, H., Satomi, Y., and Hisamitsu, T. 1963. *Proc. 6th World Petrol. Congr.*, 6: 97.

Reynolds, J.G. 1998. In *Petroleum Chemistry and Refining*, J.G. Speight (Editor). Taylor & Francis, Washington, DC, chap. 3.

Roberts, I. 1989. Preprints. *Div. Petrol. Chem. Am. Chem. Soc.*, 34: 251.

Schabron, J.F. and Speight, J.G. 1997. *Rev. Inst. Franc. Petrole*, 52: 73.

Schucker, R.C. and Keweshan, C.F. 1980. Preprints. *Div. Fuel Chem. Am. Chem. Soc.*, 25: 155.

Scott and Bridge. 1971. In *Origin and Refining of Petroleum*, Advances in Chemistry Series 103, H.G. McGrath and M.E. Charles (Editors). American Chemical Society, Washington, DC, p. 113.

Shiroto, Y., Nakata, S., Fukul, Y., and Takeuchi, C. 1983. *Ind. Eng. Chem. Process Design Dev.*, 22: 248.

Speight, J.G. 1984. In *Catalysis on the Energy Scene*, S. Kaliaguine and A. Mahay (Editors). Elsevier, Amsterdam.

Speight, J.G. 1987. Preprints. *Div. Petrol. Chem. Am. Chem. Soc.*, 32: 413.

Speight, J.G. 1992. *Proceedings of the 4th International Conference on the Stability and Handling of Liquid Fuels*, DOE/CONF-911102. U.S. Department of Energy, p. 169.

Speight, J.G. 1994. In *Asphalts and Asphaltenes*, Vol. 1, T.F. Yen and G.V. Chilingarian (Editors). Elsevier, Amsterdam, chap. 2.

Speight, J.G. 1998. In *Petroleum Chemistry and Refining*, J.G. Speight (Editor). Taylor & Francis, Washington, DC, chap. 5.

Speight, J.G. 2000. *The Desulfurization of Heavy Oils and Residua*, 2nd edition. Marcel Dekker, New York.

Speight, J.G. 2006. *The Chemistry and Technology of Petroleum*, 4th edition. Marcel Dekker, New York.

Speight, J.G. and Long, R.B. 1996. *Fuel Sci. Technol. Int.*, 14: 1.

Speight, J.G. and Moschopedis, S.E. 1979. *Fuel Process. Technol.*, 2: 295.

Storm, D.A., Decanio, S.J., Edwards, J.C., and Sheu, E.Y. 1997. *Petrol. Sci. Technol.*, 15: 77.

Storm, D.A., Decanio, S.J., and Sheu, E.Y. 1994. In *Asphaltene Particles in Fossil Fuel Exploration, Recovery, Refining, and Production Processes*, M.K. Sharma and T.F. Yen (Editors). Plenum Press, New York, p. 81.

Ternan, M. 1983. *Can. J. Chem. Eng.*, 61: 133, 689.

Tojima, M., Suhara, S., Imamura, M., and Furate, A. 1997. Preprints. *Div. Petrol. Chem. Am. Chem. Soc.*, 42: 504.

Wiehe, I.A. 1992. *Ind. Eng. Chem. Res.*, 31: 530.

Wiehe, I.A. 1994. *Energy Fuels*, 8: 536.

Wiehe, I.A. 1996. *Fuel Sci. Technol. Int.*, 14: 289.

Wolny, R.A., Green, L.A., Bendoraitis, J.G., and Alemany, L.B. 1997. Preprints. *Div. Fuel Chem. Am. Chem. Soc.*, 42: 440.

4 Thermodynamics of Hydroprocessing Reactions

Syed Ahmed Ali

CONTENTS

4.1 INTRODUCTION

Cleaner fuels are vital in combating air pollution from automobile emissions. Legislation is being enacted in many countries mandating stringent sulfur and aromatics specifications for transportation fuels. One of the most versatile options to produce cleaner fuels is hydrotreating, and its role has been pivotal. Although hydrotreating has always been an important part of refinery operations, its role has been to remove objectionable elements such as sulfur and nitrogen from products of feedstocks by reacting them with hydrogen. In the past decade, hydrotreating capacity in the world refineries has been increasing at a steady rate as a response to the following three factors: (1) increasingly stringent environmental requirements for clean-burning fuels; (2) strong demand for transportation fuels and decreased demand for heavy fuel oil; and (3) the increased share of heavy and sour crudes (Ali, 1997). As a result, hydrotreating applications in refineries include a variety of streams, such as naphtha, light and heavy gas oils, and resids. These developments have brought hydrotreating to a level of economic importance matching catalytic cracking and reforming. Newer applications of hydrotreating include fluid catalytic cracking feed pretreatment, naphtha desulfurization and olefin saturation, deep desulfurization and dearomatization of straight-run and other middle distillate streams, and catalytic reformer feedstock pretreatment to obtain ultra-low-sulfur naphtha.

Several classes of reactions occur simultaneously during the hydrotreating of petroleum fractions, such as aromatic hydrogenation, hydrodesulfurization, and hydrodenitrogenation. The thermodynamics of each of these reactions must be considered to understand the overall chemical equilibria. The subject of thermodynamics of simultaneous multiple reactions in liquid phase containing a multitude of species is extremely complex and requires experimentally determined properties (Tassios, 1993). However, it is possible to make some generalizations about each class of reaction on the basis of chemical reaction equilibrium studies made with the pure (model) compounds (Le Page, 1987).

This chapter comprehensively reviews the available experimental thermodynamic data as well as the methods to estimate the data for three main hydrotreating reactions: aromatic hydrogenation, hydrodesulfurization, and hydrodenitrogenation. Available equilibrium data for these reactions are compiled and evaluated. In the experimental part of the study, the effect of methyl group addition on the two-ring aromatic hydrogenation was investigated.

4.2 THEORETICAL INVESTIGATIONS

4.2.1 Group Contribution Methods

Estimation methods for standard enthalpies, standard free energies, entropies, and ideal gas heat capacities involve group contributions based on molecular structure of the reactants and products (Ried et al., 1987). Some earlier correlations include the methods of Anderson et al. (1944), Souders et al. (1949), Bremmer and Thomas (1948), Franklin (1949), and van Krevelen and Chermin (1951). Recent techniques assign contributions of common molecular groupings (for example, $-CH_3$, $=CH_2$, $-NH_2-$, $-COOH$). By simple additivity, one can then estimate ideal gas properties from a table of group values (Ried et al., 1987). The method of Joback employs such an approach, which lists 41 atomic and molecular groupings to obtain group contributions for standard enthalpy of formation, standard Gibbs energy of formation, and polynomial coefficients, which relate heat capacities to temperature (Ried et al., 1987; Joback, 1984). This method is applicable to hydrocarbons as well as to organic compounds containing sulfur, nitrogen, oxygen, and halogen atoms.

In the more complicated methods, which are usually more accurate, atomic or molecular groups are chosen and allowance is made for next-nearest neighbors to this atom or group. The methods of Yoneda (1979) and Benson et al. (1969), which are illustrative of this type, can estimate standard enthalpy of formation, standard entropy, and polynomial coefficients to relate heat capacities to temperature, but cannot estimate standard Gibbs energy of formation. Another disadvantage of Benson's method is the necessity of determining the symmetry number of a compound, which requires three-dimensional models of compounds. Thinh and coworkers proposed a more accurate method for estimation of standard enthalpy of formation, standard Gibbs energy of formation, and polynomial coefficients to relate heat capacities to temperature (Thinh et al., 1971; Thinh and Throng, 1976). But this method is applicable only to hydrocarbons. Hence, *Thinh's method* can be used in thermodynamic calculations of aromatic hydrogenation (AH) reactions but not in hydrodesulfurization (HDS) and hydrodenitrogenation (HDN).

For the estimation of equilibrium constants of HDS and HDN reactions, *Joback's method* is more suitable, as it provides reliable estimates of thermodynamic properties of almost all the compounds of interest. Furthermore, it is much simpler to use and no symmetry or optical isomer corrections are necessary. However, caution should be exercised while applying Joback's method to very complex molecules, as this may lead to large errors. It should be noted that the uncertainty in the standard enthalpy of formation of hydrogenated aromatic hydrocarbons is typically as high as 0.75 kJ/mol (Ried et al., 1987; Stein et al., 1977). This results in uncertainty of more than one order of magnitude in the calculated equilibrium constants. Small errors in estimating Gibbs energy of formation are amplified when calculating equilibrium constant, as they are exponentially related.

Thermodynamic properties of hydrocarbons and other organic compounds required for calculating equilibrium constants in *liquid phase* are not readily available. However, as pointed out by Shaw et al. (1977), the gas phase equilibrium constants provide good approximations for the reactions occurring in the liquid phase because enthalpies and entropies of reaction, which affect the value of the equilibrium constant most strongly, change only slightly upon a change from gas to the liquid phase.

4.2.2 Thermodynamics of Aromatic Hydrogenation

The hydrogenation reactions of aromatic hydrocarbons are reversible and, at typical hydrotreating conditions, complete conversions are not possible owing to equilibrium limitations. The hydrogenation of an aromatic species, A, is given by

$$A + nH_2 \rightleftharpoons AH \tag{4.1}$$

where AH is the hydrogenated product (a naphthene). The equilibrium concentration of the aromatic species can be approximated by the following equation:

$$Y_A/(Y_A + Y_{AH}) = 1/(1 + K_a \times P_{H_2})^n \tag{4.2}$$

where Y_A and Y_{AH} are the mole fractions of the aromatic and naphthene species, respectively, K_a is the equilibrium constant, and P_{H2} is the partial pressure of hydrogen. In the derivation of this equation it is assumed that liquid activity coefficients and fugacities for A and AH are equal, and that the hydrogen activity coefficient and the ratio of fugacity to total pressure at hydrotreating conditions are both unity.

The hydrogenation of the aromatic ring, which is stabilized by mesomerism, is more difficult than that of other unsaturated hydrocarbons. Like all hydrogenation reactions, aromatic hydrogenation is also exothermic (Jaffe, 1974). When the aromatics are hydrogenated, the conjugate π bonds of an aromatic structure are destroyed and σ C–C bonds of the naphthenes are formed. This process requires less energy per mole than the cleaving of a σ C–C bond. Consequently, the heat released during saturation is larger than that associated with cleaving σ C–C bond linkages. Because of the exothermic nature of these reactions, the extent of aromatic hydrogenation at equilibrium decreases with increase in temperature. Thus, increasing the temperature to give higher rates of other reactions, such as HDS, results in lower equilibrium conversions in aromatic hydrogenation.

4.2.2.1 Equilibrium Constants and Standard Enthalpies

Table 4.1 lists equilibrium constants and standard enthalpies of a number of representative single-ring aromatic hydrogenation reactions and the calculated heat release associated with them. These data are calculated from experimentally determined values of standard enthalpy of formation, standard Gibbs free energy of formation, and polynomial coefficients to relate heat capacity to the temperature by previous researchers. It can be noticed that the amount of heat released per mole of aromatic hydrocarbon increases proportionately with the number of hydrogen moles consumed. However, the heat release per mole of hydrogen is fairly constant of all the reactions. Hence, it can be generally concluded that aromatic hydrogenation yields 58 to 70 kJ/mol of hydrogen (Ried et al., 1987; Jaffe, 1974).

If the π C–C bonds are not conjugated, as in alkenes, the heat release is much larger. Hence, the heat release from the saturating olefinic side chains of cyclic compounds, such as styrene, is about 120 kJ/mol of hydrogen. Hydrogenations of cyclohexenes produce slightly less heat (105 to 120 kJ/mol), as shown in Table 4.1.

4.2.2.2 Benzene and Its Homologues

Experimental data for calculation of equilibrium constants are available for benzene and its homologues (Reid et al., 1987). Representative values of equilibrium constants are presented in Table 4.1. In the hydrogenation of benzene homologues, the value of the equilibrium constant decreases with an increase in both the number of side chains and the number of carbon atoms in each side chain, as shown in Figure 4.1. These trends indicate that further analysis of data may lead to useful generalized correlations. For benzene homologues (benzene, toluene,

TABLE 4.1
Equilibrium Constants and Standard Enthalpies of Representative Single-Ring Aromatic Hydrogenation Reactions

	$\log_{10} K_{eq}$ at			
Reaction	**200°C**	**300°C**	**400°C**	$\Delta H^{\circ a}$
Benzene + $3H_2 \rightleftharpoons$ cyclohexane	3.94	0.13	–2.69	–206
Toluene + $3H_2 \rightleftharpoons$ methylcyclohexane	3.54	–0.19	–2.71	–205
Ethylbenzene + $3H_2 \rightleftharpoons$ ethylcyclohexane	3.17	–0.45	–3.07	–202
Propylbenzene + $3H_2 \rightleftharpoons$ propylcyclohexane	2.91	–0.75	–3.23	–201
Cumene + $3H_2 \rightleftharpoons$ isopropylcyclohexane	4.31	0.84	–1.35	–184
n-Butylbenzene + $3H_2 \rightleftharpoons$ *n*-butylcyclohexane	2.75	–0.88	–3.32	–199
Cyclohexene + $H_2 \rightleftharpoons$ cyclohexane	5.51	3.24	1.65	–118
Styrene + $H_2 \rightleftharpoons$ ethylbenzene	6.95	4.68	3.07	–118
Styrene + $4H_2 \rightleftharpoons$ ethylcyclohexane	10.13	4.23	0.20	–319
o-Xylene + $3H_2 \rightleftharpoons$ 1,2-dimethylcyclohexane	1.96	–1.53	–3.89	–191
p-Xylene + $3H_2 \rightleftharpoons$ 1,4-dimethylcyclohexane	2.16	–1.37	–3.75	–195
m-Xylene + $3H_2 \rightleftharpoons$ 1,3-dimethylcyclohexane	2.69	–0.99	–3.47	–202
1,2,3-TMB + $3H_2 \rightleftharpoons$ 1,2,3-trimethylcyclohexane[b]	2.74	–1.03	–3.57	–206
1,2,4-TMB + $3H_2 \rightleftharpoons$ 1,2,4-trimethylcyclohexane[b]	1.68	–1.99	–4.46	–201
1,3,5-TMB + $3H_2 \rightleftharpoons$ 1,3,5-trimethylcyclohexane[b]	2.00	–1.62	–4.06	–199
1,2,3,5-tetramethylbenzene + $3H_2 \rightleftharpoons$ 1,2,3,5-tetramethylcyclohexane[b]	1.18	–2.71	–5.34	–212

[a] Standard enthalpy of reaction in kJ/mol of aromatic reactant.
[b] Estimated by Thinh's method; TMB = trimethylbenzene.

ethylbenzene, *n*-propylbenzene, and *n*-butylbenzene), the equilibrium constant for aromatic hydrogenation at different temperatures is plotted against the carbon number in Figure 4.2. In this study, attempts were made to correlate the equilibrium constant with the carbon number using different types of equations, such as linear, polynomial, logarithmic, and exponential. It was found that the logarithmic type of correlation fits the data most accurately. The generalized empirical correlation thus obtained is

$$-\log K = [-1.8399 \log T_K + 13.716] \log C_n + [21.321 \log T_K - 139.33] \quad (4.3)$$

where T_K is the temperature in degrees Kelvin and C_n is the carbon number, which can be between 6 for benzene and up to 12 for *n*-hexylbenzene. This generalized correlation is valid for the entire temperature range of hydrotreating, that is, from 200 to 400°C. The accuracy of this generalized equation is within 3 to 5%. If the alkyl group is present, the hydrogenation is thermodynamically limited. But dealkylation during hydrotreating is undesirable, as it increases hydrogen consumption and forms low molecular weight products of lesser value.

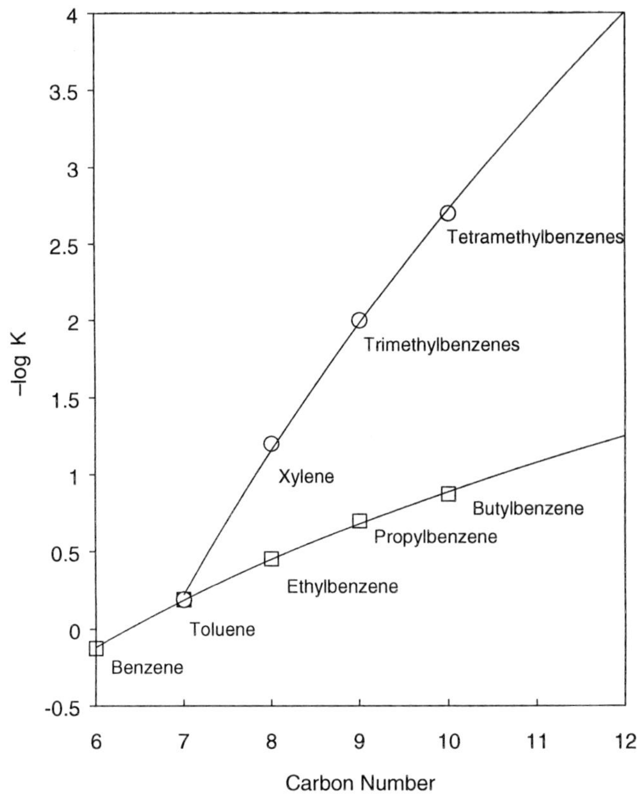

FIGURE 4.1 Equilibrium constants of benzene and its homologues at 300°C.

4.2.2.3 Multiring Aromatic Compounds

Table 4.2 lists equilibrium constants and standard enthalpies of a number of representative multiring aromatic hydrogenation reactions and the calculated heat release associated with them. For aromatic hydrocarbons containing more than one ring, hydrogenation proceeds via successive steps, with each of the hydrogenation reaction steps being reversible (Le Page, 1987; Ho, 1988; Sapre and Gates, 1982). Proposed pathways for three-fused-ring aromatics such as anthracene (Wiser et al., 1970) and phenanthrene (Lemberton and Guisnet, 1984) and four-fused-ring aromatics such as pyrene (Stephens and Kottenstette, 1985), fluorene (Lapinas et al., 1991), and fluoranthene (Lapinas et al., 1987) have been reported by different researchers. In general, the equilibrium constant is higher for the hydrogenation of the first ring than the latter rings. For example, the equilibrium constant for hydrogenation of naphthalene to tetralin at 325°C is 0.025, while that of tetralin to decalin is 0.0002.

One mole of polycyclic aromatics generates more heat than monoaromatics. For example, hydrogenation of naphthalene to decalin produces 320 kJ/mol

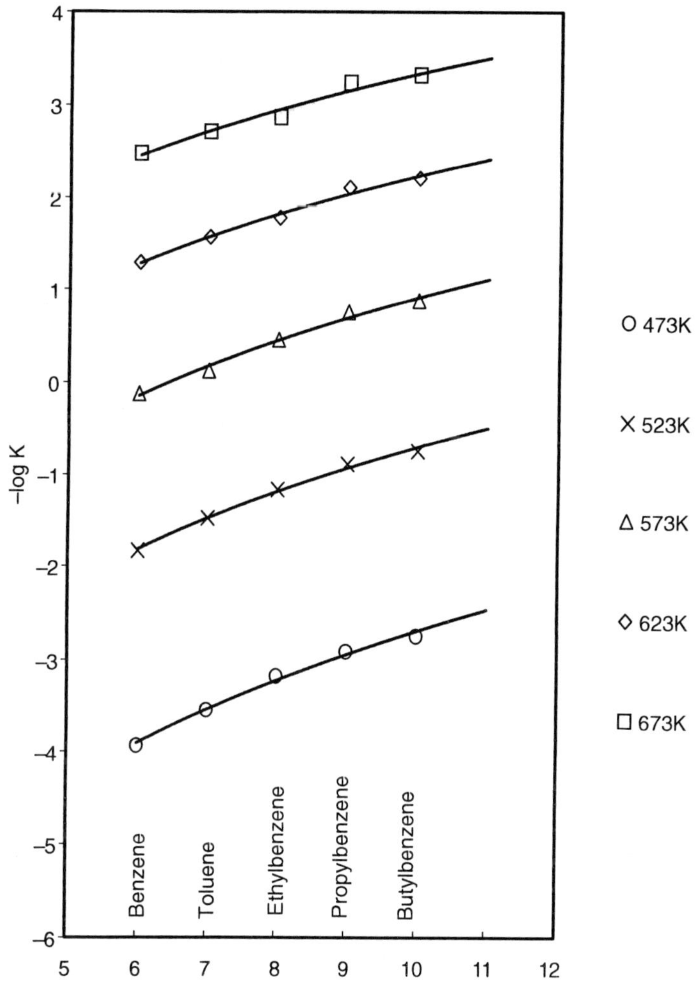

FIGURE 4.2 Equilibrium constants of benzene and its homologues at different temperatures.

of naphthalene. However, the heat release per mole of hydrogen consumed is still within the range of 58 to 67 kJ/mol. The hydrogenation of the first ring in multiring aromatics, such as napthalene to tetralin, produces about 70 kJ/mol of hydrogen consumed. On the other hand, the hydrogenation of the final ring, such as tetralin to decalin, generates only 64 kJ/mol of hydrogen consumed.

Few experimentally determined thermodynamic data are available for polyaromatic hydrocarbons and the products of their hydrogenation that allow useful calculations of equilibrium concentrations of partially hydrogenated products (Girgis and Gates, 1991). Frye and his coworkers (1962, 1969) measured the

TABLE 4.2
Equilibrium Constants and Standard Enthalpies of Representative Multiring Aromatic Hydrogenation Reactions

Reaction	$\log_{10} K_{eq}$ at 200°C	300°C	400°C	$\Delta H^{\circ a}$
Naphthalene + $2H_2$ ⇌ tetralin	1.26	−1.13	−2.80	−140
Tetralin + $3H_2$ ⇌ *trans*-decalin	0.74	−2.95	−5.56	−193
Naphthalene + $5H_2$ ⇌ *trans*-decalin	2.00	−4.08	−8.36	−333
Biphenyl + $3H_2$ ⇌ cyclohexylbenzene	3.13	−1.20	−4.25	−225
Cyclohexylbenzene + $3H_2$ ⇌ cyclohexylcyclohexane	2.47	−1.86	−4.91	−225
Biphenyl + $6H_2$ ⇌ cyclohexylcyclohexane	5.60	−3.06	−9.16	−450
Phenanthrene + H_2 ⇌ dihydrophenanthrene	−0.61	−1.57	−2.25	−50
Phenanthrene + $2H_2$ ⇌ tetrahydrophenanthrene	0.63	−1.79	−3.49	−100
Phenanthrene + $4H_2$ ⇌ octahydrophenanthrene	1.16	−3.64	−7.12	−251
Phenanthrene + $7H_2$ ⇌ perhydrophenanthrene	2.54	−6.02	−12.0	−439
Indene + H_2 ⇌ indane	4.34	2.50	1.20	−96
Indane + $3H_2$ ⇌ *cis*-hydrindane	1.24	−2.47	−5.07	−191
Acenaphthene + $2H_2$ ⇌ tetrahydroacenaphthene	0.61	−1.55	−3.07	−113
Acenaphthalene + $5H_2$ ⇌ perhydroacenaphthene	−0.65	−5.4	−8.74	−247
Fluorene + $3H_2$ ⇌ *cis*-hexahydrofluorene	0.54	−2.87	−5.27	−177
Fluorene + $6H_2$ ⇌ perhydrofluorene	−0.96	−7.54	−12.17	−351

[a] Standard enthalpy of reaction in kJ/mol of aromatic reactant.

compositions of some aromatic hydrocarbon–hydrogen mixtures in the vapor phase at several temperatures and pressures after equilibrium is attained. The hydrocarbons included biphenyl, indene, naphthalene, phenanthrene, acenaphthene, and fluorene. Representative results for multiring aromatic hydrogenations, presented in Table 4.2, show that hydrogenation equilibrium constants are less than unity at typical hydroprocessing temperatures of about 350°C for heavy feedstocks. The thermodynamic behavior becomes less favorable as the size of the aromatic molecule becomes larger.

The thermodynamic competition between hydrogenation of the first ring and final ring in three- and four-fused-ring aromatics depends on the temperature (Girgis and Gates, 1991). Since more moles of hydrogen are involved in the final ring hydrogenation than in the other rings (three moles compared to one or two for hydrogenation of phenathrene), the hydrogenation of the first ring is usually less thermodynamically favored than hydrogenation of the final ring at typical hydroreacting conditions (such as 350°C, 5 Mpa). Consequently, operation at high hydrogen partial pressure is mandatory to hydrogenate aromatic hydrocarbons to an appreciable extent. This is particularly so for reactions where the number of moles of hydrogen required to complete the ring saturation is high. This effect of pressure is in accordance with the Le Chartelier principle.

4.2.3 Thermodynamics of Hydrodesulfurization

The hydrodesulfurization of organo-sulfur compounds is exothermic. The heat of reaction varies significantly from one compound to an other. Table 4.3 lists the standard enthalpies of hydrodesulfurization of representative compounds (Gates et al., 1979; Speight, 1981). The amount of heat release increases with the number of moles of hydrogen required to desulfurize the compound. In addition, there is a marked difference between heat of reactions of different classes of sulfur compounds. For the hydrodesulfurization of mercaptans, the heat of reaction is about 55 to 70 kJ/mol of hydrogen consumed, while for thiophenes it is 63 to 66 kJ/mol of hydrogen consumed. The heat of reaction of hydrodesulfurization can increase the reactor temperature by 10 to 80°C at typical operating conditions, depending on the nature of the feedstock (Mounce and Rubin, 1971). Hence, it must be taken into consideration while designing the hydrodesulfurization reactors.

Equilibrium constants of representative hydrodesulfurization reactions of typical organo-sulfur compounds present in petroleum light distillates are calculated using the available experimental data (Ried et al., 1987) or by Joback's group contribution method (Joback, 1984), presented in Table 4.3. Notice that the

TABLE 4.3
Equilibrium Constants and Standard Enthalpies for the Hydrodesulfurization Reactions

	$\log_{10} K_{eq}$ at					
Reaction	**25°C**	**100°C**	**200°C**	**300°C**	**400°C**	**$\Delta H^{\circ a}$**
Mercaptans						
$CH_3\text{–}SH + H_2 \rightleftharpoons CH_4 + H_2S$	12.97	10.45	8.38	7.06	6.15	–72
$C_2H_5\text{–}SH + H_2 \rightleftharpoons C_2H_6 + H_2S$	10.75	8.69	6.99	5.91	5.16	–59
$C_3H_7\text{–}SH^b + H_2 \rightleftharpoons C_3H_8 + H_2S$	10.57	8.57	6.92	5.87	5.15	–57
$C_2H_5\text{–}SH^b \rightleftharpoons CH_2{=}CH_2 + H_2S$	–5.33	–2.57	–0.25	1.26	2.31	78
$C_3H_7\text{–}SH^b \rightleftharpoons CH_2{=}CH\text{–}CH_3 + H_2S$	–4.54	–2.18	–0.20	1.08	1.97	67
Thiophenes						
Thiophene + $4H_2 \rightleftharpoons n\text{-}C_4H_{10} + H_2S$	30.84	21.68	14.13	9.33	6.04	–262
3-Methylthiophene[b] + $4H_2 \rightleftharpoons$ 2-methylbutane + H_2S	30.39	21.35	13.88	9.11	5.82	–258
2-Methylthiophene[b] + $4H_2 \rightleftharpoons$ *n*-pentane + H_2S	29.27	20.35	13.33	8.77	5.66	–250
Thiophene + $2H_2 \rightleftharpoons$ tetrahydrothiophene[b]	10.51	6.47	3.17	1.12	–0.21	–116
Benzothiophenes						
Benzothiophene[b] + $3H_2 \rightleftharpoons$ ethylbenzene + H_2S	29.68	22.56	16.65	12.85	10.20	–203
Dibenzothiophene[b] + $2H_2 \rightleftharpoons$ biphenyl + H_2S	24.70	19.52	15.23	12.50	10.61	–148
Benzothiophene[b] + $H_2 \rightleftharpoons$ dihydrobenzothiophene[b]	5.25	3.22	1.55	0.49	–0.23	–58
Dibenzothiophene[b] + $3H_2 \rightleftharpoons$ hexahydrodibenzothiophene[b]	19.93	11.93	5.47	1.54	–0.98	–230

[a] Standard enthalpy of reaction in kJ/mol of organo-sulfur reactant.
[b] Thermodynamic properties of these compounds are calculated by Joback's method.

equilibrium constants for the hydrodesulfurization reactions involving conversion of sulfur compounds to saturated hydrocarbons over a wide range of temperatures are all positive. This indicates that the hydrodesulfurization reactions are essentially irreversible and can proceed to completion if hydrogen is present in stoichometric quantities under the reaction conditions employed industrially (for example, 250 to 350°C and 3 to 10 MPa). Generally, the equilibrium constants decrease with the increase in temperature, which is consistent with the exothermicity of the hydrodesulfurization reactions. The equilibrium constants approach values much less than unity only at temperatures considerably higher than those required in practice (>425°C).

Comparison of equilibrium constants of hydrodesulfurization of different classes of sulfur compounds indicates that there is a marked difference in the equilibrium constants from one class to another. However, within one class, such as mercaptans, the variation is very small and perhaps within the experimental error. The hydrodesulfurization of sulfur compounds to yield unsaturated hydrocarbons (alkenes) and hydrogen sulfide is not thermodynamically favored, as they are endothermic and will not proceed to completion at temperatures below 300°C.

Experimentally determined thermodynamics data for organo-sulfur compounds present in higher-boiling fractions (that is, multiring heterocyclics) are not widely available. The data available for dibenzothiophene hydrodesulfurization to give biphenyl (Table 4.3) indicate that it is also favored at temperatures representative of industrial practice and that the reaction is exothermic. Extrapolation of this result suggests that the hydrodesulfurization of higher molecular weight organo-sulfur compounds, such as benzonaphthothiophenes, is also thermodynamically favored. However, kinetic studies have shown that the hydrodesulfurization of multiring sulfur compounds proceeds much slower than that of single-ring compounds (Vrinat, 1983; Gates et al., 1979).

4.2.3.1 Hydrodesulfurization Reaction Pathways

Sulfur removal occurs either with or without hydrogenation of the heterocyclic ring. Givens and Venuto (1970) reported that a hydrogenation–dehydrogenation equilibrium was established between benzothiophene and 2,3-dihydrobenzothiophene during the hydrodesulfurization of the former at a rate faster than the sulfur removal. A similar result was obtained when the reactant was 2,3,7,-trimethylbenzo[*b*]thiophene. Hence, hydrogenated benzothiophenes (substituted and unsubstituted) are formed rapidly from benzothiophenes as intermediates in the hydrodesulfurization. In contrast, the hydrodesulfurization of dibenzothiophene may follow two routes: (1) hydrogenolysis of C–S bonds to give hydrogen sulfide and biphenyl and (2) hydrogenation of one of the benzoid rings followed by rapid hydrogenolysis of C–S bonds to give cyclohexyl benzene (Vrinat, 1983).

The reaction pathways involving prior hydrogenation of the heterocyclic ring are affected by the thermodynamics of reversible hydrogenation of sulfur-containing rings or benzoid rings, which is thermodynamic equilibrium limited at industrial hydrodesulfurization temperatures. For example, the equilibrium constant for

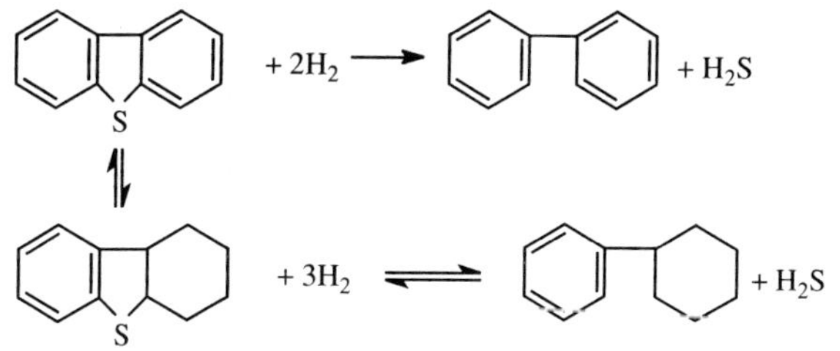

(4.4)

hydrogenation of thiophene to tetrahydrothiophene is less than unity at temperatures above 350°C (Vrinat, 1983). The cases of hydrogenation of benzothiophene to form dihydrobenzothiophene and of dibenzothiophene to form hexahydrodibenzothiophene are also similar, as shown in Table 4.3. Thus, sulfur removal pathways via hydrogenated organo-sulfur intermediates may be thermodynamically inhibited at high temperatures and low pressures because of the low equilibrium concentrations of the latter species.

4.2.3.2 Recombination Reaction

Recombination of alkenes and hydrogen sulfide was observed by Ali and Anabtawi (1995) while studying the deep desulfurization of naphtha. The product naphtha obtained at higher temperature contained more sulfur, specially as mercaptan sulfur, than the product obtained at lower temperature. It was inferred that higher temperature caused cracking, resulting in more alkenes, which recombined with the hydrogen sulfide downstream of the reactor and formed mercaptans. Analysis of sulfur compounds present in the product naphtha by gas chromatograph equipped with a flame ionization detector showed that methyl mercaptan is present (Anabtawi et al., 1995).

Thermodynamic analysis of the alkene–hydrogen sulfide recombination reaction was carried out for five reactions, as shown in Table 4.4. The experimentally

TABLE 4.4
Equilibrium Constants and Standard Enthalpies for Some Alkene–Hydrogen Sulfide Recombination Reactions

Reaction	$\log_{10} K_{eq}$ at 25°C	100°C	200°C	300°C	400°C	ΔH°[a]
$C_2H_4 + H_2S + H_2 \rightleftharpoons CH_3\text{–}SH + CH_4$	16.78	12.22	8.41	5.95	4.25	−130
$C_3H_6 + H_2S + H_2 \rightleftharpoons CH_3\text{–}SH + C_2H_6$	12.71	8.91	5.75	3.71	2.31	−108
1-$C_4H_8 + H_2S + H_2 \rightleftharpoons CH_3\text{–}SH + C_3H_8$	12.55	8.81	5.68	3.68	2.29	−107
$C_2H_4 + H_2S \rightleftharpoons C_2H_5\text{–}SH$	5.33	2.57	0.25	−1.26	−2.31	−78
$C_3H_6 + H_2S \rightleftharpoons C_3H_7\text{–}SH$	4.54	2.18	0.20	−1.08	−1.97	−67

[a] Standard enthalpy of reaction in kJ/mol.

determined thermodynamic properties of the compounds involved were reported by Reid et al. (1987). The values of standard enthalpies of reactions indicate that these reactions are exothermic. The amount of heat release is between 105 and 130 kJ/mol when methyl mercaptan is the product, and it is lower when heavier mercaptans are formed. The equilibrium constants are very high at lower temperatures, indicating the definite likelihood of these reactions in the downstream of reactors, such as in coolers, pipes, and heat exchangers. The methyl mercaptan-forming reactions are more favorable than other mercaptans. In fact, the formation of heavier mercaptans, such as propyl mercaptan, is not thermodynamically favored at temperatures above about 250°C. Hence, the thermodynamic analysis of these reactions confirms that the formation of methyl mercaptan is more favored than the formation of higher mercaptans, as observed in experimental studies.

4.2.4 Thermodynamics of Hydrodenitrogenation

The removal of undesirable nitrogen compounds from heavier petroleum fractions is best achieved by hydrodenitrogenation. Most of the nitrogen in petroleum fractions is present in heterocyclic compounds in the form of five- or six-member rings, nearly all of which are unsaturated. These compounds can be either basic or nonbasic. Pyridines and saturated heterocyclic ring compounds (indoline, hexahydrocarbazole) are generally basic, while pyrroles are not. The small quantities of nonheterocyclic nitrogen compounds present in petroleum fractions include anilines, aliphatic amines, and nitriles. These compounds are generally easier to denitrogenate by catalytic hydrogenation than the relatively unreactive heterocyclic compounds (Cocchetto and Satterfield, 1976).

Nitrogen removal from heterocyclic organo-nitrogen compounds requires hydrogenation of the ring containing the nitrogen atom before hydrogenolysis of the carbon–nitrogen bond can occur. Hydrogenation of the heteroring is required to reduce the relatively large energy of the carbon–nitrogen bonds in such rings, and thus permit more facile carbon–nitrogen bond scission. Nitrogen is then removed from the resulting amine or aniline as ammonia. The energies of carbon–nitrogen double and single bonds are 615 and 305 kJ/mol, respectively (Streitwiser and Heathcock, 1976).

The requirement that ring hydrogenation occur before nitrogen removal implies that the position of the equilibrium of the hydrogenation reactions can affect the nitrogen removal rates, if the rates of the hydrogenolysis reactions are significantly lower than the rates of hydrogenation. An unfavorable hydrogenation equilibrium results in low concentrations of hydrogenated nitrogen compounds undergoing hydrogenolysis. Consequently, hydrodenitrogenation rates will be lowered. However, high hydrogen pressure can be used to increase the equilibrium concentrations of saturated heteroring compounds (Girgis and Gates, 1991).

Previous studies have investigated the hydrodenitrogenation reaction mechanisms and reaction kinetics of different steps by conducting experiments using model compounds and actual feedstocks (Ho, 1988). Reaction mechanisms, based on experimental data, were proposed for the hydrodenitrogenation of pyrrole (Smith, 1957), pyridine (Hanlon, 1989), indole (Bhinde, 1979), quinoline (Satterfield and Cocchetto, 1981), and isoquinoline (Sonnemans et al., 1972). Some of the reaction networks, such as for quinoline, are complex and contain about 10 steps (Satterfield and Cocchetto, 1981). The reaction mechanisms are reviewed by Girgis and Gates (1991). Based on the proposed reaction mechanisms for the heterocyclic nitrogen compounds, the thermodynamic equilibrium constants can be calculated from the corresponding standard free energy changes.

However, little has been published about possible thermodynamic limitations of hydrodenitrogenation reactions, though this could have significant implications. Sonnemans et al. (1972) were the first to report the estimated thermodynamic equilibrium constants for pyridine hydrodenitrogenation:

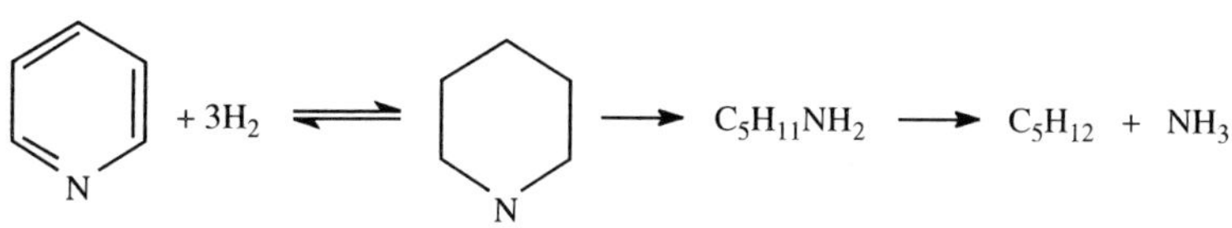

(4.5)

They found that the vapor phase hydrogenation to piperidine is reversible under representative hydrotreating operating conditions, and that the equilibrium favors pyridine at higher temperature. Satterfield and Cocchetto (1975) observed a maximum, followed by a decrease, in the extent of pyridine hydrodenitrogenation with an increase in temperature in the range of 200 to 500°C during the experiments at 1.1 MPa. This was associated with the thermodynamic equilibrium limitation. If pyridine hydrogenation in Equation 4.5 is rate limiting, the piperidine will undergo hydrogenolysis (cracking, step 2) as fast as it is formed, so that the position of the hydrogenation equilibrium does not influence the rate of hydrodenitrogenation. If, on the other hand, piperidine hydrogenolysis is rate limiting, the hydrogenation equilibrium can be established. Increased temperature shifts this equilibrium to the left (toward pyridine), decreasing the partial pressure of piperidine in the system. This can lower the rate of cracking, and therefore the rate of hydrodenitrogenation, resulting in a decrease in pyridine conversion with increasing temperature.

Illustrative values of equilibrium constants for different reactions in hydrodenitrogenation networks of some representative compounds are presented in Table 4.5. Equilibrium constants for the gas phase hydrogenation and hydrogenolysis reactions of several heterocyclic organo-nitrogen compounds have been calculated and reported in the literature (Cocchetto and Satterfield, 1976). The reactants included pyridine, quinoline, isoquinoline, acridine, pyrrole, indole, and carbazole. The equilibrium constants were calculated from the experimental values of standard Gibbs free energies of formation at 100 kPa for pyridine, pyrrole, their hydrogenated derivatives, and ammonia and are available in the literature (Reid et al., 1987). Group contribution methods were

TABLE 4.5
Equilibrium Constants and Standard Enthalpies for the Selected Hydrogenation, Hydrogenolysis, and Overall Hydrodenitrogenation Reactions

Reaction	$\log_{10} K_{eq}$ at 300°C	$\log_{10} K_{eq}$ at 400°C	ΔH°[a]
Pyrrole + $3H_2$ ⇌ pyroldine	–1.3	–2.8	–111
Pyroldine + H_2 ⇌ *n*-butylamine	2.2	1.3	–66
n-Butylamine + H_2 ⇌ *n*-butane + NH_3	9.3	8.2	–81
Pyrrole + $4H_2$ ⇌ *n*-butane + NH_3	10.0	6.1	–288
Indole + H_2 ⇌ indoline	–2.7	–3.3	–46
Indoline + H_2 ⇌ *o*-ethylaniline	4.7	3.3	–105
o-Ethylaniline+ H_2 ⇌ ethylbenzene + NH_3	5.8	5.0	–58
Indole + $3H_2$ ⇌ ethylbenzene + NH_3	7.8	5.0	–49
Carbazole + H_2 ⇌ *o*-phenylaniline	1.2	0.3	–66
o-Phenylaniline + H_2 ⇌ biphenyl + NH_3	5.8	5.0	–58
Carbazole +$2H_2$ ⇌ biphenyl + NH_3	6.8	5.1	–126
Pyridine + $3H_2$ ⇌ piperidine	–2.4	–5.1	–199
Piperidine + H_2 ⇌ *n*-pentylamine	1.3	0.8	–37
n-Pentylamine + H_2 ⇌ *n*-pentane + NH_3	9.9	8.7	–89
Pyridine + $5H_2$ ⇌ *n*-pentane + NH_3	8.9	4.4	–362
Quinoline + $2H_2$ ⇌ 1,2,3,4-tetrahydroquinoline	–1.4	–3.2	–133
Quinoline + $2H_2$ ⇌ 5,6,7,8-tetrahydroquinoline	–0.7	–3.0	–171
1,2,3,4-Tetrahydroquinoline + $3H_2$ ⇌ decahydroquinoline	–2.8	–5.4	–192
5,6,7,8-Tetrahydroquinoline + $3H_2$ ⇌ decahydroquinoline	–3.5	–5.6	–155
1,2,3,4-Tetrahydroquinoline + H_2 ⇌ *o*-propylaniline	4.3	3.0	–96
Decahydroquinoline + $2H_2$ ⇌ propylcyclohexane + NH_3	6.3	7.9	–117
o-Propylaniline+ H_2 ⇌ propylbenzene + NH_3	6.0	5.6	–29
Quinoline + $4H_2$ ⇌ propylbenzene + NH_3	7.0	3.3	–272
Isoquinoline + $2H_2$ ⇌ 1,2,3,4-tetrahydroisoquinoline	–3.0	–5.1	–155
Tetrahydroisoquinoline + H_2 ⇌ 1-ethyltoludine	3.2	2.6	–44
1-Ethyltoludine + H_2 ⇌ 1-ethyltoluene + NH_3	7.9	6.5	–103
Isoquinoline + $4H_2$ ⇌ 1-ethyltoluene + NH_3	8.1	4.1	–295

[a] Standard enthalpy of reaction in kJ/mol of organo-nitrogen reactant.

used to estimate the Gibbs free energies of formation of those compounds for which experimentally determined values are unavailable. Because the less accurate method of van Krevelen and Chermin (1951) was used for estimating the standard Gibbs free energies of the heterocyclics, errors in the equilibrium constants for ring saturation reactions are estimated to be two to three orders of magnitude, whereas those for C–N bond hydrogenolysis reactions were about one order of magnitude.

The equilibrium constants of all the reactions decrease with temperature, consistent with the fact that all the reactions are exothermic. Hydrogenolysis reactions, however, are favorable at temperatures as high as 500°C. The overall reactions are highly exothermic but favorable at temperatures up 400°C for both pyridine and pyrrole, while the equilibrium constants for the initial ring saturation steps, are favorable ($K > 1$, $\log_{10} K < 0$) above approximately 225°C.

The thermodynamics of multiring nitrogen compounds are analogous to those of single-ring compounds, that is, the equilibrium constants for the ring saturation steps are favorable only at low temperatures. Those for hydrogenolysis and overall hydrodenitrogenation reactions are favorable at temperatures up to 500°C. The exception to this generalization is the hydrodenitrogenation of carbazole (Cocchetto and Satterfield, 1976). The mechanism proposed for carbazole hydrodenitrogenation does not include an initial ring saturation step. Hence, the hydrodenitrogenation of carbazole produces biphenyl, similar to the hydrodesulfurization of benzothiophene. The equilibrium constants for hydrodenitrogenation reactions of pyridine, pyrrole, quinoline, isoquinoline, indole, and carbazole are also presented in Table 4.5.

The HDN of quinoline has been studied extensively by Satterfield and coworkers (Cocchetto and Satterfield, 1976; Satterfield et al., 1978; Cocchetto and Satterfield, 1981). Chemical equilibria among quinoline and its HDN reaction products have been experimentally determined. A detailed reaction network involving more than 10 steps was proposed. Equilibrium constants have been calculated as well as experimentally determined for the significant steps. The experimentally determined equilibrium constants are compared with those estimated by the group contribution method of van Krevelen and Chermin (1951) for hydrogenation reactions of quinoline. The agreement was between 1/2 and 1 1/2 orders of magnitude, which is within the experimental error. Considering the fact that the estimation was carried out with the less accurate group contribution method, the agreement was good. With the more recent group contribution method of Joback (1984), the accuracy of estimation did not improve much. For example, the logarithm of experimental equilibrium constant for quinoline hydrogenation to 1,2,3,4-tetrahydroquinoline was –3.3 at 420°C. By Joback's method it was estimated to be –2.1, and by van Kreleven and Chermin's method it was found to be –5.0.

4.3 EXPERIMENTAL INVESTIGATIONS

An attempt was made to determine the effect of methyl group addition on the thermodynamic equilibrium of aromatic hydrogenation. As the data for single-ring aromatic compounds were already available, the experiments were conducted for two-ring aromatic compounds, namely, the hydrogenation of naphthalene and 1-methyl naphthalene.

4.3.1 Equipment and Procedures

The experiments were performed in a 300-ml stainless steel batch autoclave reactor, which is equipped with a disperse mix agitator, thermocouple, and pressure transducer. A commercial hydrogenation catalyst was used, which was crushed to about 0.08 mm. Effectiveness factor calculations indicated that such a size was required to have negligible internal transfer resistance in the system. The catalyst required for each run was freshly sulfided with 5% hydrogen sulfide in hydrogen at 325°C for 8 h. High-purity hydrogen (99.99%), *n*-tetradecane (99% pure), naphthalene, and 1-methyl naphthalene were used. The experimental procedure consisted in loading the cleaned reactor with about 90 g of tetradecane, 10 g of reactant, and 1 g of sulfided catalyst. A stirrer speed of 500 rpm was maintained. Experiments were carried out under a hydrogen pressure of 150 kg/cm^2 at 325, 350, and 375°C, which are commercial hydrotreating conditions of diesel feedstocks.

Periodically, the liquid phase samples were taken out from the reactor from the porous stainless steel filter and analyzed by gas chromatography. Correction for the nonideality of hydrogen was made by using the activity coefficient of $1.000 + 3.30 \times 10^{-4} P_{H2}$ at all temperatures, where P_{H2} was in atmospheres. Thus, for example, calculated equilibrium constants for the hydrogenation reaction of naphthalene to tetralin were

$$K_p = \frac{(\text{moles tetralin})}{(\text{moles naphthalene})\left(P_{H_2} + 3.30 \times 10^{-4}\, P_{H_2}\right)} \tag{4.6}$$

4.3.2 Experimental Results and Discussion

The thermodynamic equilibrium of naphthalene was reported earlier by Frye and coworkers (1962, 1969). It was repeated in this study in order to gain confidence in obtaining the thermodynamic data. From Table 4.6, it can be noticed that the equilibrium constants for hydrogenation of naphthalene to tetralin obtained from this study were within ±26% of the values reported by Frye and Weitkamp (1969). Such an agreement of the results indicates that the technique applied in the present study can be reliably used to generate thermodynamic equilibrium data for other reactions. Table 4.6 also gives the values calculated by Thinh's method. It should be noted that Thinh's method is the most reliable means available for estimation of thermodynamic properties of the hydrocarbons (Reid et al., 1987). The values obtained by Thinh's method were within ±78% of those obtained by the experimental work of this study. Hence, in the absence of experimental values, Thinh's method can be used to assess the thermodynamic equilibrium within one order of magnitude.

The effect of a methyl group on the thermodynamic equilibrium is illustrated from the data obtained for 1-methyl naphthalene hydrogenation. Note that the presence of a methyl group on the two-ring aromatics decreased the thermodynamic constants by about 90% in the temperature range studied.

TABLE 4.6
Thermodynamic Equilibrium Data of Naphthalene and 1-Methyl Naphthalene Hydrogenation

Reaction/Method	$\log_{10} K_{eq}$ at 325°C	350°C	375°C
Naphthalene + $2H_2 \rightleftharpoons$ tetralin			
Thinh's method	–2.199	–2.658	–3.078
Frye's data	–1.600	–2.033	–2.433
This study	–1.707	–1.926	–2.258
Naphthalene + $3H_2 \rightleftharpoons$ decalins			
Thinh's method	–5.541	–6.695	–7.757
Frye's data	–5.295	–6.402	–7.423
This study	–5.441	–6.529	–7.207
1-Methyl naphthalene + $2H_2 \rightleftharpoons$ 1-methyl tetralin			
Thinh's method	–3.587	–3.909	–4.206
This study	–2.673	–3.068	–3.524
1-Methyl tetralin + $3H_2 \rightleftharpoons$ decalins			
Thinh's method	–7.913	–9.024	–10.049
This study	–5.992	–7.061	–8.286

Recall that a decrease of about 50% in thermodynamic equilibrium constant was reported for benzene and toluene hydrogenation to cyclohexane and methylcyclohexane, respectively (Table 4.1). The results of the present study indicate that the decrease in two-ring aromatics is much higher than in monoaromatics.

4.4 CONCLUDING REMARKS

The augmented roles of hydrotreating in the petroleum refineries demand better understanding of all the aspects of the process. Although hydrotreating is a well-established process, information about the thermodynamic aspects of hydrotreating reactions is rather limited and scattered. A comprehensive review of literature on thermodynamics of hydrotreating reactions is presented in this chapter. The areas covered include AH, HDS, and HDN, with particular emphasis on model compound data. Available experimental thermodynamic data to determine equilibrium concentrations are limited to a small number of compounds present in naphtha. Thermodynamic data for multicyclic compounds, which are present in large amounts in heavier petroleum fractions, are sparse. One of the principal reasons for the scarcity of extensive equilibrium data on multicyclic systems has been the lack of appropriate analytical techniques. The situation has changed now with the availability of very sophisticated gas chromatography–mass spectrometry (GC-MS), high-performance liquid chromatography (HPLC), and other instruments.

HDS and HDN reactions as such are not severely limited thermodynamically. However, with AH reactions it is difficult to reach completion at the process conditions normally required by hydrotreating over the conventional catalysts. Between the kinetics of hydrotreating using the conventional catalysts and the thermodynamics of AH, the available operating temperatures are within a narrow range of about 300 to 350°C. Hence, the strategy to maximize the yields of hydroaromatics by lower temperatures and moderate pressures depends on the role of the catalyst. As shown in Table 4.1 through Table 4.3, almost all hydrotreating reactions are thermodynamically favored at about 200 to 250°C and moderate pressure (3 to 5 MPa). Hence, there is a need for highly active catalysts that can facilitate significant reaction kinetics at around 200 to 250°C.

The experimental results indicate that the effect of the presence of a methyl group in two-ring aromatics is to limit the thermodynamic equilibrium conversion. The extent of this limitation was more than in monoaromatics. Hence, this study confirmed that the reported values for the single-ring AH cannot be reasonably applied to the two-ring aromatics, which are present in significant amounts in middle distillates.

There is a strong need for good thermodynamic data to determine reaction equilibria accurately, especially for the equilibrium-limited hydrogenation reactions. It would be of great value to obtain correlations of species at different temperatures and hydrogen partial pressures for representative compounds. Such correlations could be used to accurately determine hydrogen partial pressure effects on the equilibrium position at conditions representative of industrial applications. Since the hydrotreating reactions occur simultaneously, there is a need to compute multireaction equilibria. Such an attempt will elucidate the effects of one class of reaction on the thermodynamics of the other.

ACKNOWLEDGMENT

The authors acknowledge the support of the Research Institute of King Fahd University of Petroleum and Minerals.

REFERENCES

Ali, S.A. 1997. Ph.D. thesis, Hokkaido University, Sapporo, Japan.

Ali, S.A. and Anabtawi, J.A. 1995. *Oil Gas J.*, 93: 48.

Anabtawi, J.A., Alam, K., Ali, M.A., Ali, S.A., and Siddiqui, M.A.B. 1995. *Fuel*, 74:1254.

Anderson, J.W., Beyer, G.H., and Watson K.M. 1944. *Natl. Petrol. News*, 36:476.

Benson, S.W., Cruikshank, F.R., Golden, D.M., Haughen, G.R., O'Neal, H.E., Rodgers, A.S., Shaw, R., and Walsh, R. 1969. *Chem. Rev.*, 69:279.

Bhinde, M.V. 1979. Ph.D. dissertation, University of Delaware, Newark.

Bremmer, J.G.M. and Thomas, G.D. 1948. *Trans. Faraday Soc.*, 44:230.

Cocchetto, J. and Satterfield, C.N. 1976. *Ind. Eng. Chem. Pro. Des. Dev.*, 15:272.
Cocchetto, J.F. and Satterfield, C.N. 1981. *Ind. Eng. Chem. Pro. Des. Dev.*, 20:49.
Franklin, J.L. 1949. *Ind. Eng. Chem.*, 41:1070.
Frye, C.G. 1962. *J. Chem. Eng. Data*, 7:592.
Frye, C.G. and Weitkamp, A.W.J. 1969. *Chem. Eng. Data*, 14:372.
Gates, B.C., Katzer, J.R., and Schuit, G.C.A. 1979. *Chemistry of Catalytic Processes*. New York: McGraw Hill.
Girgis, M.J. and Gates, B.C. 1991. *Ind. Eng. Chem. Res.*, 30:2021.
Givens, E.N. and Venuto, P.B. 1970. Preprint. *Am. Chem. Soc. Div. Pet. Chem.*, 15:A183.
Hanlon, R.J. 1989. *Energy Fuels*, 1:424.
Ho, T.C. 1988. *Catalysis Rev. Sci. Eng.*, 30:117.
Jaffe, S.B. 1974. *Ind. Eng. Chem. Proc. Des. Dev.*, 13:34.
Joback, K.G. 1984. Sc.D. thesis, MIT, Cambridge, MA.
Lapinas, A.T., Klien, M.T., Gates, B.C., Marcis, A., and Lyons, J.E. 1987. *Ind. Eng. Chem. Res.*, 26:1026.
Lapinas, A.T., Klien, M.T., Gates, B.C., Marcis, A., and Lyons, J.E. 1991. *Ind. Eng. Chem. Res.*, 30:42.
Le Page, L.F. 1987. *Applied Heterogeneous Catalysis*. Paris: Technip.
Lemberton, J.L. and Guisnet, M. 1984. *Appl. Catalysis*, 13:181.
Mounce, E. and Rubin, R.S. 1971. *Chem. Eng. Prog.*, 87:81.
Reid, R.C., Prausnitz, J.M., and Poling, B.M. 1987. *The Properties of Gases and Liquids*. New York: McGraw Hill.
Sapre, A.V. and Gates, B.C. 1982. *Ind. Eng. Chem. Proc. Des. Dev.*, 21:86.
Satterfield, C.N. and Cocchetto, J.F. 1975. *AIChE J.*, 21:107.
Satterfield, C.N. and Cocchetto, J.F. 1981. *Ind. Eng. Chem. Prod. Des. Dev.*, 20:53.
Satterfield, C.N., Modell, M., Hites, R.A., and Declerck, C.J. 1978. *Ind. Eng. Chem. Proc. Des. Dev.*, 17:2.
Shaw, R., Golden, D.M., and Benson, S.W. 1977. *J. Phys. Chem.*, 81:1716.
Smith, H.A. 1957. In *Catalysis*, Emmet, P.H. (Ed.). New York: Reinhold, 5:231.
Sonnemans, J., Goudrian, F., and Mars, P. 1972. Presented at the Fifth International Congress of Catalysis, Paper 76. Palm Beach, FL.
Souders, M., Mathews, C.S., and Hurd, C.O. 1949. *Ind. Eng. Chem.*, 41:1037.
Speight, J.G. 1981. *The Desulfurization of Heavy Oil and Residua*. New York: Marcel Dekker.
Stein, S.F., Golden, D.M., and Benson, S.W. 1977. *J. Phys. Chem.*, 81:314.
Stephens, H.P. and Kottenstette, H.P. 1985. Preprint. *Am. Chem. Soc. Div. Fuel Chem.*, 30:345.
Streitwiser, A. and Heathcock, C. 1976. *Introduction to Organic Chemistry*. New York: Macmillan.
Tassios, D.P. 1993. *Applied Chemical Engineering Thermodynamics*. New York: Springer-Verlag.
Thinh, T.P., Duran, J.L., and Ramalho, R.S. 1971. *Ind. Eng. Chem. Proc. Des. Dev.*, 10:576.
Thinh, T.P. and Throng, T.K. 1976. *Can. J. Chem. Eng.*, 54: 344.
van Krevelen, D.W. and Chermin, H.A.G. 1951. *Chem. Eng. Sci.*, 1:66.
Vrinat, M.L. 1983. *Appl. Catalysis*, 6:137.
Wiser, W.H., Singh, S., Qader, S.A., and Hill, G.R. 1970. *Ind. Eng. Chem. Prod. Res. Dev.*, 9:350.
Yoneda, Y. 1979. *Bull. Chem. Soc. Jpn.*, 52:1297.

5 Reactors for Hydroprocessing

Jorge Ancheyta

CONTENTS

5.1 INTRODUCTION

Catalytic hydrotreating (HDT) is extensively applied in the petroleum refining industry to remove impurities, such as heteroatoms (sulfur, nitrogen, oxygen), polynuclear aromatics (PNAs), and metal-containing compounds (mainly V and Ni). The concentration of these impurities increases as the boiling point of the petroleum fraction does. S-, N-, O-, and PNA-containing compounds are found in low molecular weight feedstocks such as straight-run distillates (naphtha, kerosene, gas oil). High molecular weight feedstocks (vacuum gas oils, atmospheric and vacuum residua) contain the same impurities in higher concentrations, as well as complex V- and Ni-containing compounds and asphaltenes (Mochida and Choi, 2004).

Depending on the nature of the feed and the amount and type of the different heteroatoms, that is, different reactivities compounds, specific hydrotreating processes have been developed. The reactions occurring during hydrotreating are hydrodesulfurization (HDS), hydrodenitrogenation (HDN), hydrodeoxygenation

(HDO), hydrodearomatization (HDA), hydrodemetallization (HDM), and hydrodeasphaltenization (HDAs). In addition, the average molecular weight of the feed is lowered by hydrocracking (HDC), which can happen without a substantial loss in liquid product yield, as in the case of HDT of light distillates, or with moderate or severe reduction of molecular weight, such as in the case of heavy feeds. To accomplish the current and future stringent environmental regulations to produce the so-called clean fuels, for example, ultra-low-sulfur fuel, the extent of each reaction needs to be maximized either to obtain the final product or to prepare feeds for subsequent processes. To do that, researchers have focused their attention on the optimization of catalyst properties and composition and also on reactor and process design (Rana et al., 2007). Chapters 6 and 7 of this book are dedicated to describing the state of the art and future prospects in the area of hydrotreating catalysis, with special emphasis on heavy oils application; thus, catalysis aspects will not be discussed in the present chapter. As for reactor and process design, each process is individually optimized according to the nature and boiling range (that is, physical and chemical properties) of the feed to be hydrotreated, for which reaction conditions and reactor type and configuration are the most important features to be considered.

The severity of reaction conditions depends on the type of feed and the desired quality of the product. In general, the higher the boiling point of the feed, the higher the reaction severity. More details about the effect of reaction conditions on hydrotreating reactions' behavior will be discussed later.

In the case of reactor type and configuration, which is the main topic of this chapter, it should first be mentioned that reactors (as well as catalyst and reaction conditions) used for hydroprocessing of heavy feeds are different from those employed for hydrodesulfurization of light feeds. In general, commercial hydroprocessing reactors can be divided into three main groups: (1) fixed-bed (FBR), (2) moving-bed (MBR), and expanded- or ebullated-bed (EBR) reactors. The principles of operation of these three groups of reactors are very similar, but they differ in some technical details (Furimsky, 1998).

In the past, FBR hydrotreating reactors were exclusively utilized only for processing of light feeds, such as naphtha and middle distillates, but at present they are also used for hydroprocessing of heavier feeds, such as petroleum residua. However, when the feed contains a high amount of metals and other impurities, for example, asphaltenes, the use of FBR has to be carefully examined according to catalyst cycle life. Alternatively, MBR and EBR have demonstrated reliable operation with difficult feeds, such as vacuum residua. When hydroprocessing petroleum feeds, the life of the catalyst is crucial to retain its activity and selectivity for some time. Depending on the feed, the catalyst life may vary on the order of months or a year. It is then clear that the timescale of deactivation influences the choice of reactor (Moulijn et al., 2001).

The main objective of this chapter is to describe the major technical aspects of reactors employed for hydroprocessing of heavy oils and residua. The commercial technologies that use each type of reactor are also briefly mentioned; they are discussed with sufficient detail in Chapters 9 and 10.

5.2 REACTORS' CHARACTERISTICS

Hydroprocessing of heavy oils and residua is difficult because of the complex nature of the heteroatom-bearing molecules. Residue HDT processes are strongly influenced by the method of feed introduction, the arrangement of the catalytic beds, and the mode of operation of the reactors. Hence, proper selection and design of reactors are very important for this. Depending upon the nature of the residua to be treated, hydrotreating is generally carried out in fixed-bed, moving-bed, or ebullated-bed reactors. Sometimes a combination of different reactors is preferred (Schulman and Dickenson, 1991; Biasca et al., 2003). Figure 5.1 shows schematic representations of the reactors used for hydroprocessing of heavy oils (Morel et al., 1997).

In fixed-bed reactors, the liquid hydrocarbon trickles down through the fixed catalyst bed from the top to the bottom of the reactor. Hydrogen gas passes concurrently through the bed. A single tailor-made optimum catalyst having high hydrodesulfurization activity and low metal tolerance can handle feedstocks having less than 25 wppm metal for a cycle length of 1 year. For feedstocks containing metals in the range of 25 to 50 wppm, a dual-catalyst system is more effective. In such a system, one catalyst having a higher metal tolerance is placed in the front of the reactor, whereas the second catalyst, located in the tail end of the reactor, is generally of higher desulfurization activity (Scheffer et al., 1988; Kressmann et al., 1998). A triple-catalyst system consisting of an HDM, a balanced HDM/HDS, and a refining catalyst is generally used to handle feedstocks having metal content in the range of 100 to 150 wppm for an average cycle length of 1 year. For feeds with metal contents higher than this range, the HDM catalyst

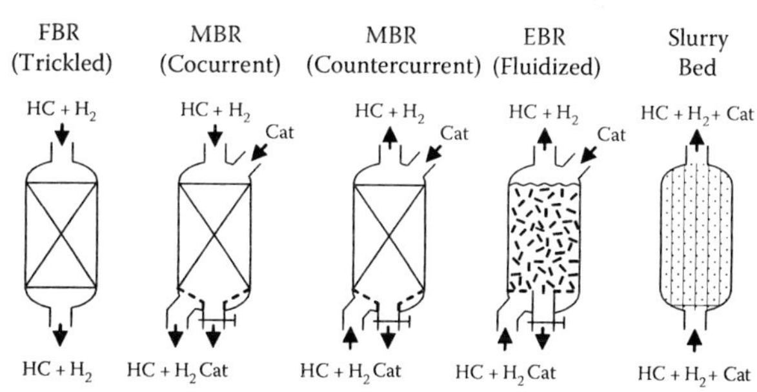

	FBR	SFB*	MBR	EBR	Slurry
Pressure, MPa	100-200	100-200	100-200	100-200	100-300
Temperature, °C	380-420	380-420	380-420	400-440	420-480
Unit LHSV, h^{-1}	0.1-0.5	0.1-0.5	0.1-0.5	0.2-1.0	0.2-1.0
Max. Conv. to 550°C-, wt%	50-70	60-70	60-70	70-80	80-95
Unit cycle length, months	6-12	Cont. opn.	12	Cont. opn.	Cont. opn.

* SFB: Swing Fixed-Bed

FIGURE 5.1 Different types of reactors used for hydroprocessing of heavy oils. (From Morel, F. et al., in *Hydrotreatment and Hydrocracking of Oil Fractions*, Froment, G.F., Delmon, B., and Grange, P., Eds., Elsevier, Amsterdam, 1997. With permission.)

quickly gets exhausted; for achieving a cycle length of 1 year, a new swing reactor fixed-bed concept has been introduced by the Institute Français du Petrole (IFP). The IFP process includes two swing fixed-bed HDM reactors that operate in a switchable mode, and the catalyst can be unloaded/reloaded without disturbing the operation of the system. The swing reactors are generally followed by various fixed-bed reactors in series containing HDS and other hydrofining catalysts.

To overcome the problems of handling feeds having metal contents higher than 150 wppm, besides the concept of fixed-bed swing reactors, another system (of moving-bed reactor) involving continuous withdrawal of deactivated catalyst and simultaneous addition of fresh catalyst has been introduced (Topsøe et al., 1996). The development of Shell bunker reactor technology is an example of such a system. In this bunker reactor system, a fresh catalyst and reacting fluids (both hydrocarbon and hydrogen) move in a cocurrent downflow mode and a special arrangement of internals inside the reactor allows withdrawal of the spent catalyst from the bottom of the reactor (Kressmann et al., 1998). In another design, as developed by Chevron in its On-Stream Catalyst Replacement (OCR) technology (Reynolds et al., 1992), the catalyst moves in a downflow mode while the reacting fluid moves countercurrently from the bottom to the top of the reactor. A similar system has been designed by the ASVAHL group (Agralwal and Wei, 1984). In all these cases, a combination of the moving-bed and fixed-bed systems is used. The moving-bed reactor containing the HDM catalyst is placed first, which is followed by other fixed-bed reactors containing HDS and hydrofining catalysts.

Another alternative for handling metal-rich feedstocks, besides the use of swing guard fixed-bed reactors and moving-bed reactors, is the use of ebullating-bed reactors. Reactants are introduced at the bottom and product is taken out from the top. The catalyst bed is kept expanded by the upward flow of reactants. The spent catalyst is continuously withdrawn and fresh catalyst is added without disturbing the operation, thereby providing a high on-stream factor.

More details about the three main reactors used for hydroprocessing are given in the following sections. The slurry phase reactor, which can also be used for hydroprocessing of heavy feeds, is also described.

5.2.1 Fixed-Bed Reactors

FBRs are the most used reactor systems in commercial hydrotreating operations. The reason is quite obvious, since they are easy and simple to operate. However, the simplicity of operation is limited to HDS of light feeds. For instance, in the case of naphtha hydrodesulfurization, the reaction is carried out in two phases (gas–solid) fixed-bed reactors since at the reaction conditions the naphtha is completely vaporized. However, for heavier fractions, that is when the distillation range of the feed increases, three phases are commonly found: hydrogen, a liquid–gas mixture of the partially vaporized feed, and the solid catalyst. This latter system is called the trickle-bed reactor (TBR), which is a reactor in which a liquid phase and a gas phase flow cocurrently downward through a fixed bed of catalyst particles while reactions take place (Rodriguez and Ancheyta, 2003). The name *trickle-bed reactor* originates from this flow pattern. The gas is the continuous phase and the liquid is the disperse phase (Quann et al., 1988). A schematic representation of the phenomena occurring in a TBR based on the three-film theory is presented in Figure 5.2 (Korsten and Hoffmann, 1996; Bhaskar et al., 2004). It is common to assume that mass transfer resistance in the gas film can be neglected and that no reaction occurs in the gas phase, so that for the reactions to occur, the hydrogen has to be transferred from the gas phase to the liquid phase, where concentration is in equilibrium with the bulk partial pressure, and then adsorbed onto the catalyst surface to react with other reactants. The gas reaction products are then transported to the gas phase while the main liquid hydrotreated reaction product is transported to the liquid phase.

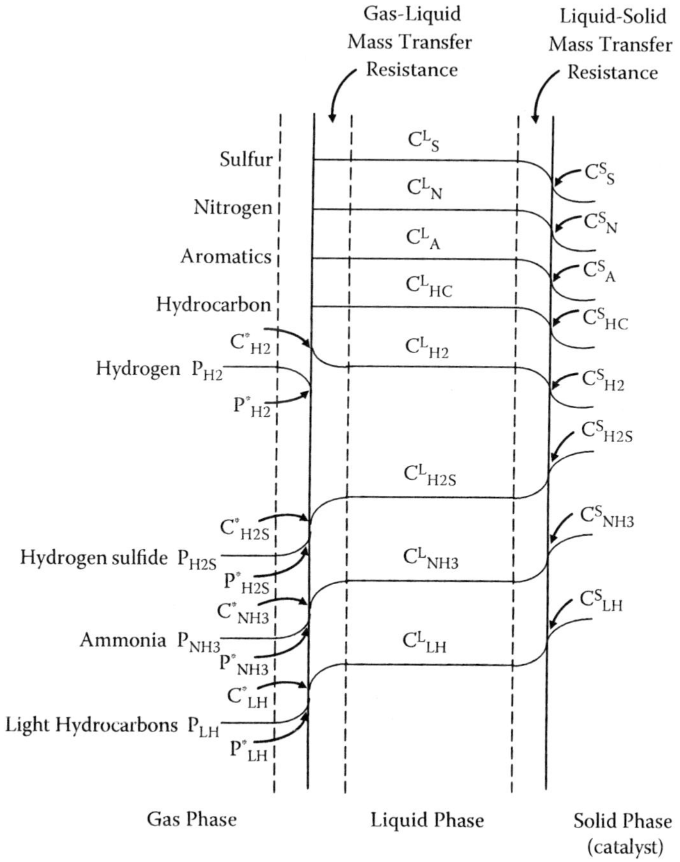

FIGURE 5.2 Concentration profiles in a hydrotreating trickle-bed reactor. (Adapted from Korsten, H. and Hoffmann, U., *AIChE J.*, 42, 1350–1357, 1996; Bhaskar, M. et al., *Ind. Eng. Chem. Res.*, 43: 6654–6669, 2004. With permission.)

TABLE 5.1
Comparison of Operating Conditions in Pilot and Commercial HDT Plants

Characteristics	Pilot Plant	Commercial Plant
Pressure, MPa	5.3	5.3
Temperature, K	613–653	633
LHSV, h^{-1}	1.0–2.5	2.5
H_2-to-oil ratio, ml ml^{-1}	356.2	356.2
Delta-T, °C	±2	15–20
Catalyst volume	75–150 mL	65.67 m^3
Liquid flow rate	75–375 mL/h	165,625 l/h
Catalytic bed length (L)	15–35 cm	9 m
Reactor inside diameter (D)	2.54 cm	3.048 m
Catalyst particle diameter (d_p), mm	2.3	2.3
L/d_p ratio	93–108	3913
D/d_p ratio	11.0	1325
Superficial liquid mass velocity (u_L), kg/m²sec	0.036–0.179	5.429
Superficial gas mass velocity (u_G), kg/m²sec	0.001–0.006	0.1859
Re in liquid phase (Re_L)	0.23–1.36	38.0
Re in gas phase (Re_G)	0.19–0.96	28.4
Pe in liquid phase (Pe_L)	0.02–0.16	0.255
Pe in gas phase (Pe_G)	3.61–5.72	0.112

Source: Ancheyta et al., *Energy Fuels,* 16: 1059–1067, 2002a. (With permission.)

Table 5.1 presents typical TBR conditions observed during operation of pilot and commercial hydrotreating plants. It is observed that superficial mass velocities and Reynolds numbers for both liquid and gas phases are always smaller in pilot reactors than in commercial reactors. For these reasons, low liquid velocities are used in small-scale reactors to match the liquid hourly space-velocity (LHSV) of commercial plants, which implies that gas–liquid and liquid–solid mass transfers are much better in commercial scale reactors. In addition, because of the lower resistance to liquid flow at the wall, the linear velocity next to the wall is greater than that found at the center of the reactor. This variation in linear velocity causes an increase in axial dispersion. The extent of this axial dispersion effect depends mainly on the bed length and conversion (Ancheyta et al., 2002a).

When processing naphtha, the main undesirable impurity is sulfur, and sulfur compounds are very easy to remove. That is the reason why only a hydrodesulfurization catalyst is loaded to an FBR. However, when processing straight-run gas oil (SRGO), the so-called refractory sulfur compounds are present (4,6-dimethylDBT and 4(or 6)-methylDBT), making deep hydrodesulfurization for ultra-low-sulfur diesel (ULSD) production hard to achieve. In addition, most of the time SRGO is blended with light cycle oil (LCO) from fluid catalytic cracking (FCC) units, and both are fed to the hydrotreater. These oil streams also contain high amounts of nitrogen and aromatics, which make the hydrotreating even more difficult, since both compete for catalytic active sites and consume high amounts of hydrogen (Ancheyta et al., 1999a). To face this problem, multibed systems with different catalysts have been proposed. Also, hydrogen is introduced between the beds as a quench because the reaction is exothermic. The heat released in light-feed HDT is relatively low, so that quenching is not necessary, and HDT units are designed with only one reactor containing a single catalyst bed. However, for heavier feeds multiple catalyst beds with cooling in between are used (Robinson and Dolbear, 2006). Multibed configuration with a hydrogen quench system is usually employed for hydrotreating of FCC feeds (a blend of heavy atmospheric gas oil and light and heavy vacuum gas oils) and heavier feedstocks.

Figure 5.3 shows an example of the widely used FBR with three catalytic beds and hydrogen quenches, in which feed distributor, quench zones, catalytic beds, and catalyst support are clearly indicated. Because the hydroprocessing reaction is always exothermic, controlling the reaction temperature is very important in order to prevent carbon deposition on the catalyst and preserve the quality of products at desired levels. The common method of controlling the reaction temperature is by combining the hot process fluids from the preceding bed with relatively cold hydrogen-rich gas before the mixture passes into the next bed, that is, between the catalytic beds, the so-called quench (Robinson and Dolbear, 2006). The problem when quenching with hydrogen is its availability in refineries. Hydrogen is also needed to keep the hydrogen-to-oil ratio (H_2/oil) along the reactors. Quenching with hydrogen takes place in some part of the reactor length and has two main functions: (1) control of reaction temperature and (2) improvement of flow distribution in the reactor bed. The quench zone is commonly a mixing chamber where the bed effluent is mixed with the hydrogen recycle stream (Muñoz et al., 2005). On the other hand, it is well known that a poor distribution design impacts on the catalyst utilization efficiency, and to minimize this problem, a liquid distributor should have high distribution element density, low sensitivity to tray out-of-levelness, low pressure drop, large spray angle, turndown flexibility, and be easy to clean (Chou, 2004).

Aiming at maximizing reactor volume utilization with superior performance, more complex reactor internal technologies have been developed. Reactor internals are extremely important for the safe, reliable, and profitable operation of a hydroprocessing unit and may have as large an effect on reactor performance as the catalyst choice. Spider Vortex™ technology, licensed by ExxonMobil, provides a better distribution of reactants and more uniform radial temperature (Figure 5.4a). Distribution of gas is done horizontally through several spokes to different parts of the quench deck. The specially designed gas–liquid redistribution equipment achieves small-scale contacting of the process gas and liquid and distributes both phases across the next catalyst bed. Radial temperature difference, that is, the maximum difference between the hottest and coldest temperatures along the horizontal thermowell at the bottom of a catalyst bed, is minimized by forcing the liquid to flow down two angled slides into a raceway, giving to the liquid some angular momentum, and the raceway gives it time to mix (Robinson and Dolbear, 2006). It has been reported that high radial delta-T (>8°C) indicates

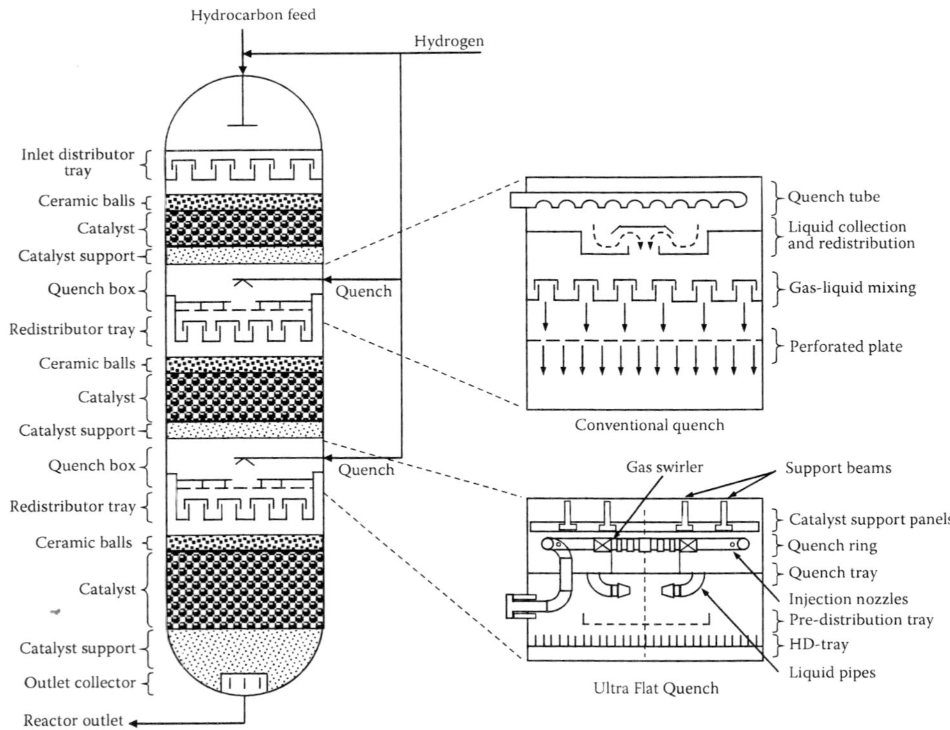

FIGURE 5.3 Fixed-bed reactor with multicatalytic bed and different quench technologies. (From Ouwerkerk, C.E.D. et al., *Petrol. Technol. Q.*, 4: 21–30, 1999; Minderhoud, J.K. et al., *Stud. Surf. Sci. Catal.*, 127: 3–20, 1999; Robinson, P.R. and Dolbear, G.E., in *Practical Advances in Petroleum Processing*, Hsu, Ch.H. and Robinson, P.R., Eds., Springer, New York, 2006, chap. 7. With permission.)

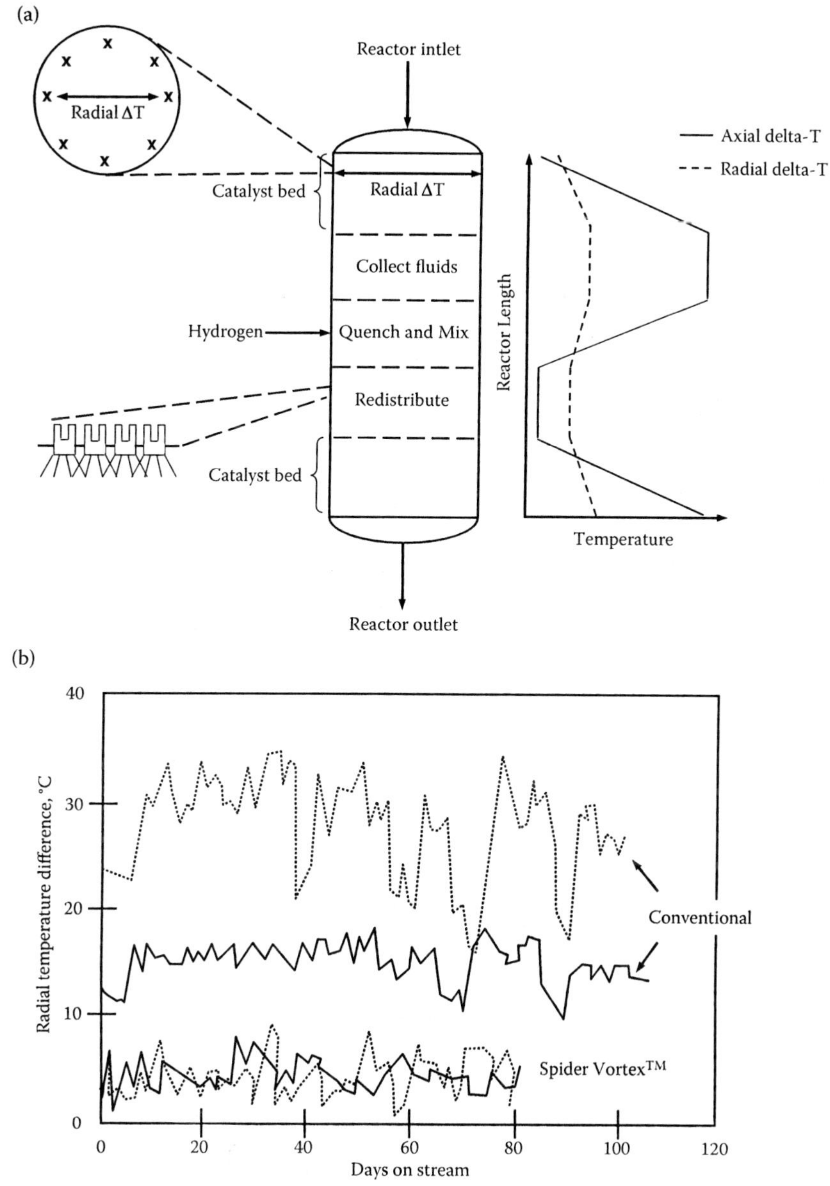

FIGURE 5.4 Spider Vortex reactor technology. (a) Fundamentals. (b) Radial temperature differences. —, catalyst bed inlet; ---, catalyst bed outlet. (From Sarli, M.S. et al., Improved Hydrocracker Temperature Control: Mobil Quench Zone Technology, paper presented at the 1993 NPRA Annual Meeting, Houston, TX, March 21–23; Hunter, M.G. et al., in *Handbook of Petroleum Refining Processes*, Meyers, R.A., Ed., McGraw-Hill, New York, chap. 7.1; EMRE, *Focus Catalysts*, 2: 5, 2006. With permission.)

poor gas/liquid flow distribution. The effects of the radial temperature maldistribution include (1) reactor runaways, (2) loss of cycle length, (3) processing limited to low-heat-release feedstocks, and (4) significant loss of product uplift (Hunter et al., 1997). Figure 5.4b shows an example of the radial temperature differences across the top and bottom of the catalyst bed. It is seen that radial temperature differences dropped from 8°C to about 3°C (top) and from 17 to 19°C to 3°C (bottom) with conventional quench design and the Spider Vortex internal design, respectively. These improvements in temperature distribution and catalyst utilization translate into better yields, longer catalyst life, and more efficient use of limited hydrogen resources (Sarli et al., 1993; Hunter et al., 1997). ExxonMobil Research and Engineering (EMRE) has announced recently that Petro-Canada has selected its commercially demonstrated Spider Vortex reactor technology to be installed in Montreal, Quebec, and Edmonton, Alberta, refineries, which will provide Petro-Canada the maximum yield of diesel product (EMRE, 2006).

The Ultra Flat Quench (UFQ) internal with a height of only 1 m, apart from providing good mixing, minimizes the height of the internal to maximize the amount of catalyst to be loaded in the reactor. Figure 5.3 also shows a schematic overview of a UFQ internal. The high performance of the UFQ has been well demonstrated in several commercial applications (Ouwerkerk et al., 1999). High-dispersion (HD) trays developed by Shell, also with proven commercial experience in hydroprocessing applications, contrary to conventional trays, pass vapor and liquid together through the nozzle, causing the acceleration of the liquid by the gas in the tube, which provokes an intimate mixing of both phases as well as almost complete distribution uniformity over the reactor cross section. HD trays have been reported to display a much better liquid distribution uniformity than conventional trays: >80% vs. 10 to 20% uniformity on top of the bed. Because of this excellent distribution at the top of the bed, there is no need for a layer of distributive packing above the catalyst bed, resulting in increased catalyst volume per reactor (Ouwerkerk et al., 1999; Minderhoud et al., 1999).

It should be mentioned that for gas phase reactors distribution trays are not necessary, but for TBR they are very important. The general benefits of using reactor internal technologies are (Sarli et al., 1993; Ouwerkerk et al., 1999; Gorshteyn et al., 2002):

1. To provide optimal conditions for catalyst performance in terms of liquid/gas distribution, inter- and intraphase mixing, quenching performance, and fouling resistance.
2. To be space efficient and easy to maintain.
3. Equal distribution of liquid over the catalyst to achieve maximum utilization of the inventory.
4. To avoid potential risks formed by stagnant zones and channeling.
5. To extend cycle length by reducing the average bed temperature or deactivation rate.
6. Improved product quality by reducing the risk of dry spots in the catalyst bed.

7. Enhanced reactant distribution and uniform radial temperatures.
8. Improved temperature control.
9. Design of large-diameter reactors.

On the other hand, in FBR (or TBR) the liquid and gas flow cocurrently downward through the catalytic bed, which has unfavorable hydrogen and hydrogen sulfide concentration profiles over the reactor, that is, high H_2S concentration at the reactor outlet, which provokes an inhibition of the removal of the last ppm sulfur compounds (Ancheyta et al., 1999b). A more suitable profile of H_2S concentration can be provided by operating the reaction in countercurrent mode, for instance, introducing the feed at the top and hydrogen at the bottom of the reactor. By this means, the possible recombination of hydrogen sulfide with olefins to form small amounts of mercaptans (rebuilding sulfur-containing molecules: $H_2C{=}CH_2 + H_2S \leftrightarrow HS{-}CH_2CH_3$) at the reactor outlet is avoided, since H_2S is removed from the top of the reactor (Babich and Moulijn, 2003). For instance, if H_2S is not removed in naphtha hydrodesulfurization units, mercaptans may cause problems for downstream catalytic reforming units. Also, the inhibition effect of H_2S in hydrotreating reactions, for example, hydrodesulfurization, can be minimized since the final deep HDS is carried out at low hydrogen sulfide concentration. The reactor employed in SynSat Technology, which combines Criterion's SynSat catalysts and ABB Lummus's reactor technologies, is an example of this approach (Langston et al., 1999). In the SynSat reactor the hydrogen sulfide is removed interstage to prevent the hydrogen sulfide formed in the first part of the reactor from passing through to the final part of the reactor. A very low H_2S partial pressure and low-temperature hydrogenation are enabled by applying a gas/liquid countercurrent operating with intake of fresh hydrogen gas in the bottom reactor. Another reactor employing this concept is that proposed by Mochida et al. (1996), in which two fractions of the feed (light and heavy fractions) react separately in upper and lower parts of the catalyst bed. The feed is introduced at about two thirds reactor height in between the top bed and a middle bed of the catalyst. Hydrogen is charged from the bottom of the reactor so that hydrogen sulfide inhibition on the heavier fraction hydrotreating can be avoided (Mochida and Choi, 2006). This reactor combines the characteristics of low-H_2S-partial-pressure final-stage hydrotreating, countercurrent operation, and catalytic distillation. One more application of this approach considers the use of a stacked bed of two catalysts, in which the final deep HDS or final deep hydrogenation step is performed at low-H_2S and high-H_2 partial pressures. Hydrocarbon feed is fed at the middle of the reactor in the downflow mode through a conventional HDS catalyst in the bottom bed, while hydrogen is fed at the top of the reactor to the top catalyst bed. The bottom bed catalyst partly desulfurizes the feed at high-H_2S partial pressure, which is then separated from the gas phase and is fed back together with hydrogen to the catalyst in the top bed, where deep HDS or deep hydrogenation is performed at low-H_2S and high-H_2 partial pressures (Sie and de Vries, 1993). A schematic representation of the reactors used in these three approaches is shown in Figure 5.5. The reduction of the hydrogen sulfide partial pressure in the last part of the reactor obtained with these approaches is also exemplified in this figure relative to a conventional downflow reactor operation.

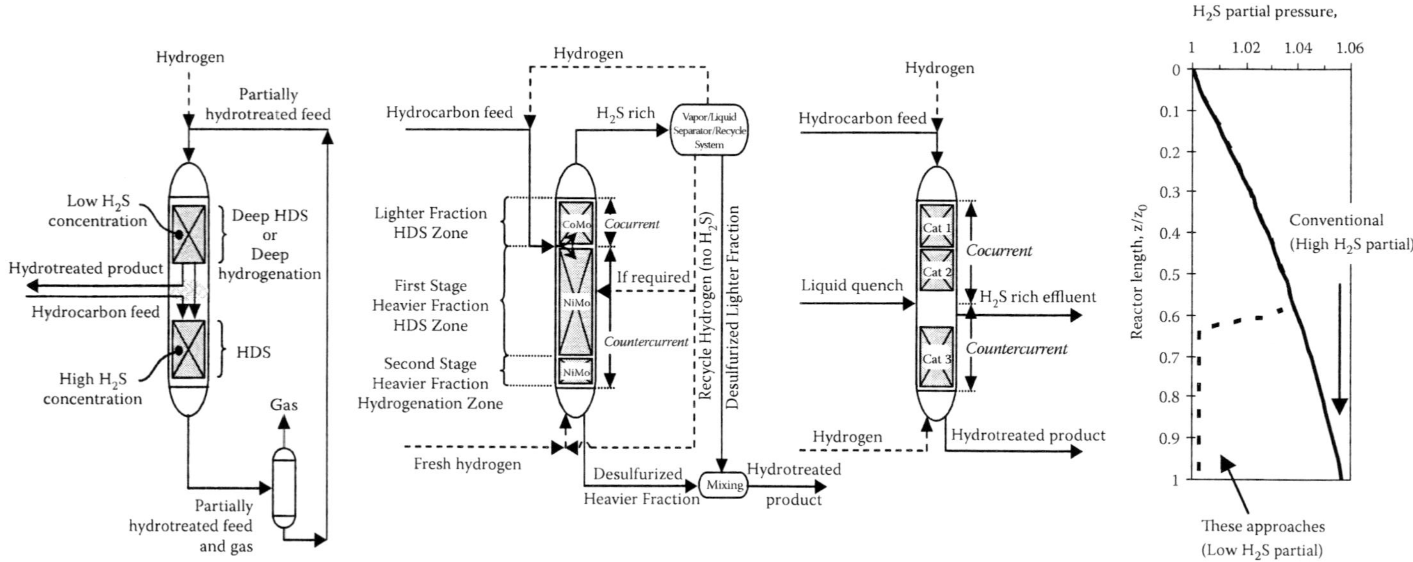

FIGURE 5.5 Different approaches of reduction of H_2S partial pressure at the final stage of hydroprocessing. (Sie, S.T. and de Vries, A.F. Hydrotreating Process. European Patent Application 0553920, 1993; Mochida et al. *Catal. Today,* 29: 185–189, 1996; Langston, J., Allen, L., and Davé, D. *Petrol. Technol. Q.*, 2: 65–69, 1999. With permission.)

During hydroprocessing of heavy oils and residua using FBR (with either single-catalyst systems or multiple-catalyst systems, in one or more reactors in series, and sometimes several parallel trains of several reactors in series), it is common to use guard materials at the top of the reactor or in a separate vessel before the main hydrotreating reactors to catch foulants. Other particulates present in the feed can also settle in the catalysts causing catalytic bed plugging. As a result, pressure drop increases and reactor performance declines, which ends up in hydrotreating plant shutdown. Pressure drop in hydroprocessing units is usually the result of the accumulation of a dense layer of particulates or the formation of gums from reactive species in the feed. In other words, delta-P develops when the particulates layer reduces the void fraction in the catalyst bed during time-on-stream. The common way to avoid this problem is to use filters to protect the catalyst beds. This solution is partial since filters are ineffective in removing complex polymeric iron sulfide gums (formed from reaction of soluble iron compounds, for example, iron porphyrins with sulfur in the feed, or by their dissolution by naphthenic acid) and particulates smaller than 5 to 20 microns, which may pass through the catalyst beds. Another solution is the use of layers of highly macroporous materials graded on the top of the FBRs, whose main objective is the protection of the catalyst bed from fouling due to the previously mentioned factors. The problem becomes more complicated when hydroprocessing heavy and extra-heavy feeds. For such cases, proper bed grading can be used to increase cycle length and prevent premature shutdown. The control of particulates is done by using large-diameter and large-void-fraction catalysts to spread out the zone of deposition of particulates. Each layer of catalyst, with particular shape and size, collects particulates within a certain size range and prevents formation of a dense layer of particulate accumulation, so that pressure drop due to gum formation is prevented. Macroporous materials and bed grading substantially reduce the pressure drop problems, as illustrated in Figure 5.6, and as a consequence extend the life of the catalyst bed and eliminate the need for frequent shutdown to skim the head part of the bed to remove crust, plugs, and agglomerates of catalyst particles, and replace fouled catalyst. The main disadvantage of using bed grading is the loss of reactor volume for loading the catalyst, since a part of the grading material can be inert. However, the use of materials with some catalytic activity mitigates these effects. Hence, finding a right balance between fouling prevention and preservation of sufficient overall catalyst activity is the key issue in optimizing bed grading (Minderhoud et al., 1999).

When the feed exhibits high amounts of metals and asphaltenes, the Quick Catalyst Replacement (QCR) reactor, developed by Royal Dutch Shell, is a good option for hydroprocessing of heavy oils and residua. The main characteristic of the QCR reactor is its conical support grids instead of the common horizontal ones (Van Zijll Langhout et al., 1980). This special design allows the use of different catalysts at the same time. For instance, to protect high-activity catalysts located at the bottom of the reactor, catalysts having large pore diameter with large metal storing capacity are used in the top section. QCR avoids the opening and closing of the reactors and shortens the cooling and heating-up periods as

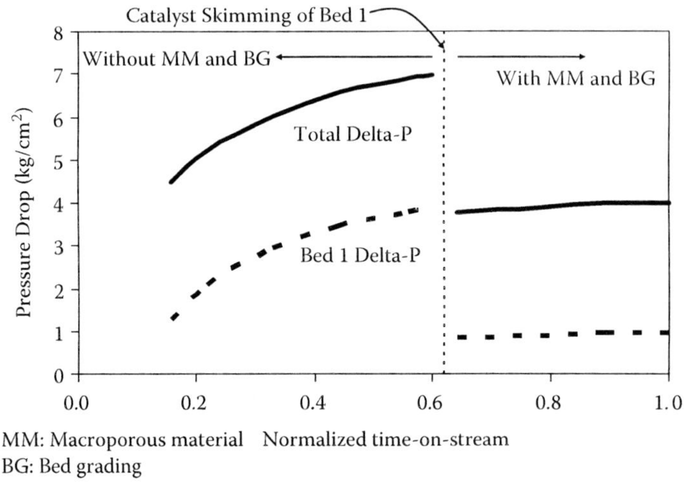

FIGURE 5.6 Common behavior of pressure drop across the catalytic bed with and without macroporous material and bed grading.

well as catalyst loading and unloading time by using hydraulic transport of catalyst/oil slurry from the storage hopper to the top of the reactor, where the catalyst and oil are separated (Figure 5.7).

A variant of downflow FBR (DF-FBR) is radial flow FBR (RF-FBR), in which the feed enters the top of the reactor and flows through the bed in a radial direction and then flows out through the base of the reactor instead of flowing downward through the catalyst bed (Figure 5.8). The following are the main advantages and disadvantages of RF-FBR over the typical DF-FBR (Speight, 2000):

Advantages:

1. Low pressure drop through the catalyst bed
2. Larger catalyst cross-sectional areas as well as a shorter bed depth

Disadvantages:

1. More chances of localized heating in the catalyst bed
2. More expensive reactor design per unit volume of catalyst bed
3. May be more difficult to remove contaminants from the bed as part of the catalyst regeneration sequence

For these reasons, RF-FBR is not recommended for hydroprocessing of heavy oils and residua, and it is limited to hydrodesulfurization of light petroleum distillates, such as naphtha and kerosene.

In the same situation is the flooded-bed or upflow reactor (UFR), in which gas and liquid are fed cocurrently upward through the catalyst bed. This latter configuration is rarely used in commercial practice, where downflow TBR prevails. Under this category is the upflow reactor licensed by Chevron Lummus

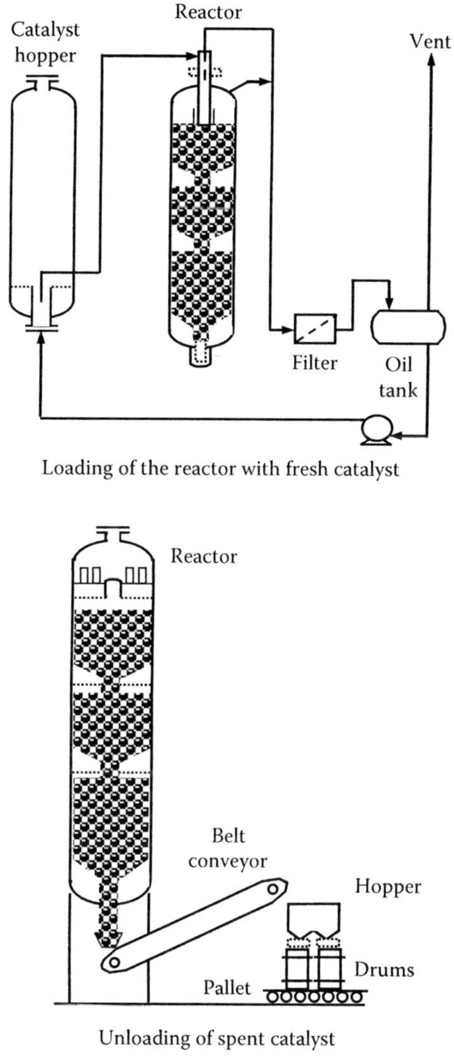

FIGURE 5.7 Loading of fresh catalyst and unloading of spent catalyst operations in QCR system. (From Van Zijll Langhout, W.C. et al., *Oil Gas J.*, 78: 120–126, 1980. With permission.)

Global LLC, which is a guard-bed reactor that is added before an FBR hydroprocessing unit. Its main advantages and characteristics are:

1. Increased capacity or improved product quality from an existing FBR residua hydroprocessing unit.
2. When adding a UFR, feed throughput is increased and heavier and more contaminated feeds can be processed.
3. Low plugging tendency due to the slightly expanded catalyst bed.

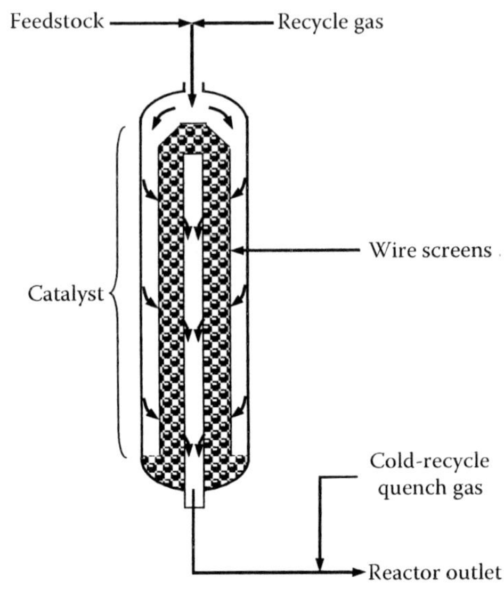

FIGURE 5.8 Radial reactor, RF-FBR. (From Speight, J.G., *The Desulfurization of Heavy Oils and Residua*, 2nd ed., Marcel Dekker, New York, 2000. With permission.)

4. Low pressure drop across the reactor loop compared with FBR.
5. Opportunity to increase capacity of the system.
6. Longer life of downstream fixed-bed catalyst.

Among the different reactors based on the fixed-bed system (DF-FBR, FBR with reduced H_2S partial pressure in the last part of the reactor, RF-FBR, UFR), the downflow fixed-bed reactor is the most used for hydroprocessing. The flow pattern in this reactor behaves as in the TBR, and its advantages and disadvantages are summarized as follows (Gianetto and Specchia, 1992; Al-Dahhan et al., 1997):

Advantages:

1. Liquid flow behaves mainly as piston flow, very little back-mixing
2. Low catalyst loss, no catalyst attrition
3. No moving parts
4. Flexibility of operation at high pressure and temperature
5. Larger reactor sizes
6. Low investment and operating costs
7. High catalyst loading per unit volume of the liquid and low energy dissipation rate

Disadvantages:

1. Lower catalyst effectiveness because of the use of large catalyst particle size
2. Long diffusion distance
3. High pressure drop

4. Inappropriate catalyst wetting with low liquid flow rates
5. Possibility of liquid maldistribution, which may give rise to hot spots and reactor runaway
6. Short catalyst cycle life due to the reactor's impracticality for reactions with rapid-deactivation catalysts. This can be solved by using a guard-bed reactor to minimize the metal and coke deposition on the downstream reactors.

The problem of short catalyst cycle life, which is the main reason for declining the use of FBR when feeds have a high amount of metals, may be solved by the association of adequate HDM and HDS catalysts as well as appropriate selection of reaction conditions, which contribute to increase strongly the performance of new processes for residue refining (Ancheyta et al., 2006).

5.2.2 Moving-Bed Reactors

Differently than in an FBR, in an MBR the catalyst goes in downflow through the reactor by gravitational forces. The fresh catalyst enters at the top of the reactor and the deactivated catalyst leaves the reactor at the bottom, while the hydrocarbon goes in either counter- or cocurrent flow through the reactor. With this moving-bed system, the catalyst can be replaced either continuously or discontinuously (Gosselink, 1998).

The main differences and advantages of MBR over FBR are:

1. In MBR the catalysts are homogeneously loaded by metal sulfides, while in FBR a profile exists over the catalyst bed (see Figure 5.24).
2. Higher catalyst utilization in MBR.
3. When using more than one reactor, for example, MBRs combined with FBRs, this configuration enables on-stream replacement of one catalyst system by another, thus increasing the flexibility of the unit.
4. Although the catalysts used in MBR are chemically quite similar to those of FBR, their mechanical strengths and shapes (that is, strong, attrition-resistant catalyst) should meet the more demanding situation, for example, severe grinding and abrasion.
5. In MBR, the top layer of the moving bed consists of fresh catalyst, and metals deposited on the top move downward with the catalyst and are released at the bottom.
6. Tolerance to metals and other contaminants is much greater in MBR than in FBR.

One example of application of MBR for hydroprocessing of heavy oils and residua is the bunker reactor used in the HYCON process developed by Shell (Van Gineken et al., 1975; Scheffer et al., 1988). This process enables easy catalyst replacement (to remove or add portions of catalyst) without interrupting operation by means of valves of lock-hoppers. The catalyst and heavy oil are fed in cocurrent flow; the fresh catalyst enters at the top of the reactor, while deactivated catalyst is removed from the bottom (Figure 5.9). A special design of the reactor internals

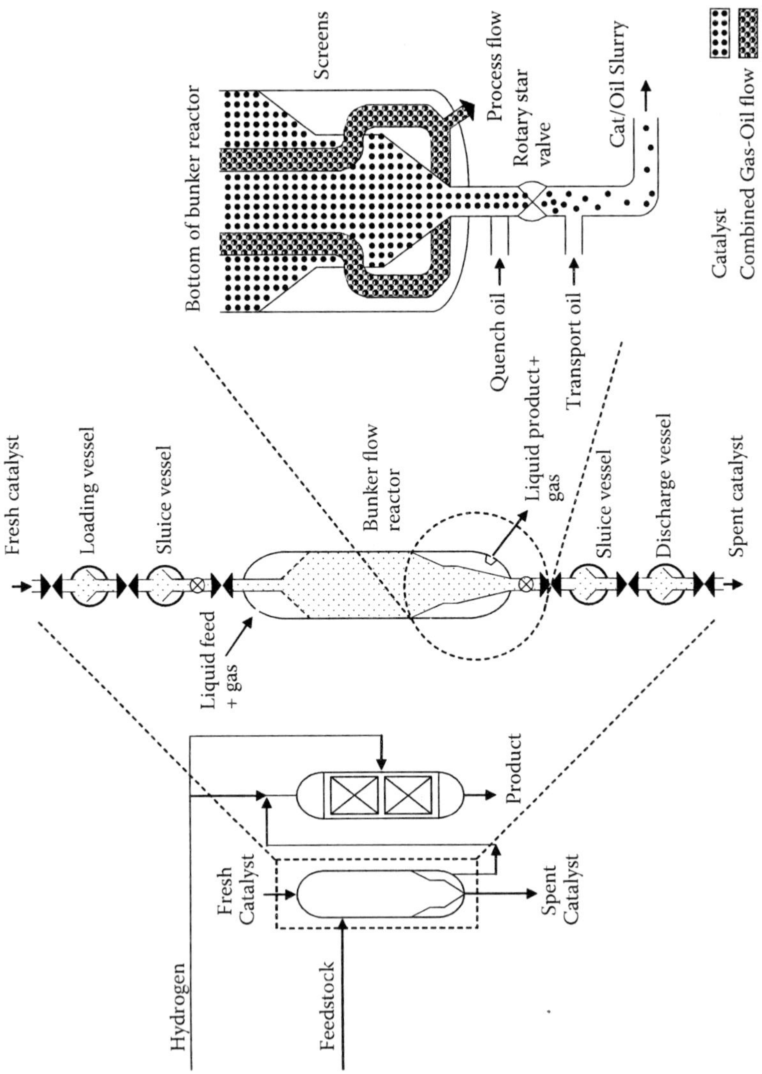

FIGURE 5.9 Bunker reactor. (From Van Gineken, A.J.J. et al., *Oil Gas J.*, 73: 59–63, 1975. With permission.)

ensures that the catalyst can move down in plug flow. Special screens separate the catalysts from the process fluids before leaving the reactor. Catalyst replacement frequency depends, of course, on the deactivation rate and is done at a rate to ensure a total plant runtime of at least a year, which depends mainly on the metal contaminants in the feed. During catalyst replacement, in which addition and withdrawal are performed via the sluice system at the top and bottom of the reactor, the catalyst moves slowly compared with the linear velocity of the feed. In this way, the bunker reactor technology combines the advantages of plug flow FBR operation with easy catalyst replacement. For feeds containing a high amount of metals, metal sulfide can be better accommodated onto the catalyst in a bunker flow reactor than in other reactor systems. Operating conditions and catalyst addition/withdrawal rate can be also adjusted to ensure that the catalyst taken out is completely spent, while still retaining an acceptable average activity in the reactor (Sie, 2001).

The On-Stream Catalyst Replacement (OCR) process is another option for hydroprocessing heavy oils and residua with a significant amount of metals. OCR is a moving-bed reactor operating in a countercurrent mode at high temperature and pressure. Fresh catalyst is added at the top of the reactor and the feed into the bottom, and both move through the reactor in a countercurrent flow, causing the dirtiest feed (that is, the feed with the highest content of impurities) to contact the oldest catalyst first. The fresh catalyst can be added at the top of the reactor and the spent catalyst removed from the OCR reactor while the unit is on-stream (Figure 5.10). An OCR moving-bed reactor can be incorporated to the processing scheme either before or after FBRs. Its main application is before FBRs, so that heavier feeds with higher levels of contaminants can be processed, while maintaining constant product quality and economical FBR run lengths (Scheuerman et al., 1993).

Another alternative of moving systems, with feed and catalyst flowing countercurrent, is the MBR used in the Hyvahl-M process (Euzen, 1991). This technology is part of the following series of processes for residua hydrotreating licensed by IFP/Asvahl (Billon et al., 1991):

1. Hyvahl-F: Uses FBR, in which liquid and gas flow cocurrently downstream in a trickle flow regime. Its main application is for processing distillate fractions and some atmospheric residua. For vacuum residua and heavy oils, it may require frequent catalyst replacements, which may be complex and uneconomical. Optimization of catalyst properties, space-velocity, and maximum reaction temperature may extend the life of the catalyst, depending mainly on metals content in the feed.
2. Hyvahl-S: Also called Hyvahl-F-swing reactor; uses two guard reactors in swing arrangement and switchable operation, which are FBRs with simple internals (Figure 5.11). This configuration allows fast switching of the guard reactor in operation with a deactivated catalyst to the other guard reactor with a fresh catalyst, without shutting down the plant.
3. Hyvahl-M: Employs countercurrent MBRs and is recommended for feeds containing a high amount of metals and asphaltenes. It requires

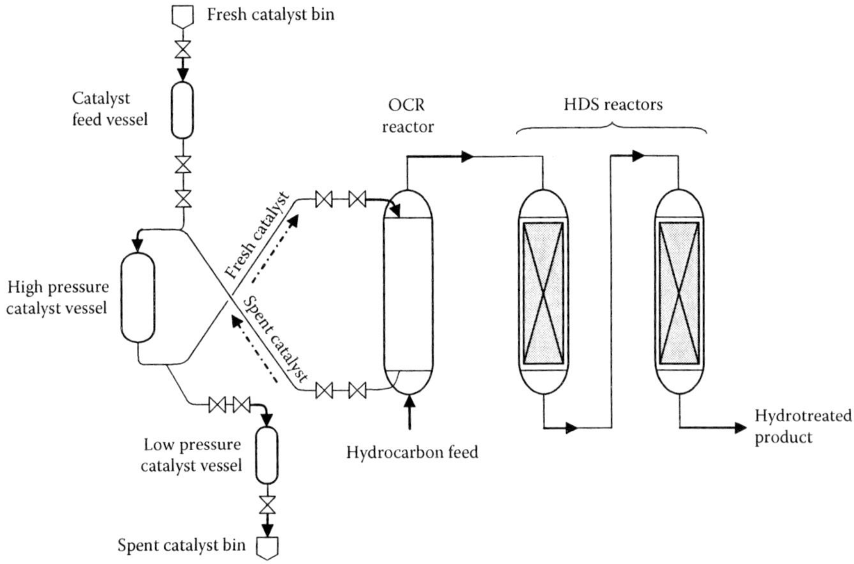

FIGURE 5.10 On-Stream Catalyst Replacement (OCR) reactor. (From Scheuerman, G.L. et al., *Fuel. Process. Technol.*, 35: 39–54, 1993. With permission.)

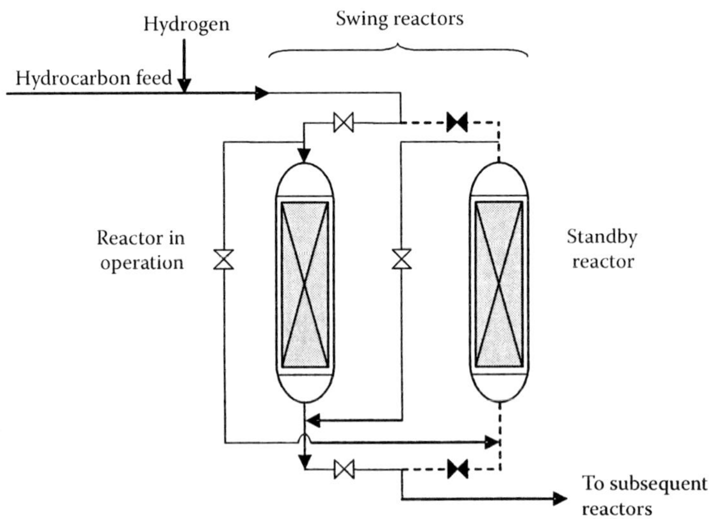

FIGURE 5.11 Swing reactors used in the Hyvahl-S process. (From Billon, A. et al., Hyvahl F versus Hyvahl M, Swing Reactor or Moving Bed, paper presented at the 1991 NPRA Annual Meeting, San Antonio, TX, March 17–19. With permission.)

special equipment and procedures for safe and effective catalyst transfer into and out of the high-pressure unit, similar to the OCR reactor shown in Figure 5.10. Catalyst is taken at atmospheric pressure and transferred to a reactor operating under hydrogen pressure, and then the catalyst is taken from the reactor at the high operating conditions and discharged to the atmosphere.

In general, MBR catalyst replacement is commonly a batch operation, which is typically done once or twice a week. Catalyst transfer, that is, adding and removing of the catalyst, is the most critical section. The countercurrent mode of operation of MBRs seems to be the best configuration since the spent catalyst contacts the fresh feed at the MBR bottom while the fresh catalyst reacts with an almost already hydrodemetallized feed at the MBR top, resulting in lower catalyst consumption (Morel et al., 1997).

5.2.3 Ebullated-Bed Reactors

Similarly to MBRs, to handle problematic heavy feeds with a high amount of metals and asphaltenes, such as vacuum residua, EBRs are used in process technologies to overcome some of the deficiencies of FBRs. H-Oil, T-Star (an extension of H-Oil), and LC-Fining processes are the commercialized technologies that use EBR. These hydroprocessing technologies possess very similar features (process parameters and reactor design), but are different in mechanical details. Figure 5.12 shows both EBRs employed in H-Oil and LC-Fining technologies (Daniel et al., 1988).

In EBR, hydrocarbon feed and hydrogen are fed upflow through a catalyst bed, expanding and back-mixing the bed, minimizing bed plugging, and consequently reducing pressure drop problems. The mixture of the gas (makeup and recycle hydrogen) and liquid (feed and recycle oils) reactants enters the reactor plenum chamber and is well mixed through the specially designed gas/liquid mixer, spargers, and catalyst support grid plate. A homogeneous environment is created to hydrotreat and hydrocrack the heavy feedstocks (Kam et al., 1999). Product quality is constantly maintained at a high level by intermittent catalyst addition and withdrawal. Reactor features include on-stream catalyst addition and withdrawal, thereby eliminating the need to shut down for catalyst replacement.

Hydroprocessing EBR is a three-phase system — gas, liquid, and solid (catalyst) — in which oil is separated from the catalyst at the top of the reactor and recirculated to the bottom of the bed to mix with the new feed. The large liquid recycle causes the reactor to behave as a continuous-stirred-tank reactor. EBR is provided with an ebullating pump, located externally for the H-Oil reactor and internally for the LC-Fining reactor (Figure 5.12), to maintain liquid circulation within the reactor. This liquid circulation is what maintains the reactor at essentially isothermal conditions, so that there is no need for quenches within the reactor. The liquid recycle rate can be adjusted by varying the ebullition pump speed. The unconverted heavy oils are recirculated back to the reactor with a

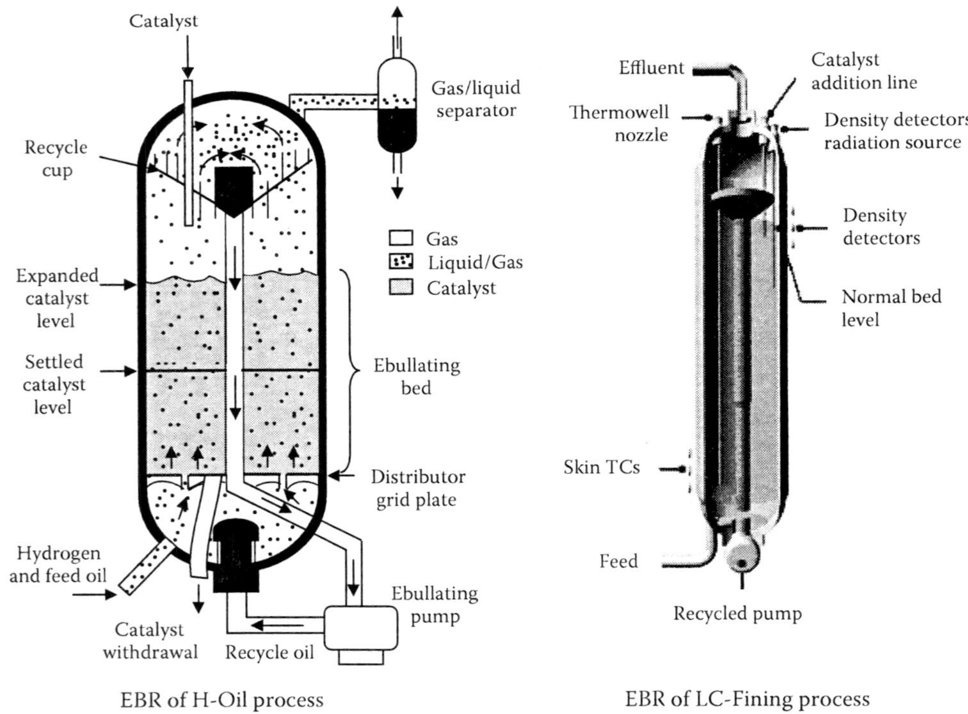

FIGURE 5.12 H-Oil and LC-Fining ebullated-bed reactors (EBRs). (From Daniel, M. et al., Amocos LC-Fining Residue Hydrocracker Yield and Performance Correlations from a Commercial Unit, paper presented at the 1988 NPRA Annual Meeting, San Francisco, CA, March 15–17. With permission.)

small amount of diluent to improve fluidity and thus overall conversion. The fluidization of the catalyst also results in solid back-mixing, which implies that the catalyst removed also contains fresh catalyst particles. That is why this mixture of spent catalyst and fresh catalyst removed from the reactor is also called equilibrium spent catalyst, or only equilibrium catalyst. Fresh catalyst is added to the top of the reactor and spent catalyst is withdrawn from the bottom of the reactor. The inventory of catalyst in the reactor is maintained at the desired level by adjusting the catalyst addition rate equal to the withdrawal rate plus any losses. The catalyst replacement rate can be adjusted to suit feed properties or product slate and quality requirements. A common daily catalyst addition rate is about 90 to 100 g catalyst/barrels of liquid feed.

Catalyst bed expansion, that is, the level of the expanded catalyst bed, is controlled by adjusting the circulating ebullating pump flow. Bed expansion is measured with a nuclear density detector. To give an example of bed expansion determination, Figure 5.13 shows a simplified scheme of a pilot EBR (900 mL catalyst volume, 29 mm diameter, 434 mm height) in which bed density measurements along the reactor length are presented (Ruiz et al., 2005). At pilot-scale the EBR is first loaded with a light petroleum fraction without catalyst (for example gas oil). Bed density is then determined for this oil fraction, which corresponds to the base density line. During operation catalyst bed density measurements (operation density line) are compared with the base density line, and finally, bed expansion is determined by this comparison, that is, the limit of bed expansion is determined when the two lines approach. In the given example catalyst bed expansion is about 45%.

EBRs are commonly designed to have an expanded catalyst bed of 130 to 150% of the settled catalyst bed, which has been demonstrated to be the bed expansion for achieving uniform fluidization and good contacting between hydrogen, oil, and catalyst. Other common design parameters of EBR are heat of reaction of 45 BTU/SCF of chemical hydrogen consumption, superficial gas velocity of 0.052 m/sec, and hydrogen consumption of 990 SCF/Bbl.

The main advantages and disadvantages of an ebullated-bed hydroprocessing reactor are (Quann et al., 1988; Furimsky, 1998; Gosselink, 1998; Sie, 2001; Babich and Moulijn, 2003; Ruiz et al., 2004a, 2004b, 2005):

Advantages:

1. Very flexible operation (high and low conversion modes).
2. Capability to periodically withdraw or add the catalyst to the reactor without interrupting the operation to maintain the necessary catalyst activity. This characteristic of EBR increases the operating factor of the process unit and decreases refinery maintenance costs associated with catalyst dumping and reloading, as compared with FBR.
3. Its design (catalyst bed expansion of 30 to 50% by using an ebullition pump) ensures ample free space between particles, allowing entrained solids to pass through the bed without accumulation, plugging, or increased pressure drop.

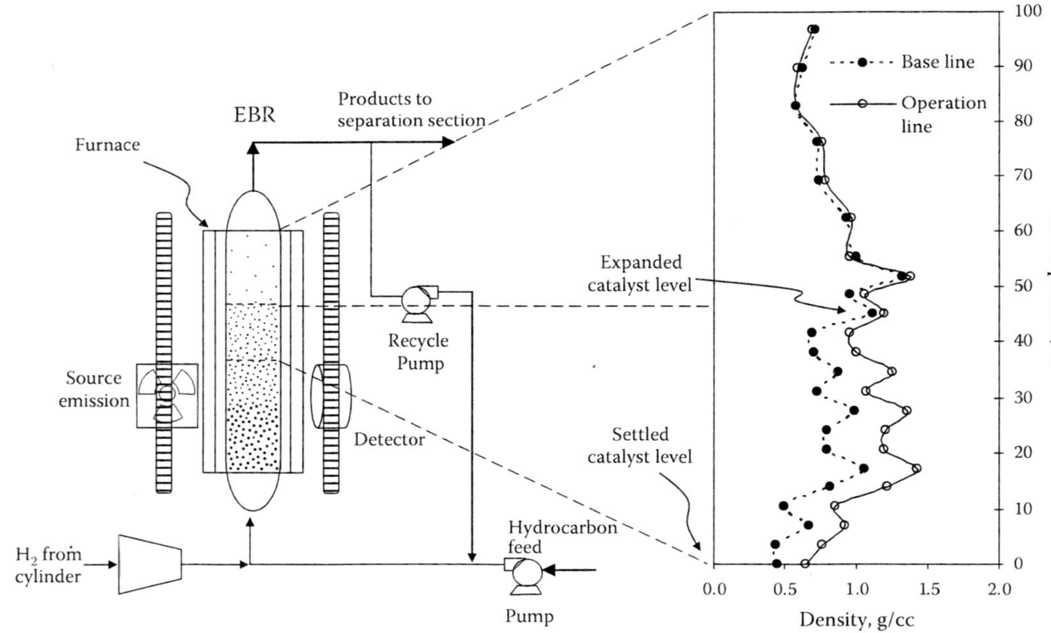

FIGURE 5.13 Catalyst bed expansion in a pilot EBR. (From Ruiz, R.S. et al., *Catal. Today*, 109: 205–213, 2005. With permission.)

4. Increase of reaction rate by the use of smaller-diameter catalyst particles due to significantly diminished diffusion limitations, that is, higher effectiveness factors. This smaller size of the catalyst also makes it less susceptible to pore mouth plugging by metal deposits.
5. Good heat transfer so that overheating of the catalyst bed is minimized and less coke forms.
6. The average activity of the catalyst inventory reaches a steady state, which results in constant product properties during the cycle length. This is a major difference compared with FBR, in which reaction selectivities are changing during the run in response to increases in operating temperature necessary to compensate declining catalyst activity.
7. Operation nearly isothermal (uniform temperature) since the EBR behaves as a well-mixed reactor, which easily dissipates the exothermic heat of reaction and enables higher operating temperatures, and thus higher conversion levels.

Disadvantages:

1. Absence of plug-flow regime, which is kinetically more favorable than a well-mixed regime. This can be partially improved by setting several EBRs in series.
2. Catalyst attrition and erosion, which means that the catalyst has to be mechanically stable and resist attrition.
3. The smaller catalyst particle size and the lower catalyst holdup (higher void fraction) require a higher reactor volume than FBR and MBR.
4. Catalyst consumption rate is quite high.
5. Stagnant zones may be developed and careful monitoring is required to prevent growth of such zones, which can lead to unstable and runaway conditions.
6. Sediment formation.
7. EBR scale-up and design are more difficult than for other reactors since it requires much more information, for example, data of feedstock composition, catalyst properties, catalysis and chemical kinetics aspects, hydrodynamic phenomena, heat transfer at the catalyst pellet and bed scales, and so forth.

The following significant improvements in EBR have been announced by the two major licensers (Axens, H-Oil; Lummus, LC-Fining) (Phillips and Lie, 2002; Kressmann et al., 2004):

1. Second-generation catalyst technology
2. Catalyst rejuvenation to reuse spent catalyst
3. New reactor design to increase single-train capacity to around 50,000 Bbl/day
4. Interstage separation for a two-EBR H-Oil unit

With interstage separation in the two EBRs, that is, an additional vessel between EBRs for separating the effluent of the first EBR into vapor and liquid products (Figure 5.14), offloading of the first EBR gas results in improved reaction kinetics in the second EBR (Kressmann et al., 2004). This configuration is best applied when EBR is limited by shipping, weight, or plot restrictions, for revamping of an existing unit with the objective of increased feed capacity, or for high-conversion applications where the hydrogen consumption is high. Interstage separation has been reported to have the following benefits (Kressmann et al., 2004):

1. The amount of gas holdup in the second EBR is reduced and the liquid holdup is increased, which enables greater conversion of the heaviest part of the feed.
2. The second EBR size can be reduced to decrease the unit investment.
3. Optimization of the recycle gas rates to each EBR, providing an operating cost savings.

When operating hydroprocessing EBRs at a high level of conversion (>50%), the main problem is sediment formation, which leads to deposits on the reactor internal parts and downstream vessels (heat exchangers, separators,

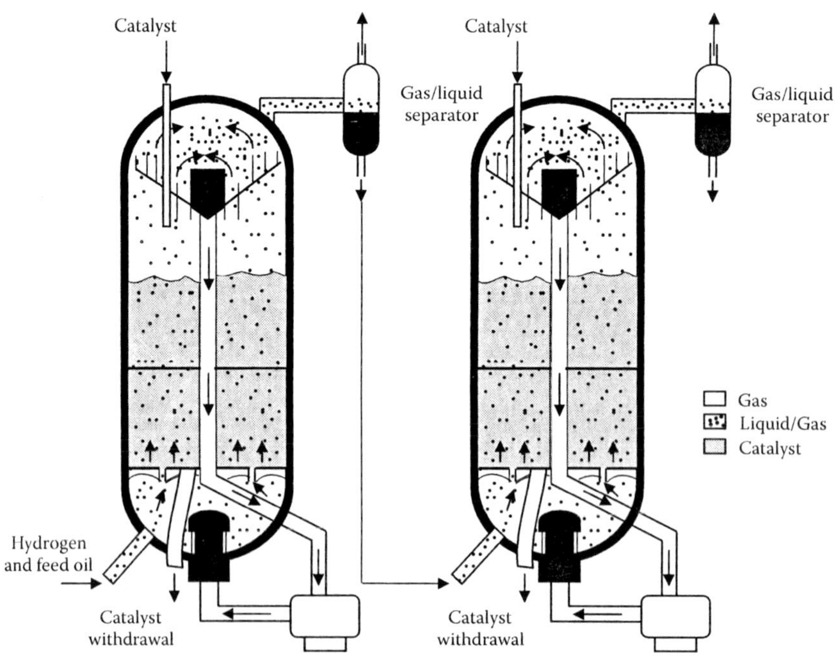

FIGURE 5.14 Interstage separation in H-Oil EBRs. (From Kressmann, S. et al., A New Generation of Hydroconversion and Hydrodesulfurization Catalysts, paper presented at the 14th Annual Symposium Catalysis in Petroleum Refining and Petrochemicals. Dhahran, Saudi Arabia, December 5–6, 2004. With permission.)

fractionating towers) and transfer lines. Sedimentation also causes EBR operability problems, catalyst deactivation, increase of catalyst consumption, and decrease in residual fuel stability, which eventually cause shutdown of commercial units (Morel et al., 1997). For this reason, refiners commonly limit sediments formation at values of 0.8 to 1.0 wt% to guarantee continuous operation of commercial hydrocracking units. Since sediments formation is a major problem in hydroprocessing of heavy oil and residua not only when using EBR but also for the other reactors, it deserves special attention. In the following paragraphs the origin, causes, and approaches to minimize sediments formation are summarized.

Solids formation is attributed in the literature to occur by the following reasons (Storm et al., 1997; Inoue et al., 1998; Speight, 2001; Marroquin et al., 2007):

1. As a result of changes in the relative amounts of the lower-boiling hydrocarbons and higher molecular weight polar species.
2. At the high temperature in the reactor the liquids split into a light aliphatic phase and a heavy aromatic phase. This separation of the aromatic material is supposed to be the initial step of solid deposition.
3. Reductions either of solubility of asphaltenes in the nonasphaltene fraction or of the ability of the nonasphaltenic fraction to solubilize the asphaltenes, both due to their chemical transformation during hydroprocessing of heavy feeds. More specifically, hydrocracking of the resin fraction (solvent) for large colloidal asphaltenes proceeds faster than the conversion of asphaltenes. The solvent becomes less effective under the severe conditions, causing the precipitation of asphaltenes. This is further enhanced by dealkylation of the asphaltenes, which also decreases solubility.

It has been shown that at temperatures lower than 420°C, only dealkylation of side alkyl chains is observed, and at higher temperatures, hydrocracking of the asphaltene molecule is prominent. Thus, asphaltenes' composition and structural parameters change significantly depending on hydroprocessing reaction conditions (Ancheyta et al., 2003a), and consequently, sediments formation is also influenced by the severity of reaction. For instance, during hydrocracking of a vacuum residue of Arabian light oil (bp > 550°C) with a commercial $CoMo/Al_2O_3$ catalyst in a two-stage fixed-bed microreactor, reaction temperatures of 395, 405, and 418°C are considered the conditions for no formation, beginning of formation, and some formation of sediments, respectively (Mochida et al., 1989). Marroquin et al. (2007) also found similar temperatures for different stages of sediment formation at pilot scale in a single FBR for hydrotreating of heavy crude oils. Of course, these temperature values are highly dependent on the type of catalyst and feed. What is clear is that there is an intimate dependency between the temperature where hydrocracking of asphaltenes is severe and sediments formation.

The limit of 50% conversion to avoid unmanageable sediment formation has been clearly demonstrated with experimental data of sediments and insoluble toluene vs. conversion at typical hydroprocessing conditions, as shown in Figure 5.15 (Inoue et al., 1998; Fukuyama et al., 2004). Characterization of sediments collected from different zones of an H-Oil plant has shown that all deposits contain appreciable amounts of metals coming from either the catalyst (Ni, Mo, Al) or the feed (V, Ni, Fe). Other metals (Mn, Cu, W, and Fe partially) are also present, which are corrosion products (Table 5.2). It was also found that the farther from the reactor, the lower the metal content in the deposit. The chemical composition of sediments also reveals that its content of insoluble organic material increases in the direction from the reactor toward the vacuum column, indicating that the insoluble organics are generated by the asphaltenes precipitated from the hydrotreated feed. The inorganics content in the sediments decreases toward the end of the installation, since the majority of the inorganic components come from the catalyst and therefore settle in the installation zone closest to the reactor (Gawel and Baginska, 2004).

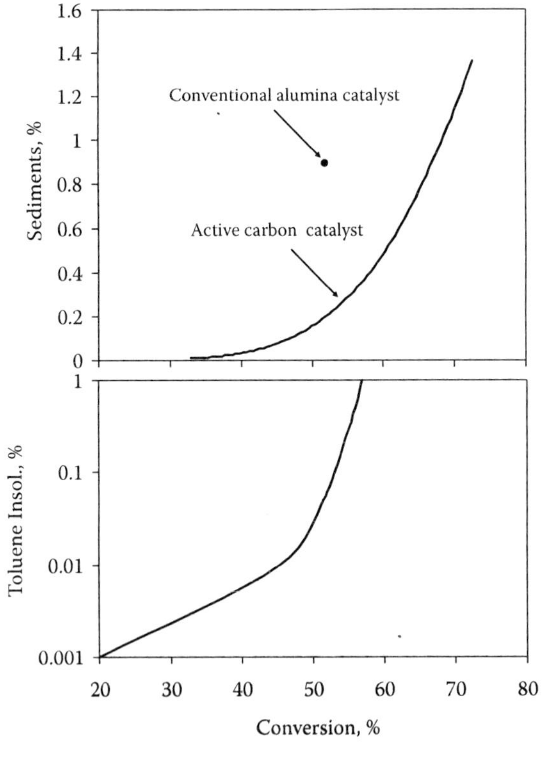

FIGURE 5.15 Sediment formation as a function of conversion. (From Inoue, S. et al., *Catal. Surv. Jpn.*, 2: 87–97, 1998; Fukuyama, H. et al., *Catal. Today*, 98: 207–215, 2004. With permission.)

TABLE 5.2
Metal Content and Chemical Composition of Sediments from an H-Oil Plant

	Metal Content, ppm			Composition, wt%	
Deposit	**Ni**	**V**	**Fe**	**Insoluble Organics**	**Inorganics**
S	7279	9616	16,528	28.02	31.98
D1	3175	7294	6,374	35.89	27.39
D2	995	5610	3,905	58.10	6.92

Note: S = sludge precipitated from a sample taken from the filter between the H-Oil reactor and the atmospheric column; D1 = solid deposit collected from the filter after the atmospheric column; D2 = solid deposit collected from the filter after the vacuum column.

Source: Gawel, I. and Baginska, K., *Pap. Am. Chem. Soc. Div. Petrol. Chem.*, 49: 265–267, 2004. (With permission.)

To minimize sediment formation, some aromatics-rich gas oil streams, such as HCGO and LCO (heavy-cycle gas oil and light-cycle oil) produced in FCC units, have been used as diluents to the vacuum residue fed to the H-Oil process instead of the heavy-vacuum gas oil (HVGO) produced in the same plant and recycled to the inlet of the H-Oil reactor. No appreciable improvement on sulfur, metals, nitrogen, and other impurities removals was reported. However, sediments formation is substantially reduced when HCGO and LCO are utilized as diluents instead of HVGO, as can be seen in Figure 5.16. This reduction in sediment formation was attributed to (Marafi et al., 2005):

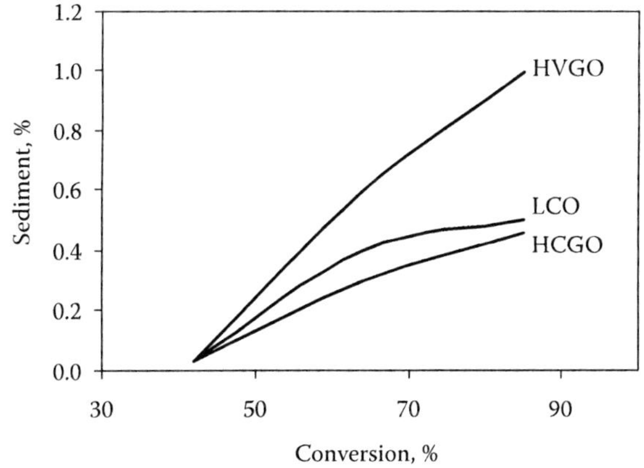

FIGURE 5.16 Effect of diluents on sediment formation. (From Marafi, M. et al., *Petrol. Sci. Technol.*, 23: 899–908, 2005. With permission.)

1. Reduction in liquid viscosity and surface tension of the dilute feed, which causes improvements in flow dynamics.
2. Diluents may associate the asphaltene aggregates and reduce their size, enhancing their diffusion into the pore and increasing the reaction rate.
3. The high aromatic content (>70%) in the diluents may help to disperse the scarcely soluble asphaltenes and improve the compatibility between them and oil fractions and control the onset of asphaltenes precipitation.

5.2.4 Slurry Phase Reactors

A slurry phase reactor (SPR) can also be used for hydroprocessing of feeds with very high metals content to obtain lower-boiling products using a single reactor. SPR-based technologies combine the advantages of the carbon rejection technologies in terms of flexibility with the high performances peculiar to the hydrogen addition processes (Panariti et al., 2000). SPR achieves a similar intimate contacting of oil and catalyst and may operate with a lower degree of back-mixing than EBR. Different from in FBR and EBR, in SPR a small amount of finely divided powder is used (typically from 0.1 to 3.0 wt%), which can be an additive or a catalyst (or catalyst precursor). The catalyst is mixed with the feed (heavy oil), and both are fed upward with hydrogen through an empty reactor vessel. The SPR is free of internal equipment and operates in a three-phase mode. The solid additive particles are suspended in the primary liquid hydrocarbon phase, through which the hydrogen and product gases flow rapidly in bubble form. Since the oil and catalyst flow cocurrently, the mixture approaches plug flow behavior (Quann et al., 1988; Speight, 2000).

In an SPR the fresh catalyst is slurried with the heavy oil prior to entering the reactor, and when the reaction finishes, the spent catalyst leaves the SPR together with the heavy fraction and remains in the unconverted residue in a benign form (Furimsky, 1998). Inside the reactor, the liquid–powder mixture behaves as a single phase (homogeneous phase) due to the small size of the catalyst/additive particles. Figure 5.17 shows a typical simplified process scheme for catalyst (or additive) addition to an SPR. It has been reported that the powder mostly serves as a site where the small amount of coke can deposit, so as to keep walls, valves, and heat exchangers clean, and thus maintain good operability (Schulman and Dickenson, 1991). In other words, the use of a selected catalyst dispersed into the feed inhibits coke formation.

The following are the main advantages and disadvantages of SPRs (Morel et al., 1997; Furimsky, 1998; Speight, 2000; Robinson and Dolbear, 2006):

Advantages:

1. SPRs use disposable catalysts with low cost. These slurry catalysts possess high surface area and micron-scale particle size, which means high catalyst utilization.
2. The catalyst or additive used in SPR has a larger number of external pores on the surface than the extruded catalyst used in FBR, EBR, and MBR. This difference makes a slurry catalyst less likely to have metal contaminants plugging up these pores, because of their size.

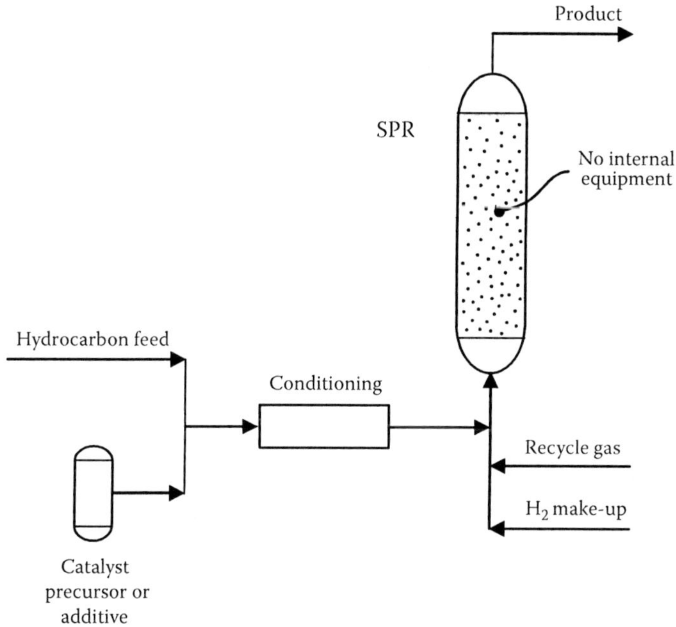

FIGURE 5.17 Schematic representation of a hydroprocessing SPR.

3. Spent hydroprocessing catalysts from SPRs are not hazardous wastes and do not require special handling procedures such as those recovered from processes using FBR, MBR, or EBR.
4. SPRs can be used for hydroprocessing of very poor quality feedstocks.
5. Good external mass transfer.
6. The smaller size of catalysts used in SPR compared with catalysts employed in other reactors results in smaller distances between particles and less time for a reactant molecule to find an active catalyst site.
7. The effectiveness of the dual-role additive (hydrogenation and suppression of coke formation) used in SPRs allows the use of operating temperatures that give high conversion in a single-stage reactor.
8. Residue conversion higher than 90% (sometimes higher than 95% and very close to 100%) can be obtained at space-velocities much higher than those used for FBR, MBR, or EBR.
9. Because of its particular features (empty reactor without internal equipment) the volume of the SPR is maximized.
10. The homogeneous phase operation of SPR provides a thermally stable operation (uniform temperature) with no possibility of temperature runaway.

Disadvantages:

1. Product quality is very poor, for example, essentially all of the feed metals remain in the unconverted feed.
2. The poor quality of the unconverted pitch (for example, high metals and sulfur contents) makes it almost impossible to be used as a fuel unless it is blended with other streams. This problem has kept SPR-based technologies from gaining industry-wide acceptance.
3. Catalyst particle size must be strictly designed to achieve excellent dispersion conditions in order to obtain a high level of conversion.
4. SPR has to be carefully designed to maintain a mixed three-phase slurry of heavy oil, fine power catalyst, and hydrogen, and to promote effective contact.
5. The effectiveness of the hydroprocessing SPR is highly dependent upon catalyst selection, since a slurry catalyst with poor activity will result in coke formation and a biphasic, incompatible product. That means that before commercial use a slurry catalyst has to be subject to different coke suppression and product compatibility tests, among others. Problems of product stability can be prevented by the use of more active catalysts in higher concentration.
6. The inexpensive additives, for instance, Fe-based compounds, or carbonaceous material to control coke formation, show very low activity toward hydrogenation reactions compared with transition metals (molybdenum, nickel, and so on). These low-cost materials can be used in a once-through mode. However, very high hydrogen pressures are necessary to compensate the scarce catalytic performance. These restrictions have limited commercialization of hydroprocessing SPR due to the high investment, operating costs, and severity of the process.
7. SPR-based hydroprocessing technologies are still at demonstration scale, and the system has to be demonstrated at a larger scale for refiners to consider this a commercially competitive alternative.

5.3 PROCESS VARIABLES

There are four process variables frequently reported in the literature to be the most important in hydrotreating operations:

1. Total pressure and hydrogen partial pressure
2. Reaction temperature
3. Hydrogen-to-oil ratio and recycle gas rate
4. Space-velocity and fresh feed rate

In general, for a specific feedstock and catalyst, the degree of impurities removal and conversion increases with the increasing severity of the reaction, that is, increasing pressure, temperature, or H_2-to-oil ratio, and decreasing space-velocity. In the following sections, the effects of these variables on hydroprocessing reactions and the global behavior of hydroprocessing reactors are described.

5.3.1 Total Pressure and Hydrogen Partial Pressure

The performance of any hydroprocessing reactor and process is limited by the hydrogen partial pressure at the inlet to the reactor. The higher the hydrogen partial pressure, the better the hydroprocessing reactor performance. The overall effect of increasing the partial pressure of the hydrogen is to increase the extent of the conversion (Speight, 2000). This has been extensively confirmed by studies conducted with model compounds for HDS, HDN, HDA, and so forth, reactions as well as with real feeds (light distillates, middle distillates, heavy oils, and so on) at microscale, bench scale, and pilot plants. Figure 5.18 shows an example of the effect of hydrogen partial pressure on saturation of polyaromatic hydrocarbons (PAHs) and sulfur removal during hydrotreating of straight-run gas oil and light cycle oil, respectively (Bingham and Christensen, 2000; Chen et al., 2003). In both cases, the combined effect of reaction temperature with H_2 partial pressure is also presented. It is clearly seen that PAHs react quite readily, but their conversion is thermodynamically limited, and neither increasing the temperature nor increasing the H_2 partial pressure reduces the PAH content in the product to values lower than 2 wt%.

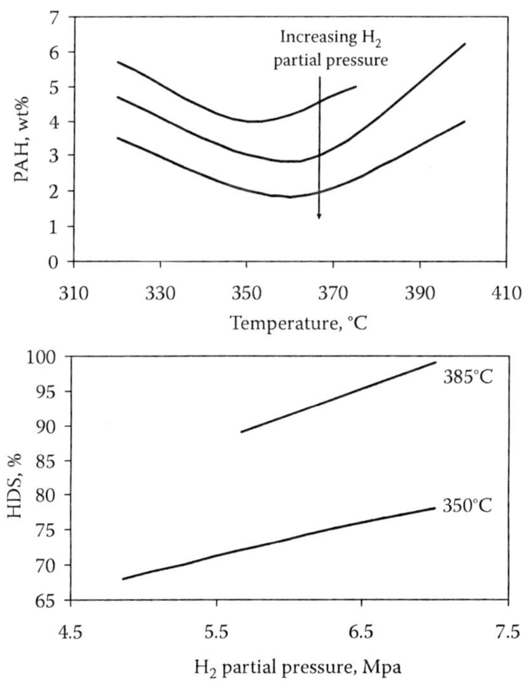

FIGURE 5.18 Effect of H_2 partial pressure on impurities removal. (From Bingham, F.E. and Christensen, P., Revamping HDS Units to Meet High Quality Diesel Specifications, paper presented at the Asian Pacific Refining Technology Conference, Kuala Lumpur, Malaysia, March 8–10, 2000; Chen, J. et al., *Petrol. Sci. Technol.*, 21: 911–935, 2003. With permission.)

The presence of heteroatom compounds with different reactivities in a hydroprocessing feed makes hydrodesulfurization of refractory multiring sulfur compounds very difficult and a rather demanding hydrogen process, whose pathway goes through prehydrogenation of one of the aromatic rings. Therefore, high-H_2 partial pressure is required; otherwise (Ho, 2003):

1. The HDN rate may be so slow that nitrogen compounds block off virtually all active sites that are available for HDS.
2. The HDS rate of refractory sulfur compounds may be limited by a thermodynamically mandated low hydrogenation rate.
3. The catalyst surface may be starved of adsorbed hydrogen.

The total pressure of a hydroprocessing reactor is fixed by the design and is controlled by the pressure that is maintained at the high-pressure separator (HPS), while hydrogen partial pressure is determined by multiplying total pressure (at the inlet of the reactor) by the hydrogen purity of the recycle gas. Hydrogen partial pressure and other variables are determined by type of feedstock, feed quality, and process objectives (desired levels of impurity removals, conversion, and so on). Operating the reactor at higher hydrogen partial pressures yields the following main benefits (Mehra and Al-Abdulal, 2005; Gruia, 2006):

1. Longer catalyst cycle life
2. Capability for processing heavier feeds
3. Higher throughput capability
4. Higher conversion capability
5. Better distillate quality
6. Purge gas elimination

A hydroprocessing reactor must be operated at H_2 partial pressure very close to the design value; otherwise, the catalyst deactivation rate will substantially be increased, and consequently, catalyst cycle life will be reduced due to excessive condensation reactions that form coke. Catalyst activity will also be negatively affected by operating the hydroprocessing reactor at low-H_2 partial pressure. However, equipment limitations (reactors, heaters, exchangers, vessels) restrict the operation at pressure close to or a little bit higher than the design value, and the unique option to increase hydrogen partial pressure is to increase the purity of the recycle gas, which can be achieved by (Gruia, 2006):

1. Increasing the H_2 purity of the makeup hydrogen
2. Venting gas off the HPS
3. Reducing the temperature at the HPS

In commercial operation, hydrogen partial pressure is mainly obtained by feeding the proper amount of makeup gas. More details about purity of hydrogen in the recycle gas and its effect on hydroprocessing reactor behavior will be given below.

The increase of catalyst activity for achieving higher impurities removal and conversion rates would require significant modifications in hydroprocessing reactor operation, primarily through the use of higher pressure, and also by increasing hydrogen rate and purity, reducing space-velocity, and proper selection of catalyst. Pressure requirements would depend, of course, on the feedstock quality and on product quality target of each refinery. For instance, it has been suggested that in order to produce diesel with less than 30 ppm sulfur, new high-pressure hydrotreaters would be required, operating at pressures between 7.6 and 8.3 MPa (NPC, 2000). Also, if all aromatics need to be hydrogenated, a higher pressure is needed in the reactor compared with the conventional operating mode. The level of pressure required for such product specifications will be limited by cost and availability of the technology.

At higher-H_2 partial pressures, sulfur (and other impurities) removal is easier; however, reactors become more expensive. Also, at higher pressure, hydrogen consumption of the unit increases, which can become a significant cost factor for the refinery. The minimum pressure required typically goes up with the required severity of the unit, that is, the heavier the feedstock, the lower the levels of sulfur required in the product. For new units, higher-H_2 partial pressure can be achieved by higher total pressure plant design, but the increase is not as pronounced as that achieved by increasing H_2 purity, since with the former approach H_2S partial pressure is also increased, whereas an increase in the recycle gas purity does not affect (or slightly reduces) the H_2S partial pressure (Bingham and Christensen, 2000).

5.3.2 Reaction Temperature

In combination with hydrogen partial pressure, the reactor temperature generally determines the types of components that can be removed from the hydrocarbon feed and also establishes the working life of the catalyst. In general, an increase in reaction temperature will substantially increase reaction rates and, consequently, removal of impurities, as is exemplified in Figure 5.19 for hydrodesulfurization of

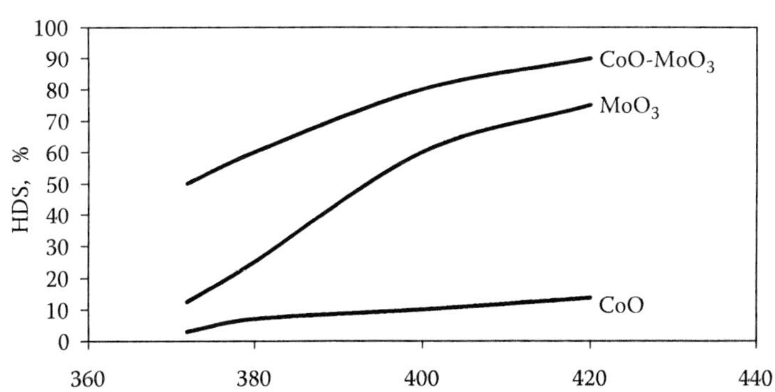

FIGURE 5.19 Effect of reaction temperature on HDS of an atmospheric residuum. Feed: 3.95 wt% sulfur, 117 wppm Ni + V. (From Takahashi, T. et al., *Catal. Today*, 104: 76–85, 2005. With permission.)

an atmospheric residuum with different catalysts (Takahashi et al., 2005). However, at temperatures above 410°C thermal cracking of the hydrocarbon constituents becomes more prominent, which can lead to formation of considerable amounts of low molecular weight hydrocarbon liquids and gases, and also to catalyst deactivation, much more quickly than at lower temperatures. Thermal cracking also produces olefins, which when hydrogenated release heat, increasing the temperatures further, and thermal cracking rates go up (hot spots). Finally, this condition inside the reactor provokes temperatures higher than the safe upper limits for the reactor walls (Speight, 2000).

There are some charts, based on commercial experiences reported in the literature, which help us to know the required temperature to reach a certain level of impurity removal or conversion. For hydroprocessing of residua, Figure 5.20 shows the required temperature increase or decrease to achieve the desired sulfur content in the product. For instance, if the actual sulfur content in the product is 0.60 wt% and the desired sulfur level is 0.55 wt%, the correction in reactor temperature would be

$$Deviation = \left(\frac{0.60 - 0.55}{0.55} \right) \times 100 = 9.1\%$$

From Figure 5.20, 9.1% deviation corresponds to a 6°F temperature correction. In this case, since the actual sulfur is higher than the desired sulfur content, the temperature correction must be an increase. On the contrary, if the actual sulfur content in the product is 0.45 wt% and the desired sulfur content level is 0.55 wt%, the correction in reactor temperature would be

$$Deviation = \left(\frac{0.55 - 0.45}{0.55} \right) \times 100 = 18.2\%$$

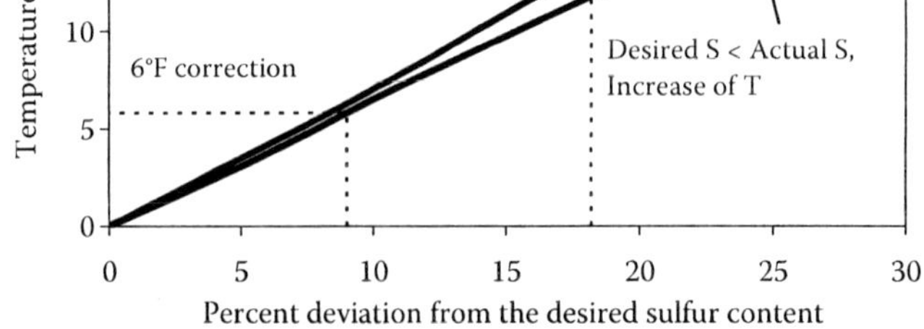

FIGURE 5.20 Typical correction temperature chart as a function of sulfur in the product during hydroprocessing of residua.

For this case, the temperature correction is 13°F. Given that the actual sulfur is lower than the desired sulfur level, the temperature correction must be a decrease.

Since hydroprocessing reactions are mostly exothermic, the reactor temperature will increase as the reactants proceed through the catalyst beds. This means that outlet reactor temperature will be higher than inlet reactor temperature. This increase in reactor temperature is not linear, as can be seen in Figure 5.21 (Rodriguez and Ancheyta, 2004; Mederos et al., 2006). In other words, for determining the average temperature of the reactor, an arithmetic average of temperatures, that is, (inlet temperature + outlet temperature)/2, will not be the more accurate representative reactor temperature. To properly monitor the reactor temperature, it is necessary to observe the temperature at the inlet, outlet, and axially (and radially) throughout the catalyst bed. A temperature profile can be constructed by plotting the catalyst temperature vs. weight percent of the catalyst. For this reason, the weight average bed temperature (*WABT*) is typically used as a measure of reactor temperature. Commercial reactors commonly have various temperature indicators (TIs) located in different zones of the catalytic beds. Most of the time TIs are placed at the inlet and outlet of the beds; then for determining the global *WABT*, the average $WABT_i$ of each catalytic bed is first determined as follows (Stefanidis et al., 2005):

$$WABT_i = \frac{1}{3}T_i^{in} + \frac{2}{3}T_i^{out} = \frac{T_i^{in} + 2T_i^{out}}{3} \tag{5.1}$$

and the global *WABT* is calculated with the following equation:

$$WABT = \sum_{i=1}^{N} (WABT_i)(Wc_i\%) \tag{5.2}$$

where N is the number of catalyst beds, T_i^{in} and T_i^{out} are the inlet and outlet temperatures in each catalytic bed, respectively, and $Wc_i\%$ is the weight percent of catalyst in each bed with respect to the total.

If there are not sufficient TIs along the length of the reactor (only at the inlet and outlet of the reactor) or simply for rapid calculation, *WABT* is sometimes determined by

$$WABT = T_{in} + \frac{1}{2}\Delta T = T_{out} - \frac{1}{2}\Delta T = \frac{T_{in} + T_{out}}{2} \tag{5.3}$$

Equation 5.3 is the common arithmetic average, which, as can be seen in Figure 5.21, induces higher error. Another, better approach is to employ Equation 5.1 for the complete reactor:

$$WABT = \frac{1}{3}T_{in} + \frac{2}{3}T_{out} = \frac{T_{in} + 2T_{out}}{3} = T_{in} + \frac{2}{3}\Delta T = T_{out} - \frac{1}{3}\Delta T \tag{5.4}$$

$$\Delta T = T_{out} - T_{in} \tag{5.5}$$

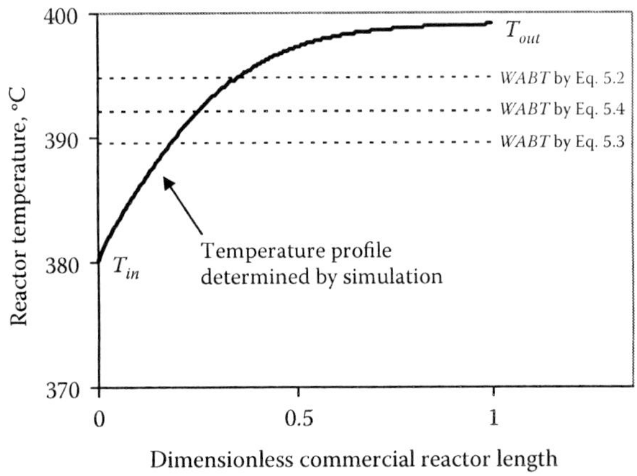

FIGURE 5.21 Example of axial temperature profile along a hydroprocessing reactor. (From Rodriguez, M.A. and Ancheyta, J., *Energy Fuels*, 18: 789–794, 2004; Mederos, F.S. et al., *Energy Fuels*, 20: 936–945, 2006. With permission.)

In this case, T_{in} and T_{out} are the inlet and outlet temperatures in the reactor, and not in each catalytic bed, as in Equation 5.1.

Equation 5.4 takes into consideration the trend of the temperature increase and considers that in the last part of the reactor (2/3 catalytic bed) the temperature is at T_{out}, while the first part of the reactor (1/3 catalytic bed) the prevailing temperature value is T_{in}. Figure 5.21 clearly shows that Equation 5.4 gives a *WABT* closer to the real average value determined by simulation with 200 beds using Equation 5.2, compared with *WABT* as determined by Equation 5.3. For the calculation of *WABT* using Equations 5.3 and 5.4, only inlet and outlet reactor temperatures were employed. If complete data of temperature are available for different catalyst beds, that is, TIs at different reactor axial positions, Equation 5.2 is the most accurate approach.

Figure 5.22 summarizes the common ways for calculating *WABT* depending on the available data of temperature indicators along the reactor length. Comparisons of *WABT* indicate considerable differences among the different approaches, with *WABT* calculated with Equation 5.2 the best approach. For practical purposes, *WABT* obtained by Equation 5.2 is expressed as a linear equation as a function of TI values, which for the case illustrated in Figure 5.22 gives

$$\begin{aligned} WABT = 0.0333TI_1 + 0.1333TI_2 + 0.1833TI_3 + 0.1500TI_4 \\ + 0.2333TI_5 + 0.2667TI_6 \end{aligned} \tag{5.6}$$

During hydroprocessing of heavy oils and residua the temperature has to be increased to compensate for catalyst deactivation in order to maintain the desired conversion level. The initial temperature is known as the start-of-run (SOR) temperature ($WABT_{SOR}$) and the final temperature as the end-of-run (EOR)

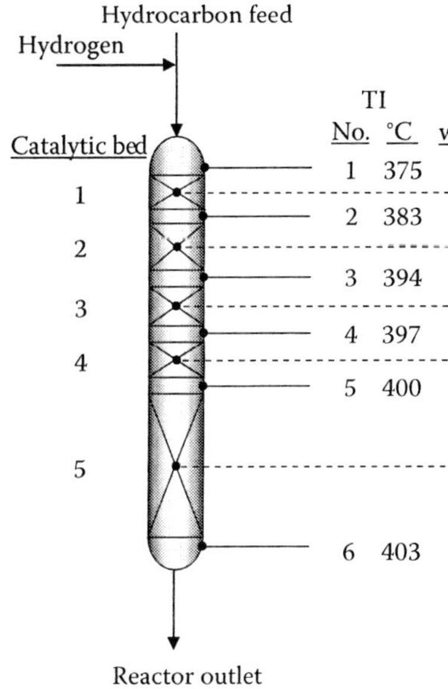

TI No.	TI °C	% Catalyst weight (Wc_i%)	$WABT_i$(°C) Eq. (1)	($WABT_i$) (Wc_i%)
1	375			
		10	380	38.00
2	383			
		20	390	78.00
3	394			
		15	396	59.40
4	397			
		15	399	59.85
5	400			
		40	402	160.80
6	403			
		100		396

$$WABT = 396°C \qquad (Eq. 5.2)$$

$$WABT = \frac{375 + 403}{2} = 389°C \qquad (Eq. 5.3)$$

$$WABT = \frac{375 + 2(403)}{3} = 393°C \qquad (Eq. 5.4)$$

FIGURE 5.22 Different approaches for calculation of *WABT*. (From Stefanidis, G.D. et al., *Fuel Proc. Technol.*, 86: 1761–1775, 2005; Gruia, A., in *Practical Advances in Petroleum Processing*, Hsu, Ch.H. and Robinson, P.R., Eds., Springer, New York, 2006, chap. 7. With permission.)

temperature ($WABT_{EOR}$). Values of these two temperatures depend mainly on the feed type. Typically ($WABT_{EOR} - WABT_{SOR}$) = 30°C. The temperature increase in many cases is very small to prevent catalyst coking, about 1 to 2°C per month. For feed containing a relatively high amount of metals this increase in temperature is required more frequently. When *WABT* reaches a value close to the designed maximum, the catalyst has to be replaced.

Figure 5.23 presents typical results of the effect of reaction temperature and time-on-stream on product quality during hydrotreating of heavy oils. Decay of catalyst activity is clearly observed from this figure. The well-known rapid catalyst deactivation period during the first hours of the run (0 to 100 h) is also observed. Asphaltenes content in product considerably increases in this period, after which it is maintained at a constant value. Being the major coke precursor, this behavior in asphaltenes content indicates that coke is the main source of catalyst deactivation during the initial period of time-on-stream. Ramsbottom carbon (RBC) follows a trend similar to that of asphaltenes, which corroborates this deactivation mechanism since RBC also indicates the tendency of the feed to form coke. On the other hand, metals (Ni and V) follow a different behavior compared with asphaltenes and RBC. The contents of Ni and V in product show a linear increase

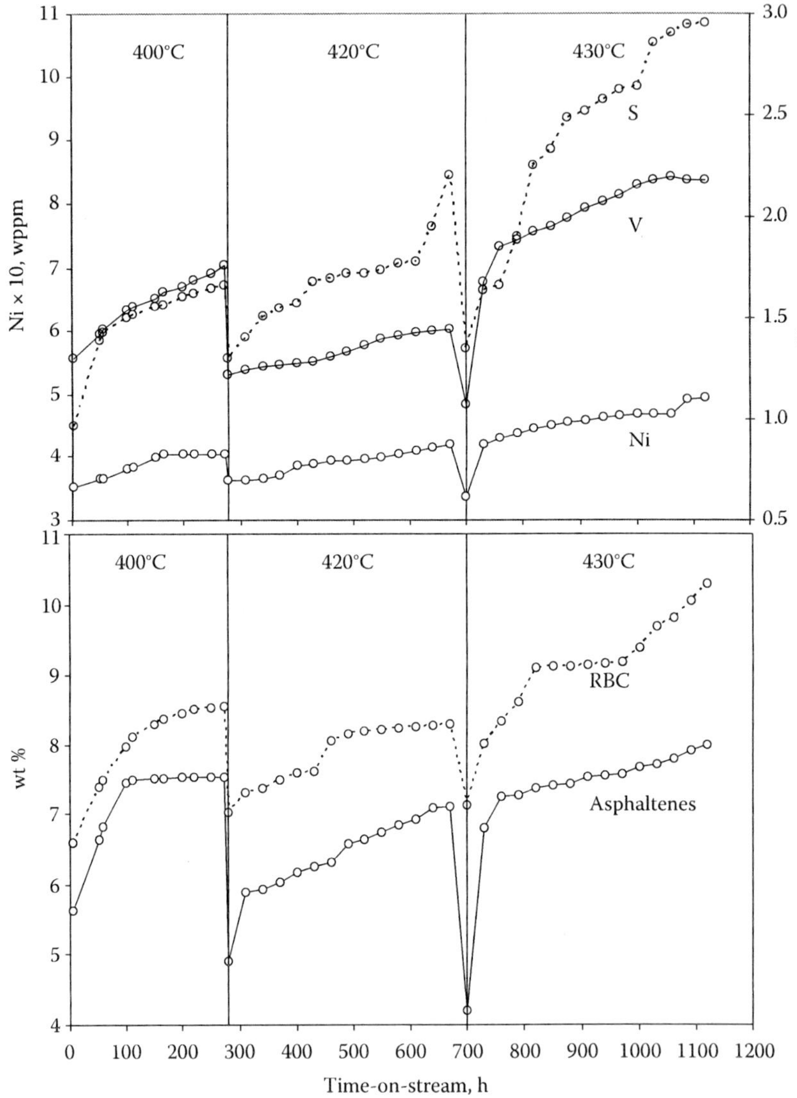

FIGURE 5.23 Effect of temperature and time-on-stream on impurities removal during hydroprocessing of Maya crude. (From Ancheyta, J. et al., *Energy Fuels*, 17: 462–467, 2003. With permission.)

with run of operation. Sulfur content incremented at a similar rate as vanadium content. V removal was higher during all times-on-stream at 420°C than at 400°C, which is a consequence of the higher hydrocracking of asphaltenes activity commonly observed at elevated temperature. Also, removal of vanadium was much faster than nickel removal, indicating that the vanadium moieties are the most reactive. At this temperature level, those metal-containing compounds located at

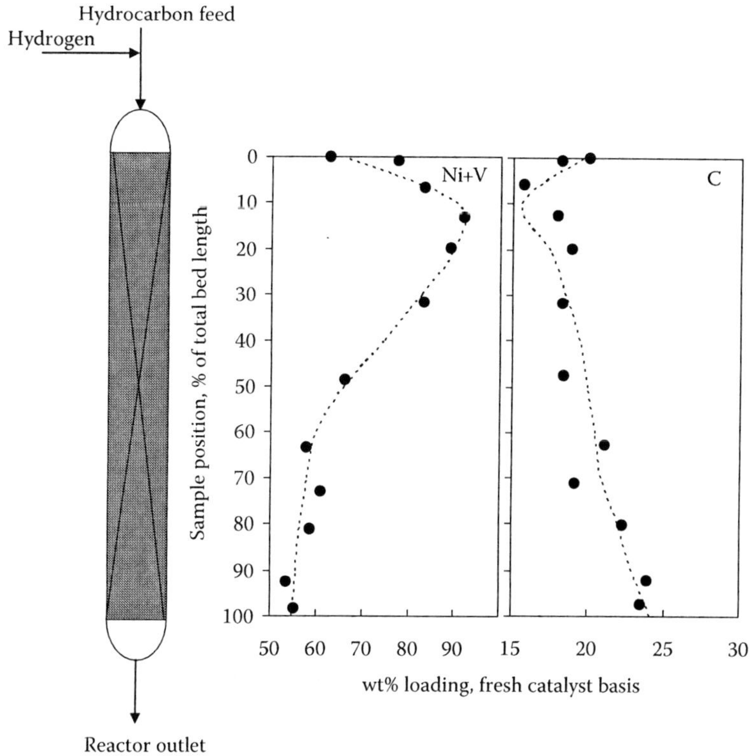

FIGURE 5.24 Metals and carbon concentration profiles on spent catalyst employed for hydroprocessing of residuum. (From Fleisch, T.H. et al., *J. Catal.*, 86: 147–157, 1984. With permission.)

the internal part of the asphaltene molecule are released, and hence can be easily removed. With run of operation, the asphaltenes content continued to more or less have the same values as the first temperature, which confirms that coke deposition is mainly present at short times-on-stream. On the contrary, nickel and vanadium contents are higher than at previous temperature stages.

Characterization of spent catalysts taken at different positions of a fixed catalytic bed during resid hydroprocessing (Figure 5.24) indicates that maximum deposition of metals occurs in the first part of the bed (15 to 20% from top to bottom), while carbon deposition is relatively uniform at the beginning of the bed and steadily increases down the catalyst bed. This behavior in metals and carbon depositions has been attributed to the following (Fleisch et al., 1984):

1. A consecutive mechanism of HDM, where metal-bearing asphaltenes are first cracked to form resins that subsequently demetallate.
2. The top part of the catalytic bed could function primarily as an asphaltene-cracking section.

3. As new metal-bearing resins are generated, the HDM rate increases until the supply becomes depleted and the exponential decline in deposition occurs.
4. Nonisothermality or unusual deactivation behavior near the top of the bed.

5.3.3 H_2-to-Oil Ratio and Recycle Gas Rate

The choice of gas rate is governed by economic considerations. Recycle is used to maintain the hydrogen partial pressure and the physical contact of the hydrogen with the catalyst and hydrocarbon to ensure adequate conversion and impurities removal while minimizing carbon deposition. Increasing the H_2 partial pressure reduces the reactor start-of-run temperature as well as the rate of catalyst deactivation. Above a certain gas rate, the change in hydrogen partial pressure will be relatively small. In general, higher gas rates than necessary incur extra heating and cooling rates, which may outweigh other advantages.

Increasing the recycle gas rate increases the hydrogen-to-oil (H_2/oil) ratio in the reactor. This increased ratio acts in much the same manner as increased hydrogen partial pressure. The H_2/oil ratio is determined by

$$H_2/\text{oil} = \frac{\text{Total gas to the reactor, SCF}/day}{\text{Total feed to the reactor, Bbl}/day}\left[=\right]\frac{SCF}{Bbl} \tag{5.7}$$

In addition to affecting the hydrogen partial pressure, the gas rate is important, as it acts to strip volatile products from the reactor liquids, and thus affects the concentration of various components in the reactive liquid phase. Similarly to H_2 partial pressure, the H_2/oil ratio must be maintained at the design value, and any reduction of it below the design minimum will have adverse effects on the catalyst life.

In the case of EBR, excessive gas rates affect the nature of the ebullating bed, can result in catalyst carryover into the downcomer feeding the ebullating pumps, and increase the concentration of heavy materials in the reactor liquid phase by stripping volatile products, which tends to increase the conversion level but can lead to loss of liquid in the reactor or to an excessive concentration of very heavy materials.

Figure 5.25 presents all the streams involved in the hydrogen circuit of a hydroprocessing plant. The reactor effluent stream is separated in a HPS into liquid hydrotreated products and noncondensable H_2-rich gases (78 to 83 mol% H_2 plus CH_4, C_2H_6, C_3H_8, butanes, pentanes, and H_2S). Hydrogen sulfide, either formed via HDS or present in the reactor feed, is commonly separated with an amine contactor to improve H_2 purity, but lighter hydrocarbon gases are still present in the recycle gases stream. Part of these gases (10 to 15%) is purged to the fuel gas system or to a hydrogen purification process (for example, pressure swing adsorption, or PSA) from which about 20% is lost to the fuel gas system. The other portion of gases leaving the amine contactor is compressed and recycled back to the top of the reactor or used for quenching (80 to 85 mol% H_2). The separated H_2-rich gases from the hydrogen purification process are mixed with the makeup hydrogen and also recycled back to the top of the reactor. Depending

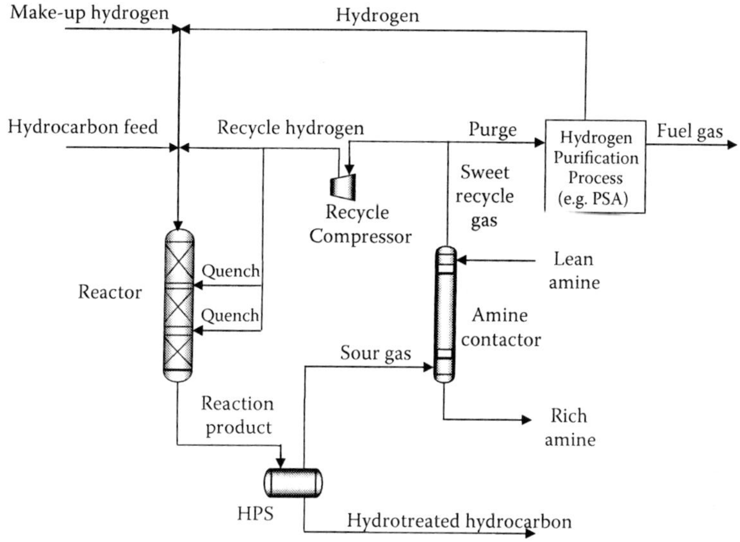

FIGURE 5.25 Hydrogen circuit in a hydroprocessing unit.

upon the source of makeup hydrogen, it is typically available at 96 to 99.9 mol% H_2 purity. On the other hand, the separated hydrotreated liquids flow from the HPS to the low-pressure separator (LPS) and eventually to the fractionation section for further processing (Mehra and Al-Abdulal, 2005).

With high-sulfur feedstocks the level of H_2S can build up to high values, reducing purity of the recycle gas stream and also the hydrogen partial pressure and inhibiting the HDS reaction. Thus, the recycle gas from the HPS is generally water-washed to remove ammonia, preventing the formation of ammonium sulfide, which might form blockages in the reactor effluent cooler, and then is sent to the sour water plant for removal of H_2S. If a recycle gas scrubber is not available, the reactor temperature must be increased to offset the H_2S inhibition, whose effect is greater at higher total reactor pressure.

Makeup hydrogen is very important to compensate hydrogen consumption during hydroprocessing reactions, which is recognized as an important operation factor in refineries from the viewpoint of hydrogen supply investment and energy cost reduction. The usual manner for calculating hydrogen consumption is by means of experimental data either by a hydrogen balance in gas streams or with hydrogen content in the liquid feed and products. Another quick approach is to employ the typical hydrogen consumptions reported in the literature (Edgar, 1993; Speight, 1999). Table 5.3 shows an example of this calculation, in which comparisons with experimental values obtained at pilot scale are presented. The difference among both values, experimental and calculated, is low, which is surely within experimental error. Hydrogen consumption during hydroprocessing is dependent upon the feedstock properties and impurity removals and conversion level. The heavier feed requires substantially more addition of hydrogen to attain a fixed level of upgrading.

TABLE 5.3
Example of Calculation of Hydrogen Consumption

	Feed	Product	Contribution to H_2 Consumption, SCF/Bbl
Sulfur, wt%	6.21	2.19	392
Nitrogen, wt%	0.66	0.38	91
Hydrocracking,[a] vol%	—	7.5	188
HDA of polyaromatics,[b] wt%	25.1	11.7	362
Subtotal	—	—	1033
Metals (Ni + V), wppm	696	260	11.86[c]
Total			1155
Experimental H_2 consumption			1082

Typical H_2 Consumptions, SCF/Bbl (Edgar, 1993)

HDS	95–100 per each 1 wt%
HDN	300–350 per each 1 wt%
HDC	25 per each 1 vol%
HDA	27 per each 1 vol%

Correction in H_2 Consumption by Metals, % (Speight, 1999)

Ni + V, ppm	Correction, %	Ni + V, ppm	Correction, %
0–100	–2	700	12
200	1	800	16
300	2.5	900	21
400	4	1000	28
500	6.5	1100	38
600	9	1200	50

[a] Determined as liquid volumetric expansion.
[b] As hydrodeasphaltenization.
[c] Interpolated from data at the bottom of the table.

5.3.4 Space-Velocity and Fresh Feed Rate

Space-velocity is a process variable normally used to relate the amount of catalyst to the amount of feed and is calculated on a volume basis (liquid hourly space-velocity, or LHSV) or weight basis (weight hourly space-velocity, or WHSV). LHSV and WHSV are calculated as follows:

$$\text{LHSV} = \frac{\text{Total volumetric feed flow rate to the reactor}}{\text{Total catalyst volume}} \left[=\right] h^{-1} \quad (5.8)$$

$$\text{WHSV} = \frac{\text{Total mass feed flow rate to the reactor}}{\text{Total catalyst weight}} \left[=\right] h^{-1} \quad (5.9)$$

LHSV is most commonly used as the process variable in hydroprocessing operations.

Increasing the space-velocity (higher feed rate for a given amount of catalyst) requires a higher reactor temperature to maintain the same impurity removal and conversion levels and results in an increase in the deactivation rate, thus reducing catalyst life.

With a fixed reactor size the residence time is inversely proportional to the fresh feed rate and also to the space-velocity. When the feed rate exceeds the design value, the conversion level will often have to be lower than the design level.

Figure 5.26 shows the influence of LHSV on sulfur, asphaltenes, nitrogen, and metals (Ni + V) contents in products obtained at two temperatures during hydroprocessing of Maya crude. It is seen from the figure that a decrease in

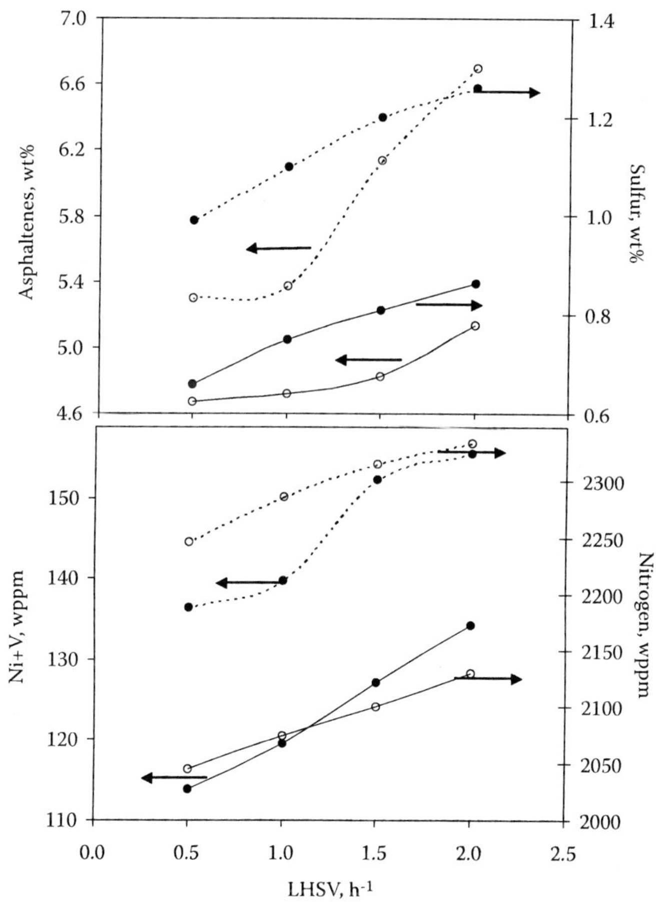

FIGURE 5.26 Effect of LHSV on impurities removal during hydroprocessing of Maya crude: (---) 360°C, (—) 400°C. (From Ancheyta, J. et al., *Appl. Catal. A*, 233: 159–170, 2002. With permission.)

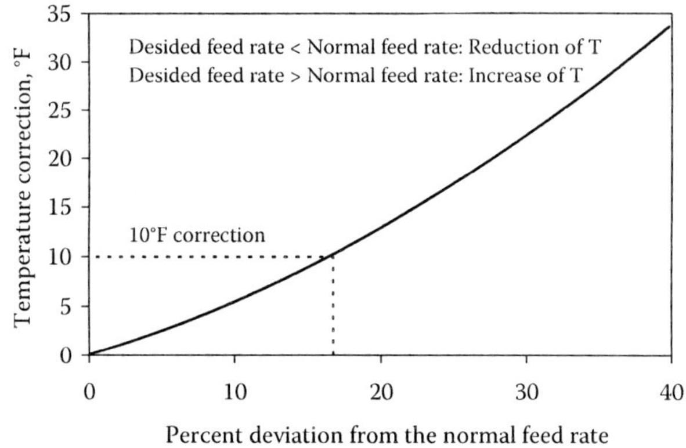

FIGURE 5.27 Typical correction temperature chart as a function of feed rate during hydroprocessing of residua.

LHSV results in diminished asphaltenes, sulfur, nitrogen, and Ni + V contents in the product. Sulfur and nitrogen contents are approximately proportional to space-velocity, while asphaltenes and metals do not totally follow this proportionality, especially at low space-velocity (0.5 to 1.0 h^{-1}). While sulfur and nitrogen continue to be removed at approximately a constant reaction rate when LHSV is reduced from 1.0 to 0.5 h^{-1}, there are no significant changes in metal and asphaltene removal with the same reduction in space-velocity.

Similarly to reaction temperature, there are some charts based on commercial experiences that help us to know the required temperature when a change in feed rate is observed without affecting the desired sulfur content in the product. Figure 5.27 shows an example of these charts. For instance, if the actual feed rate is 30,000 Bbl/day and it drops to 25,000 Bbl/day, the correction in reactor temperature would be

$$Deviation = \left(\frac{30{,}000 - 25{,}000}{30{,}000} \right) \times 100 = 16.7\%$$

From Figure 5.27, 16.7% deviation corresponds to a 10°F temperature correction. Since the desired feed rate is lower than the actual feed rate, that is, desired LHSV < actual LHSV, the temperature correction must be a decrease.

REFERENCES

Al-Dahhan, M.H., Larachi, F., Dudukovic, M.P., and Laurent, A. 1997. High pressure trickle-bed reactors: a review. *Ind. Eng. Chem. Res.* 36:3292–3314.

Agralwal, R.M. and Wei, J. 1984. Hydrodemetallation of nickel and vanadium porphyrins. 1. Intrinsic kinetics. *Ind. Eng. Chem. Proc. Des. Dev.* 23:505–514.

Ancheyta, J., Aguilar, E., Salazar, D., Betancourt, G., and Leiva, M. 1999a. Hydrotreating of straight run gas oil-light cycle oil blends. *Appl. Catal. A* 180:195–205.

Ancheyta, J., Aguilar, E., Salazar, D., Marroquin, G., Quiroz, G., and Leiva, M. 1999b. Effect of hydrogen sulfide on the hydrotreating of middle distillates over $Co/MoAl_2O_3$ catalyst. *Appl. Catal. A* 183:265–272.

Ancheyta, J., Betancourt, G., Centeno, G., and Marroquin, G. 2003b. Catalyst deactivation during hydroprocessing of Maya heavy crude oil. II. Effect of reaction temperature during time-on-stream. *Energy Fuels* 17:462–467.

Ancheyta, J., Betancourt, G., Marroquín, G., Centeno, G., Alonso, F., and Muñoz, J.A. 2006. Process for the Catalytic Hydrotreatment of Heavy Hydrocarbons of Petroleum. U.S. Patent, pending.

Ancheyta, J., Betancourt, G., Marroquin, G., Centeno, G., Castañeda, L.C., Alonso, F., Muñoz, J.A., Gómez, M.T., and Rayo, P. 2002b. Hydroprocessing of Maya heavy crude oil in two reaction stages. *Appl. Catal. A* 233:159–170.

Ancheyta, J., Centeno, G., Trejo, F., and Marroquin, G. 2003a. Changes in asphaltene properties during hydrotreating of heavy crudes. *Energy Fuels* 17:1233–1238.

Ancheyta, J., Marroquin, G., Angeles, M.J., Macías, M.J., Pitault, I., Forissier, M., and Morales, R.D. 2002a. Some experimental observations of mass transfer limitations in a hydrotreating trickle-bed pilot reactor. *Energy Fuels* 16:1059–1067.

Babich, I.V. and Moulijn, J.A. 2003. Science and technology of novel processes for dep desulfurization of oil refinery streams: a review. *Fuel*, 82:607–631.

Bhaskar, M., Valavarasu, G., Sairam, B., Balaraman, K.S., and Balu, K. 2004. Three-phase reactor model to simulate the performance of pilot-plant and industrial trickle-bed reactors sustaining hydrotreating reactors. *Ind. Eng. Chem. Res.* 43:6654–6669.

Biasca, F.E., Dickenson, R.L., Chang, E. Johnson, H.E., Bailey, R.T., and Simbeck, D.R. 2003. Upgrading Heavy Crude Oils and Residues to Transportation Fuels: Technology, Economics, and Outlook. SFA Pacific, Inc., Phase 7, California.

Billon, A., Peries, J.P., Espeillan, M., and des Courieres, T. 1991. Hyvahl F versus Hyvahl M, Swing Reactor or Moving Bed. Paper presented at the 1991 NPRA Annual Meeting, San Antonio, TX, March 17–19.

Bingham, F.E. and Christensen, P. 2000. Revamping HDS Units to Meet High Quality Diesel Specifications. Paper presented at the Asian Pacific Refining Technology Conference, Kuala Lumpur, Malaysia, March 8–10.

Chen, J., Te, M., Yang, H., and Ring, Z. 2003. Hydrodesulfurization of dibenzotiophenic compounds in a light cycle oil. *Petrol. Sci. Technol.* 21:911–935.

Chou, T. 2004. Causes of fouling in hydroprocessing units. *Petrol. Technol. Q.* 3:1–5.

Daniel, M., Lerman, D.B., and Peck, L.B. 1988. Amoco's LC-Fining Residue Hydrocracker Yield and Performance Correlations from a Commercial Unit. Paper presented at the 1988 NPRA Annual Meeting, San Francisco, CA, March 15–17.

EMRE. 2006. ExxonMobil's dewaxing technology selected for two Petro-Canada. *Focus Catalysts* 2:5.

Edgar, M.D. 1993. Hydrotreating, Q&A. Paper presented at the 1993 NPRA Annual Meeting, San Antonio, TX, March.

Euzen, J.P. 1991. Moving-bed process for residue hydrotreating. *Rev. Inst. Fr. Petrol* 46:517–527.

Fleisch, T.H., Meyers, B.L., Hall, J.B., and Ott, G.L. 1984. Multitechnique analysis of a deactivated resid demetallation catalyst. *J. Catal.* 86:147–157.

Fukuyama, H., Terai, S., Uchida, M., Cano, J.L., and Ancheyta, J. 2004. Active carbon catalyst for heavy oil upgrading, *Catal. Today* 98:207–215.

Furimsky, E. 1998. Selection of catalysts and reactors for hydroprocessing. *Appl. Catal. A* 171:177–206.

Gawel, I. and Baginska, K. 2004. Characterization of deposits collected from residue hydrotreatment installation. Preprint. *Pap. Am. Chem. Soc. Div. Petrol. Chem.* 49:265–267.

Gianetto, A. and Specchia, V. 1992. Trickle-bed reactors: state of art and perspectives. *Chem. Eng. Sci.* 47:3197–3213.

Gorshteyn, A., Kamiesnky, P., Davis, T., Novak, W., and Lee, M. 2002. ExxonMobil Catalytic Dewaxing: A Commercial Proven Technology. Paper presented at the 2nd Russian Refining Technology Conference, Moscow, September 26–27.

Gosselink, J.W. 1998. Sulfide catalysts in refineries. *Cattech* 2:127–144.

Gruia, A. 2006. Recent advances in hydrocracking. In *Practical Advances in Petroleum Processing*, Hsu, Ch.H. and Robinson, P.R. (Editors). Springer-Verlag, New York, chap. 7.

Ho, T.C. 2003. Hydrodesulfurization with RuS_2 at low hydrogen pressures. *Catal. Lett.* 89:21–25.

Hunter, M.G., Pappal, D.A., and Pesek, C.L. 1997. MAK moderate-pressure hydrocracking. In *Handbook of Petroleum Refining Processes*, Meyers, R.A. (Editor). McGraw-Hill, New York, chap. 7.1.

Inoue, S., Asaoka, S., and Nakamura, M. 1998. Recent trends of industrial catalyst for resid hydroprocessing in Japan. *Catal. Surv. Jpn.* 2:87–97.

Kam, E.K.T., Al-Mashan, M.-H., and Al-Zami, H. 1999. The mixing aspects of NiMo and CoMo hydrotreating catalysts in ebullated-bed reactors. *Catal. Today* 48:229–236.

Korsten, H. and Hoffmann, U. 1996. Three-phase reactor model for hydrotreating in pilot trickle-bed reactors. *AIChE J.* 42:1350–1357.

Kressmann, S., Guilaume, D., Roy, M., and Plain, C. 2004. A New Generation of Hydroconversion and Hydrodesulfurization Catalysts. Paper presented at the 14th Annual Symposium Catalysis in Petroleum Refining and Petrochemicals, Dhahran, Saudi Arabia, December 5–6.

Kressmann, S., Morel, F., Harlé, V., and Kasztelan, S. 1998. Recent developments in fixed-bed catalytic residue upgrading. *Catal. Today* 43:203–215.

Langston, J., Allen, L., and Davé, D. 1999. Technologies to achieve 2000 diesel specifications. *Petrol. Technol. Q.* 2:65.

Marafi, M., Al-Barood, A., and Stanislaus, A. 2005. Effect of diluents in controlling sediment formation during catalytic hydrocracking of Kuwait vacuum residue. *Petrol. Sci. Technol.* 23:899–908.

Marroquin, G., Ancheyta, J., Vera, I., and Díaz, J.A.I. 2007. Sediments formation during fixed-bed hydrotreating of heavy feedstocks. *Catal. Today*, submitted.

Mederos, F.S., Rodriguez, M.A., Ancheyta, J., and Arce, E. 2006. Dynamic modeling and simulation of catalytic hydrotreating reactors. *Energy Fuels* 20:936–945.

Mehra, Y.R. and Al-Abdulal, A.H. 2005. Hydrogen Purification in Hydroprocessing (HPH^{SM} Technology). Paper presented at the 103rd NPRA Annual Meeting, San Francisco, March 13–15.

Minderhoud, J.K., van Veen, J.A.R., and Hagan, A.P. 1999. Hydrocracking in the year 2000: a strong interaction between technology development and market requirements. *Stud. Surf. Sci. Catal.* 127:3–20.

Mochida, I. and Choi, K. 2004. An overview of hydrodesulfurization and hydrogenation. *J. Jpn. Petrol. Inst.* 47:145–163.

Mochida, I. and Choi, K. 2006. Current progress in catalysts and catalysis for hydrotreating. In *Practical Advances in Petroleum Processing*, Hsu, Ch.H. and Robinson, P.R. (Editors). Springer-Verlag, New York, chap. 9.

Mochida, I., Sakanishi, K., Ma, X., Nagao, S., and Isoda, T. 1996. Deep hydrodesulfurization of diesel fuel: design of reaction process and catalysts. *Catal. Today* 29:185–189.

Mochida, I., Zhao, X., Sakanishi, K., Yamamoto, S., Takashima, H., and Uemura, S. 1989. Structure and properties of sludges produced in the catalytic hydrocracking of vacuum residue. *Ind. Eng. Chem. Res.* 28:418–421.

Morel, F., Kressmann, S., Harlé, V., and Kasztelan, S. 1997. Processes and catalysts for hydrocracking of heavy oil and residues. In *Hydrotreatment and Hydrocracking of Oil Fractions*, Froment, G.F., Delmon, B., and Grange, P. (Editors). Elsevier, Amsterdam.

Moulijn, J.A., van Diepen, A.E., and Kapteijn, F. 2001. Catalyst deactivation: is it predictable? What to do? *Appl. Catal. A* 212:3–16.

Muñoz, J.A.D., Alvarez, A., Ancheyta, J., Rodríguez, M.A., and Marroquín, G. 2005. Process heat integration of a heavy crude hydrotreatment plant. *Catal. Today* 109:214–218.

National Petroleum Council (56 NPC). 2000. *U.S. Petroleum Refining: Assuring the Adequacy and Affordability of Cleaner Fuels*, June, chap. 7, pp. 132–133.

Ouwerkerk, C.E.D., Bratland, E.S., Hagan, A.P., Kikkert, B.L.J.P., and Zonnevylle, M.C. 1999. Performance optimisation of fixed bed processes. *Petrol. Technol. Q.* 4:21–30.

Panariti, N., del Bianco, A., del Piero, G., and Marchionna, M. 2000. Petroleum residue upgrading with dispersed catalysts. Part 1. Catalysts activity and selectivity. *Appl. Catal. A* 204:203–213.

Phillips, G. and Lie, F. 2002. Advances in Residue Upgrading Technologies Offer Refiners Cost-Effective Options for Zero Fuel Oil Production. Paper presented at the 2002 European Refining Technology Conference, Paris, November.

Quann, R.J., Ware, R.A., Hung, Ch., and Wei, J. 1988. Catalytic hydrodemetallation of petroleum. *Adv. Chem. Eng.* 14:95–259.

Rana, M.S., Samano, V., Ancheyta, J., and Diaz, J.I.A. 2007. A review of recent advances of processing technologies for upgrading of heavy oils and residua. *Fuel*, 86: 1216–1231.

Reynolds, B.E., Bachtel, R.W., and Yagi, K. 1992. Chevron's Onstream Catalyst Replacement (OCR™) Provides Enhanced Flexibility to Residue Hydrotreaters. Paper presented at the 1992 NPRA Annual Meeting, New Orleans, March 22–24.

Robinson, P.R. and Dolbear, G.E. 2006. Hydrotreating and hydrocracking: fundamentals. In *Practical Advances in Petroleum Processing*, Hsu, Ch.H. and Robinson, P.R. (Editors). Springer, New York, chap. 7.

Rodriguez, M.A. and Ancheyta, J. 2004. Modeling of hydrodesulfurization (HDS), hydrodenitrogenation (HDN), and the hydrogenation of aromatics (HDA) in a vacuum gas oil hydrotreater. *Energy Fuels* 18:789–794.

Ruiz, R.S., Alonso, F., and Ancheyta, J. 2004a. Minimum fluidization velocity and bed expansion characteristics of hydrotreating catalysts in ebullated-bed systems. *Energy Fuels* 18:1149–1155.

Ruiz, R.S., Alonso, F., and Ancheyta, J. 2004b. Effect of high pressure operation on overall phase holdups in ebullated-bed reactors. *Catal. Today* 98:265–271.

Ruiz, R.S., Alonso, F., and Ancheyta, J. 2005. Pressure and temperature effects on the hydrodynamic characteristics of ebullated-bed systems. *Catal. Today* 109:205–213.

Sarli, M.S., McGovern, S.J., Lewis, D.W., and Sinder, P.W. 1993. Improved Hydrocracker Temperature Control: Mobil Quench Zone Technology. Paper presented at the 1993 NPRA Annual Meeting, Houston, TX, March 21–23.

Scheffer, B., van Koten, M.A., Robschlager, K.W., and de Boks, F.C. 1988. The shell residue hydroconversion process: development and achievements. *Catal. Today* 43:217–224.

Scheuerman, G.L., Johnson, D.R., Reynolds, B.E., Bachtel, R.W., and Threlkel, R.S. 1993. Advances in Chevron RDS technology for heavy oil upgrading flexibility. *Fuel Process. Technol.* 35:39–54.

Schulman, B.L. and Dickenson, R.L. 1991. Upgrading heavy crudes: a wide range of excellent technologies now available. In *UNITAR 5th International Conference*, Caracas, Venezuela, August, pp. 105–113.

Sie, S.T. 2001. Consequences of catalyst deactivation for process design and operation. *Appl. Catal. A* 212:129–151.

Sie, S.T. and de Vries, A.F. 1993. Hydrotreating Process. European Patent Application 0553920.

Speight, J.G. 1999. *The Chemistry and Technology of Petroleum*, 3rd ed. Marcel Dekker, New York.

Speight, J.G. 2000. *The Desulfurization of Heavy Oils and Residua*, 2nd ed. Marcel Dekker, New York.

Speight, J.G. 2001. *Handbook of Petroleum Analysis*. John Wiley & Sons, New York.

Stefanidis, G.D., Bellos, G.D., and Papayannakos, N.G. 2005. An improved weighted average reactor temperature estimation for simulation of adiabatic industrial hydrotreaters. *Fuel Proc. Technol.* 86:1761–1775.

Storm, D.A., Decanio, S.J., Edwards, J.C., and Sheu, E.Y. 1997. Sediment formation during heavy oil upgrading. *Petrol. Sci. Technol.* 15:77–102.

Takahashi, T., Higashi, H., and Kai, T. 2005. Development of a new hydrodemetallization catalyst for deep desulfurization of atmospheric residue and the effect of reaction temperature on catalyst deactivation. *Catal. Today* 104:76–85.

Topsøe, H., Clausen, S., and Massoth, F.E. 1996. *Hydrotreating Catalysis Science and Technology*. Springer-Verlag, Berlin.

Van Gineken, A.J.J., Van Kessel, M.M., Pronk, K.M.A., and Renstrom, G. 1975. Shell process desulfurizes resids. *Oil Gas J.*, 73: 59–63.

Van Zijll Langhout, W.C., Ouwerkerk, C., and Pronk, K.M.A. 1980. New process hydrotreats metal-rich feedstocks. *Oil Gas J.*, 78: 120–126.

6 Characteristics of Heavy Oil Hydroprocessing Catalysts

Jorge Ramírez, Mohan S. Rana, and Jorge Ancheyta

CONTENTS

6.1 INTRODUCTION

During the past years the development and use of technologies for hydroprocessing of heavy oils has been slow because of the availability of light sweet crudes, mainly from the Middle East, Nigeria, and the North Sea. Additionally, the small price differential between light and heavy oils has oriented refiners to the use of feedstocks based on light oils or mixtures with a high light/heavy oil ratio. Refining companies have been under pressure to achieve a short payback period for investment, so funds have been used, where this can be achieved, in the upstream processing of oil. However, the world's reserves of sweet crude oil resources are expected to decline. At the same time, an increasing proportion of the oil reserves in some producing countries is of heavy crudes, which contain a lower amount of distillates and more residues. So, it is foreseen that in the future, some refineries will increasingly replace light crudes with heavier oils, like those produced today in countries like Venezuela and Mexico.

Today we observe an increase in the rate of decline of the size of the residual fuel oil market, associated with the more strict legislation regarding the sulfur content of fuel oil and a growing demand for middle distillates. The imbalance created by the increasing demand for distillate products and the declining one for heavy fuel oil can only be solved by increasing the upgrading of heavy and residual oils. Moreover, the high oil prices observed today and the declining supplies of light crude oil favor investments in residue and heavy oil conversion and upgrading. The availability of heavy oils and the price differential between heavy and light oils, which is substantial although not stable, gives an incentive for direct processing of heavy petroleum crude oil (Figure 6.1).

Petroleum is a very complex mixture consisting of hydrocarbons of different sizes and compounds containing sulfur, nitrogen, and oxygen, in addition to metals like vanadium, nickel, and iron. About 35% or more of the oil components are in the gasoline boiling range. Additionally, oil contains varying proportions of higher-boiling compounds up to the nonvolatile compounds present in lubricants and asphalts (Schabron and Speight, 1997). Moreover, the properties of petroleum are not uniform all over the world, and its composition depends on

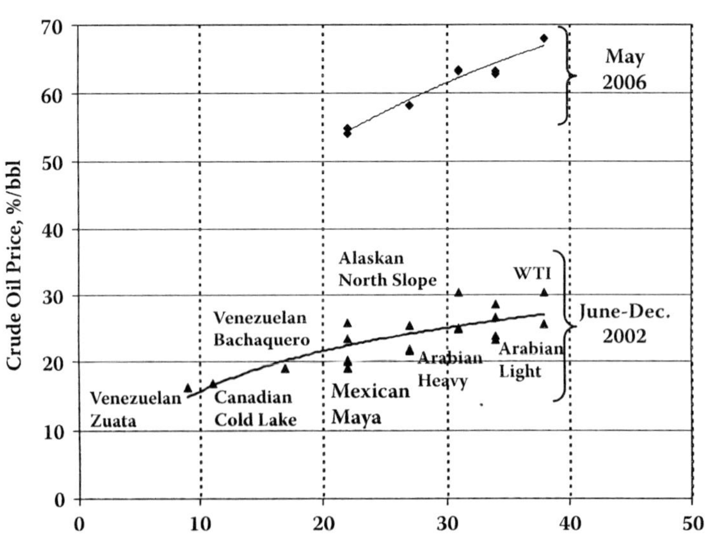

FIGURE 6.1 Relationship between crude oil price and API gravity.

the characteristics of the extraction site, on the manner of production, and on the life of the reservoir.

Because of the variability in the quality of petroleum feeds, the configuration and complexity of refinery processes can vary substantially according to the strategy of each refinery to meet the new goals of fuel specification. So, it appears that there is no generic technology solution, and each refinery must look at its own unique requirements and integration opportunities.

To select adequate methods of processing heavy oils and residua, it is necessary to define the feedstock in as much detail as possible. However, the chemical composition of heavy oils and residua is more complex than that of light oils, and their precise chemical and physical constitution is not well understood. This is due, in part, to the lack of acceptable analytical procedures capable of dealing with high concentrations of metals, sulfur, nitrogen, and oxygen compounds, as well as with the high proportions of asphaltenes and resins (50% or more) present in heavy feeds. It has been found that the properties of residua determined by conventional methods used for lighter cuts are not good predictors of their behavior in upgrading processes (Dawson et al., 1989).

With the increasingly heavy feedstock available for processing, production of the required high-quality transportation fuels has to be achieved by either residue upgrading or additional heavy crude processing (Weiss and Schmalfeld, 1999). Therefore, the development of upgrading technology for these heavy feeds with low API gravity is becoming more important for refineries. However, the growing use of heavy oils (less than 20° API) as refinery feedstocks entails unique additional difficulties related to the presence of high viscosity at operational temperatures and high concentration of elements such as sulfur, nitrogen, asphaltenes, and heavy metals (nickel and vanadium) (Speight, 1991). An account of the catalysis research performed to meet the future refining needs is given in Absi-Halabi et al. (1997).

Residue oil upgrading is the conversion of no longer distillable oil residues into distillates by increasing the H/C ratio. This may be done by the removal of carbon (coking and deasphalting) or hydrogen addition (catalytic hydrotreating and hydrocracking). Treatment of distillation residua is necessary to increase the liquid yield per unit of crude. For this, several commercial processes, such as visbreaking and delayed fluid and flexicoking, based on carbon rejection, are offered commercially (Speight, 1991; MacKetta, 1992). Coking is a severe thermal conversion of the feed at temperatures between 480 and 550°C and vapor phase residence times of 20 seconds or more, providing for a significant degree of cracking and dehydrogenation of the feed, which makes subsequent processing more cumbersome and produces low-value by-products such as gas and coke. On the other hand, hydrocracking processes, based on the addition of hydrogen, can also be used for upgrading heavy crudes and residua (Scherzer and Gruia, 1996; Schuetze and Hofmann, 1997; Dickenson et al., 1997). In this type of process, residual oils are hydrogenated at high pressures and temperatures in the liquid phase because such oils can no longer be evaporated.

Of the available possibilities for treatment of residua, hydrogen addition processes lead to high hydrogen consumption but higher liquid yields. These processes, which provide the feedstock for the subsequent fluid catalytic cracking (FCC) process, require the use of well-designed catalysts capable of dealing with the high concentrations of metals and asphaltenes present in the feedstock. Moreover, the multifunctional catalysts used for hydrocracking processes become poisoned by coke deposition and the heavy metals present in the feed create a hazardous waste, which has to be disposed of properly and with safety (Furimsky, 1996). A high catalyst demetallization function is necessary because vanadium destroys the zeolitic catalyst used in the subsequent FCC process. Besides, the concentration of nitrogen compounds must be taken to a minimum to avoid poisoning of the catalyst acid sites in this and the subsequent FCC process. Although in the hydrocracking process the amount of metals is not as critical as in FCC, the elimination of nitrogen compounds is determinant to avoid poisoning of the catalyst acid sites.

Finally, one may well recall that although a good match between the properties of feed and catalyst has to be achieved to obtain high hydroprocessing conversions, catalyst life, and stability, the final selection of a catalyst for a particular process has to be based on price and performance (van Kessel et al., 1987). Below we discuss some of the important features that must be taken into account when processing heavy feeds.

6.2 CHARACTERISTICS OF HEAVY PETROLEUM FEEDS

Heavy petroleum feeds such as vacuum (VR) and atmospheric (AR) residua as well as bitumen from tar sands are composed primarily of heavy hydrocarbons (asphaltenes and resins) and metals, predominantly in the form of porphyrins. The amount of VR and AR varies according to the crude source. The differences between light and heavy feedstocks are clearly stressed in Table 6.1, where the variabilities in yields of atmospheric (>345°C) and vacuum (>565°C) residua from several crudes are shown. Heavy crudes produce more than twice the amount of vacuum residue than light ones.

Information on the properties of heavy feeds is available in the literature (Suzuki et al., 1982; Gray, 1994; Wiehe and Liang, 1996; Speight, 1998, 1999; Hauser et al., 2005). Viscosity, which depends on the amount of >350°C residue and the presence of metals and asphaltenes, dictates some of the main problems in the hydroprocessing of heavy feeds. At ambient conditions a vacuum residue is in a semisolid state and heating is necessary to achieve a viscosity low enough to allow adequate pumping. In the case of residua, to avoid reheating, hydroprocessing is integrated after distillation. For heavy crudes upgrading has to be performed on site to achieve enough pumpability for pipelining (Ancheyta et al., 2005a). Blending the heavy crude or residue with a lighter fraction can be an option to improve pumpability. However, attention has to be paid to the compatibility of the mixture to avoid the formation of sediments (Rayo et al., 2004).

TABLE 6.1
Crude Oil Properties and Yields of Atmospheric (>345°C) and Vacuum (>565°C) Residues*

Crude	Composition of Whole Crude, wt% S	N (total)	Metals (Ni+V)	Yield of AR (>345°C)	Yield of VR (>565°C)
Alaskan North	1.1	—	—	51.5	21.4
Bachaquero	2.4	0.43	453	70.2	38.0
Cold Lake	4.11	—	—	83.1	50.0
Kuwait Export	2.52	—	—	45.9	21.8
Maya	3.31	0.36	325	56.4	31.2
North Slope	1.06	—	—	52.6	18.0
Santa Barbara	0.59	0.08	10	—	—
Zuata	4.16	0.69	553	84.9	51.0

* Data taken from various literature sources.

As mentioned before, petroleum is a complex liquid mixture whose chemical composition and properties vary depending on their type and origin. In general, heavy feeds can be described as colloidal solutions consisting of three main fractions with increasing order of molecular weight: oils, resins, and asphaltenes. The latter occupy the core of micelles, with resins adsorbed on their surface acting as dispersing agents. The amount of each fraction varies significantly with the type of feed. Some differences in the content of asphaltenes, resins, and oils between light and heavy feeds are shown in Table 6.2. Moreover, Table 6.3 shows that important differences exist between light and heavy oils in the relevant parameters for hydroprocessing.

TABLE 6.2
Properties and Amount of Asphaltenes, Resins, and Oils in Light, Heavy Crudes, and Residue*

	API Gravity	Asphaltenes, wt%	Resins, wt%	Oils, wt%	Contaminants S, wt%	N, wt%	Metals (Ni+V), ppm
Extra light	> 50	0–< 2	0.05–3	—	0.02–0.2	0.0–0.01	< 10
Light crude	22–32	<0.1–12	3–22	67–97	0.05–4.0	0.02–0.5	10–200
Heavy crude	10-22	11–25	14–39	24–64	0.1–5.0	0.2–0.8	50–500
Extra heavy	< 10	15–40	—	—	0.8–6.0	0.1–1.3	200–600
Residue	—	15–30	25–40	<49	—	—	100– >1000

*Data taken from various literature sources.

TABLE 6.3
Properties of Crude Oils*

Properties	Boscan	Athabasca	Maya Crude	Arabian Heavy	Arabian Light
Sulfur, wt%	5.5	3.9	3.52	2.85	1.79
Nitrogen, wt%	0.52	0.41	0.32	0.2	0.1
Ni + V, ppm	340	253	325	66	22
Asphaltene, wt%	11.9	17.1	12.7	—	—
Carbon Conradson, wt%	10.4	18.5	10.8	7	3
API gravity	10.4	7.1	22	27.9	33.4

*Data taken from various literature sources.

Heavy feeds contain aggregates of resins and asphaltenes dissolved in the oil fraction held together by weak physical interactions. With resins being less polar than asphaltenes but more polar than oil, equilibrium between the micelles and the surrounding oil is established, leading to homogeneity and stability of the colloidal system. If the amount of resins decreases, the asphaltenes coagulate, forming sediments. Evidently this fact is important during hydroprocessing of heavy feeds since asphaltenes need to be eliminated at about the same rate as resins to maintain the stability of the colloidal system and avoid the deposition of asphaltenes on the surface of the catalyst (Ancheyta et al., 2005a). Clearly, tuning hydrogenating and hydrocracking functionalities of the catalyst is of outmost importance to achieve long catalyst life during the hydrotreating of heavy feeds.

Direct hydrotreating of heavy oil to produce lighter synthetic crude oil (SCO) has been addressed as a means to provide reasonable feedstock for FCC and hydrocracking (Ancheyta et al., 2005b). Although this alternative would require larger equipment to process the whole crude volume rather than only the residual fraction, some advantages are worth mentioning: a better-priced synthetic crude oil would be produced, the problems of processing heavy residua would diminish, and residue transportation would be avoided when processing is not done at the site where the residue is produced.

One of the main problems during the hydroprocessing of heavy feeds is the precipitation of asphaltenes on the catalyst surface. Asphaltenes act as coke precursors, causing catalyst deactivation. They are responsible for the formation of deposits on the refinery equipment, which needs to be taken out of service for deposit removal, with the consequent cost increase. Additionally, frequent replacement of the catalyst due to loss of activity caused by coke and metal deposition is an important imbalance to the economics of the process.

Asphaltenes are complex polar structures with polyaromatic character containing metals (mostly Ni and V) that can not be properly defined according to their chemical properties. Instead, they are defined according to their solubility.

Asphaltenes are the compounds that precipitate from petroleum by addition of at least 40 volumes of *n*-heptane or *n*-pentane for each volume of crude. Asphaltenes precipitated with *n*-heptane have a lower H/C ratio than those precipitated with n-pentane. In contrast, asphaltenes obtained with n-heptane are more polar, have a greater molecular weight, and display higher N/C, O/C, and S/C ratios than those obtained with *n*-heptane. As shown in Table 6.2, depending on the crude source, the percent content of asphaltenes can vary widely from light (0.1 to 12 wt%) to heavy (11 to 45 wt%) crude.

Asphaltenes from heavy crudes are constituted by condensed aromatic nuclei carrying alkyl groups, alicyclic systems, and heteroelements (Dickie and Yen, 1967; Tynan and Yen, 1970). Asphaltene molecules are grouped together in systems of up to five or six sheets, which are surrounded by the so-called maltenes (all those structures different from asphaltenes that are soluble in n-heptane). The exact structure of asphaltenes is difficult to obtain, and several structures have been proposed for the asphaltenes present in different crudes (Speight and Moschopedis, 1979, 1981; Speight, 1981; Beaton and Bertolacini, 1991). The length of the alkylic chains in asphaltenes has been the subject of different studies. Mojelsky et al. (1992) found chains of 3 or 4 carbon atoms, while Speight (1981) and Savage and Klein (1989) found alkylic chains of up to 30 carbon atoms. Other studies on the structure of asphaltenes were performed by Bestougeff and Byramjee (1994), Shirokoff et al. (1997), Miller et al. (1998), and Mullins and Groenzin (1999).

An asphaltene molecule may be 4 to 5 nm in diameter, which is too large to pass through a micropore or even some mesopores in the catalyst (Larson and Beuther, 1966). Metals in the asphaltene aggregates are believed to be present as organometallic compounds associated with the asphaltene sheets, making the asphaltene molecule heavier than its original structure, as shown in Figure 6.2. The characterization studies on Ni and V ethioporphyrins indicate that they have diameters of around 1.6 nm (Fleischer, 1963; Fleisch et al., 1984).

Usually the tendency toward coke formation during hydroprocessing increases with the content of asphaltenes. Coke formation will poison the catalyst. So a high content of asphaltenes creates a refining problem that is not easy to solve. To avoid coke formation, the catalyst needs to have a high hydrogenating functionality. However, hydrogenation of the aromatic rings in the maltene fraction will reduce its aromatic character, making it more paraffinic and causing instability of the colloidal system, leading to precipitation of asphaltenes on the surface of the catalyst.

From the hydroprocessing point of view, the presence of a large concentration of metals is of great importance in the upgrading of heavy feeds (Goulon et al., 1984; Jacobsen et al., 1987). Although several metals such as Ni, V, Fe, Ti, and others are present in heavy feeds, Ni and V are predominant, with the latter being present in higher concentrations, varying from a few to several thousand ppm. Clay-like mineral matter is also found, and alkali metals in the form of chlorides and bromides are also present (Dekkers, 1999). It is considered that Ni and V are in the form of porphyrins of various types (Goulon et al., 1984; Jacobsen et al., 1987)

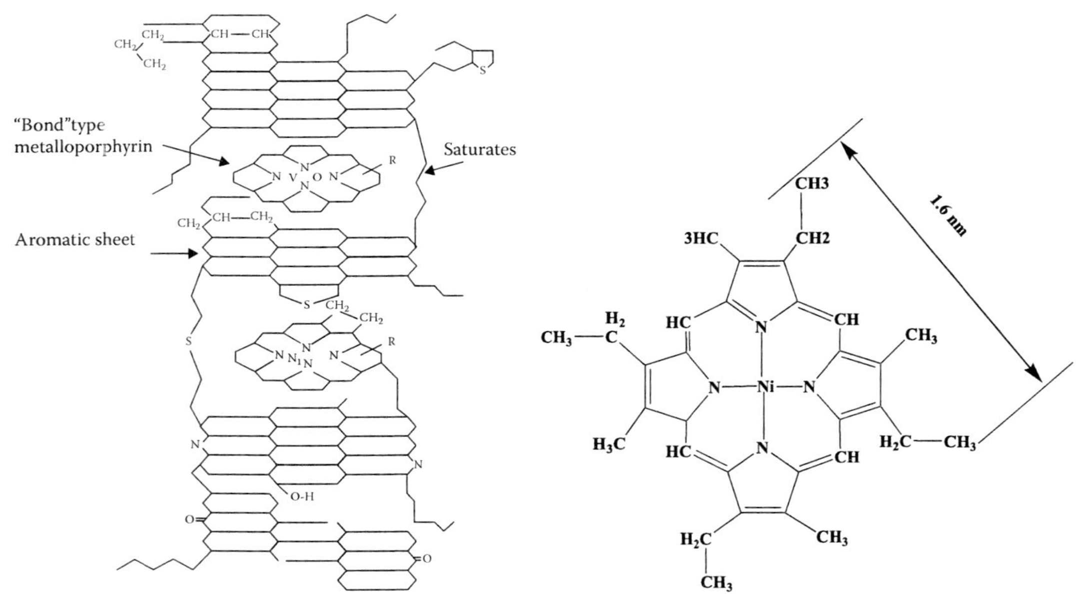

(a) (b)

FIGURE 6.2 Hypothetical asphaltene molecule and its interaction with organometallic compounds. ((a) Adapted from Speight, J.G., in *Asphaltenes and Asphalts*, Vol. 1, *Developments in Petroleum Science*, Yen, T.F. and Chilingarian, G.V., Eds., Elsevier Science, Amsterdam, 1994, chap. 2. (b) Adapted from Fleischer, E.B., *J. Am. Chem. Soc.*, 85: 146–148, 1963. With permission.)

and that they are associated with the asphaltene aggregates in a porphyrin-like form by strong noncovalent interactions (Sakanishi et al., 1997; Ancheyta et al., 2005a). However, it has been suggested that porphyrins account for only about half of the concentration of V and Ni in petroleum (Mitchell, 1990). The kinetics of Ni and V elimination have been studied in the past, and the kinetics of Ni elimination over a Mo/Al_2O_3-TiO_2 catalyst using Ni 5,10,15,20-tetraphenyl porphyrin (Ni-TPP) as a model molecule have been recently described by García-López et al. (2005). It was found that Ni-TPP reacts through a sequential mechanism. In the first step, Ni-TPP is reversibly hydrogenated to Ni 5,10,15,20-tetraphenylclorin (Ni-TPC); this compound is then hydrogenated to Ni 5,10,15,20-tetraphenylbacterioclorin (Ni-TPiB). Finally, Ni-TPiB reacts via a series of fast reactions ending in demetallation and ring fragmentation. The mechanism is similar to others, with the difference that the controlling step is the first hydrogenation of Ni-TPP instead of the previously reported hydrogenolysis step. This change in controlling step was ascribed to the acidity of the Al_2O_3-TiO_2 support.

Increased catalyst deactivation by metal deposition is a major problem during the hydroprocessing of heavy feeds for the elimination of sulfur (hydrodesulfurization) or cracking to distillates. Compared to the hydrotreatment of light feeds, where catalysts can remain active for several years, in the case of heavy feeds catalyst life can be several weeks or months, as Figure 6.3 shows.

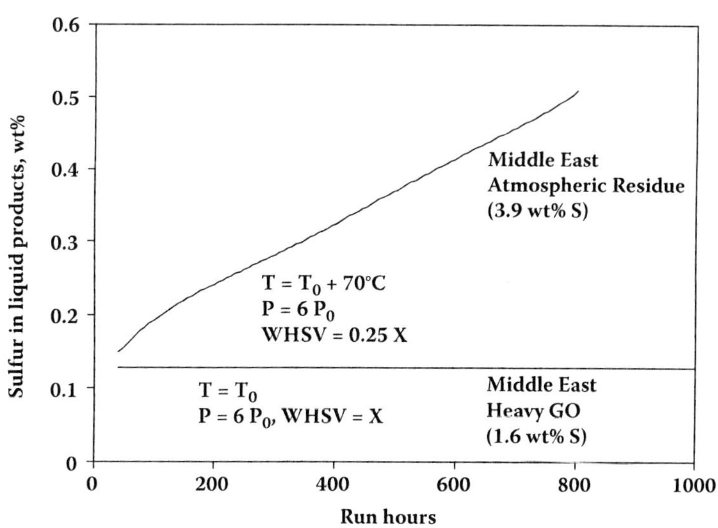

FIGURE 6.3 Comparison of catalyst activity and stability of a conventional HDS catalyst in the processing of distillate and residual feeds from the same Middle East crude oil. (Reprinted from Scheuerman, G.L. et al., *Fuel Process. Technol.*, 35: 39–54, 1993. With permission.)

The difference between the deactivation caused by coke formation and that of metal deposits is that the deactivation by coke is reversible while that caused by metals is not. So, in spite of the metal concentration in the feed being generally much lower than that of carbonaceous matter, deactivation by metals is much more serious. As it happens, contrary to coke deactivation, that produced by metal deposits does not stop when the original active sites on the catalyst surface are covered with deposited Ni and V (Sie, 2001). The reason for this is that during hydroprocessing Ni and V sulfides (Ni_3S_2 and V_2S_3) are formed due to the presence of H_2S, and they autocatalyze the demetallization reaction (Sie, 1980). As Figure 6.4 shows, the relative life of a catalyst is strongly dependent on the metal content of the feed, which can vary widely according to the feed source. Since the metal content largely determines the rate of catalyst deactivation, specific catalyst combinations and reactor technologies are necessary to process each feedstock. Additionally, it must be taken into account that after the distillation of heavy crude oil, AR and VR become richer in asphaltenes and metals (Table 6.4), and therefore in sulfur molecules refractory to hydrodesulfurization (HDS). In general the content of aromatics in petroleum fractions increases with boiling point, while that of paraffins decreases; in fact, at about 510°C almost 50% of the hydrocarbons are aromatics containing more than 80% polyaromatics or asphaltenes, molecules that are deficient in hydrogen, as shown in Figure 6.5.

The high viscosity of heavy feeds is another important parameter for the processability of these feeds because it has to be low enough to allow adequate

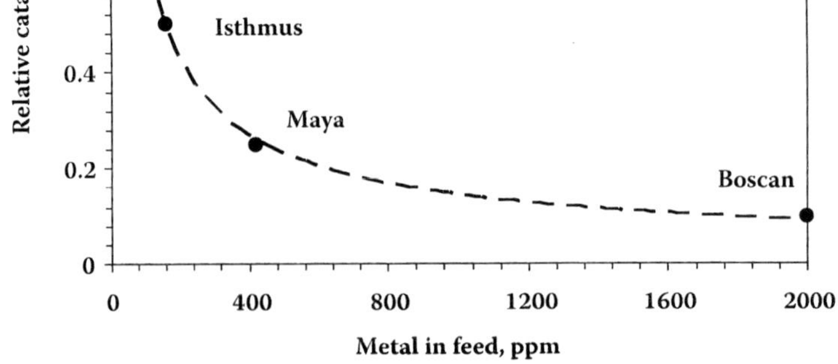

FIGURE 6.4 Effect of metal content on catalyst life for different feeds. (Reprinted from Sie, S.T., *Appl. Catal. A Gen.*, 212: 129–151, 2001. With permission.)

TABLE 6.4
Origin and Properties of Different Residua

	Crude Oil	Ni+V (ppmw)	S (wt%)	Residue Yield (vol% of crude) AR, 343°C+	VR, 565°C+
South America	Venezuela, Bachaquero	509	3.0	70.2	38.0
North America	Mexico, Maya	620	4.7	56.4	31.2
	Alaska, North Slope	71	1.8	51.5	21.5
	Canada, Athabasca	374	5.4	85.3	51.4
	Canada, Cold Lake	333	5.0	83.7	45.0
	California, Hondo	489	5.8	67.2	44.3
Europe	North Sea, Ekofisk	6	0.4	25.2	13.2
Middle East	Arabian, Safaniya	125	4.3	53.8	23.2
	Iranian	197	2.6	47.6	—
	Kuwait, Export	75	4.1	49.5	21.8

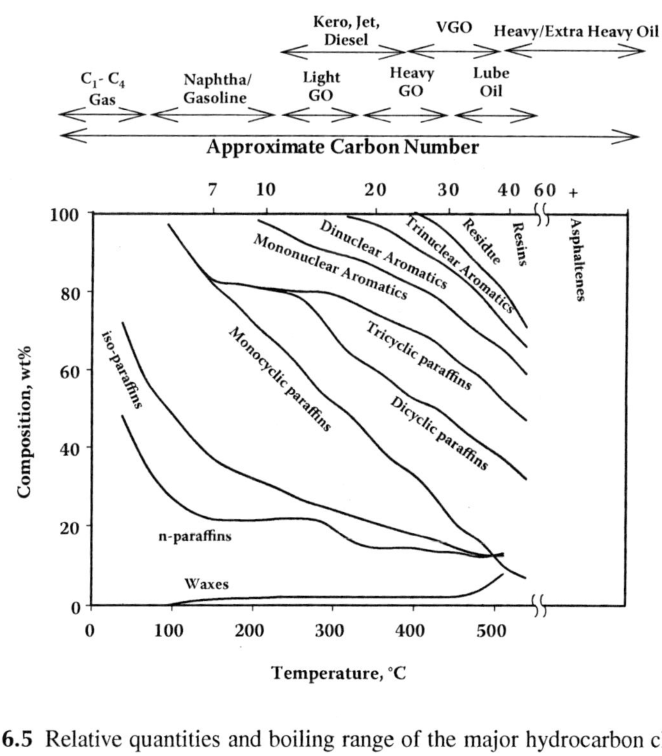

FIGURE 6.5 Relative quantities and boiling range of the major hydrocarbon class in the crude of the Ponca City field, OK. (Adapted from Venuto, P.B. and Habib, E.T., Jr., *Catal. Rev. Sci. Eng.*, 18: 1–150, 1978. With permission.)

TABLE 6.5
Properties of Whole Crude Oils*

Crude Source	API Gravity	Paraffins, vol%	Naphthenes, vol%	Aromatics, vol%	Sulfur, wt%
		Light Crude Oils			
Saudi Light	34	63	18	19	2.0
South Lousiana	35	79	45	19	0.0
Beryl	37	47	34	19	0.4
North Sea Brent	37	50	34	16	0.4
Nigerian Light					
Lost Hill Light		Nonaromatic, 50%		50	0.9
U.S. Mid-Continental Sweet	40				0.4
		Midrange Crude Oil			
Venezuela Light	30	52	34	14	30
Kuwait	31	63	20	24	2.4
U.S. West Texas Sour	32	46	32	22	1.9
		Heavy Crude Oil			
Prudhoe Bay	28	27	36	28	0.9
Saudi Heavy	28	60	20	15	2.1
Venezuela Heavy	24	35	53	12	2.3
Belridge Heavy		Nonaromatic, 37%			

* From data from several literature sources.

pumpability pipelining. Heavy crudes have high viscosity and low API gravity, as Table 6.5 shows. A general classification of crude oils based on these physical properties is shown in Figure 6.6. As mentioned before, to avoid reheating and pumping problems, in the refinery the distillation unit is integrated with the catalytic reactors for hydroprocessing residua and other heavy feeds. For pumping at lower temperature, dilution is necessary; however, sludge formation can be a problem if the mixture lies off the compatibility zone (Rayo et al., 2004; Takatsuka et al., 1989).

6.3 THE CATALYST

6.3.1 The Catalyst Support

Crude oils are mixtures of a virtually infinite number of different hydrocarbon molecules, such as paraffins, naphthenes, asphaltenes, and aromatics, together with molecules containing heteroatoms such as sulfur, nitrogen, vanadium, and nickel.

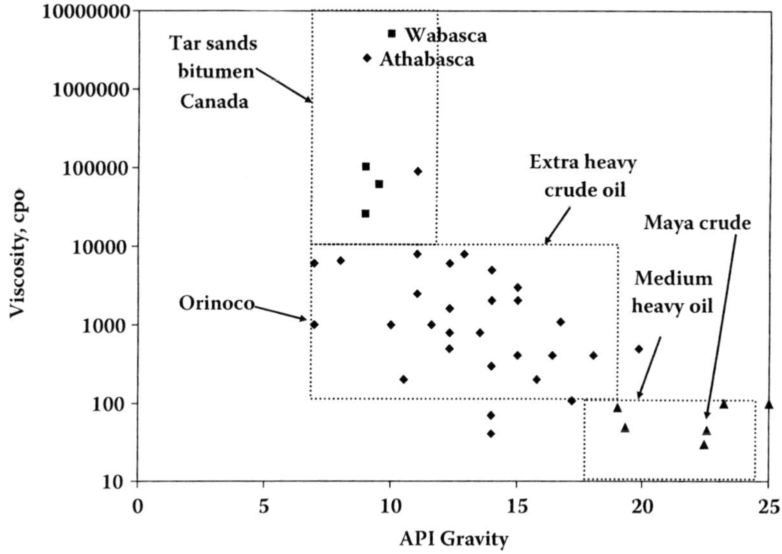

FIGURE 6.6 Heavy oil classification based on physical properties.

As shown before, there are significant differences between light and heavy crudes. The latter contain more metals and asphaltenes. To transform such heavy molecules, catalysts need a wide range of pores to allow the diffusion of such large molecules to the catalytic sites. However, increasing catalyst porosity leads to a reduction in surface area, and consequently a decrease in specific activity. For light feeds where diffusional limitations are not a big problem, surface area is critical. In contrast, for heavy feeds, adequate porosity is essential since in this case it is necessary to achieve a large metal-retaining capacity to extend the catalyst life, which decreases rapidly due to coking and metal deposition. So, in the case of heavy feeds, the textural properties of the catalyst could be of greater importance than surface area and chemical composition of the surface (given an acceptable catalyst activity).

It is a general practice in heterogeneous catalysis to disperse the active components on the surface of a support with adequate textural properties in order to increase the efficiency of utilization of active components. Therefore, for hydrotreating and hydrocracking of heavy feeds, the catalyst support plays an important role (Ward, 1983; Le Page et al., 1992; Speight, 2004). Because of this, there is a continuous effort to develop improved support formulations through suitable methods of preparation, leading to adequate surface composition and textural properties that allow adequate diffusion of the feed molecules to take full advantage of the deposited active phase. In a review covering the effects of support on the activity and selectivity of hydrotreating catalysts, Luck (1991) stated the inherent support properties that fulfill the various technical and economical criteria that the ideal hydrotreating catalyst must have:

- Stabilization of group VI and VIII oxides in highly dispersed or microcrystalline phases without interactions forming inactive compounds.
- Stabilization of the corresponding sulfides in highly dispersed phases.
- High purity or at least absence of any negative interference of adventitious impurities with the active phase.
- No parasitic reactions of the support with the feedstock to be treated.
- Positive contribution of the active sites of the carrier to the catalyst performance.
- Easily tailored pore structure and specific surface area.
- Thermal stability under reaction and regeneration conditions.
- Easy to form into the desired shape (pellets, extrudates, and so on) with a good mechanical strength.
- Low cost.
- No adverse effect from metal recovery from used catalysts.

Although the importance of each of these factors varies according to the type of hydrotreating (HDT) process, alumina is by far the most widely used support because it combines virtually all the above characteristics. A comparison of the properties of alternative carriers was also presented in the same report.

A thorough review of the scientific and patent literature on the methods of catalyst preparation is vast and is therefore outside the scope of this work. Some of the methods of catalyst preparation have been reviewed in detail elsewhere (Sanfilipo, 1994; Eartl et al., 1999). For hydroprocessing catalysts, alumina has remained the most used support because, as we mentioned before, it fulfills the various technical and economical criteria that the ideal hydrotreating catalyst support must have (Luck, 1991). A large variety of aluminas exist (see Figure 6.7). However, -alumina, which is the most versatile and extensively used catalyst support, can be prepared by thermal decomposition of gibbsite or boehmite, or by precipitation of colloidal gels (Luck, 1991). Gamma-alumina has a high surface area, is stable at the temperatures used in hydroprocessing, is easily formed in spheres or extrudates, and is reasonably inexpensive. The microporosity, surface area, pore distribution, and pore volume of alumina can be finely tuned, each aspect almost separately, through the physical parameters of the oxyhydroxide crystallites. Figure 6.8 shows clearly the surface area–pore volume region that aluminas can cover. The acid–base character of alumina plays an important role during the impregnation of the active phase precursor. The acid properties of alumina are due to incompletely coordinated aluminum cations (Lewis acid centers).

The alumina support can be easily prepared in the laboratory or may be obtained from commercial companies in sufficiently large quantities. Usually, preparation of alumina in the laboratory is carried out with aluminum nitrate (Rana et al., 2004a). Aluminum sulfate has also been used for several studies,

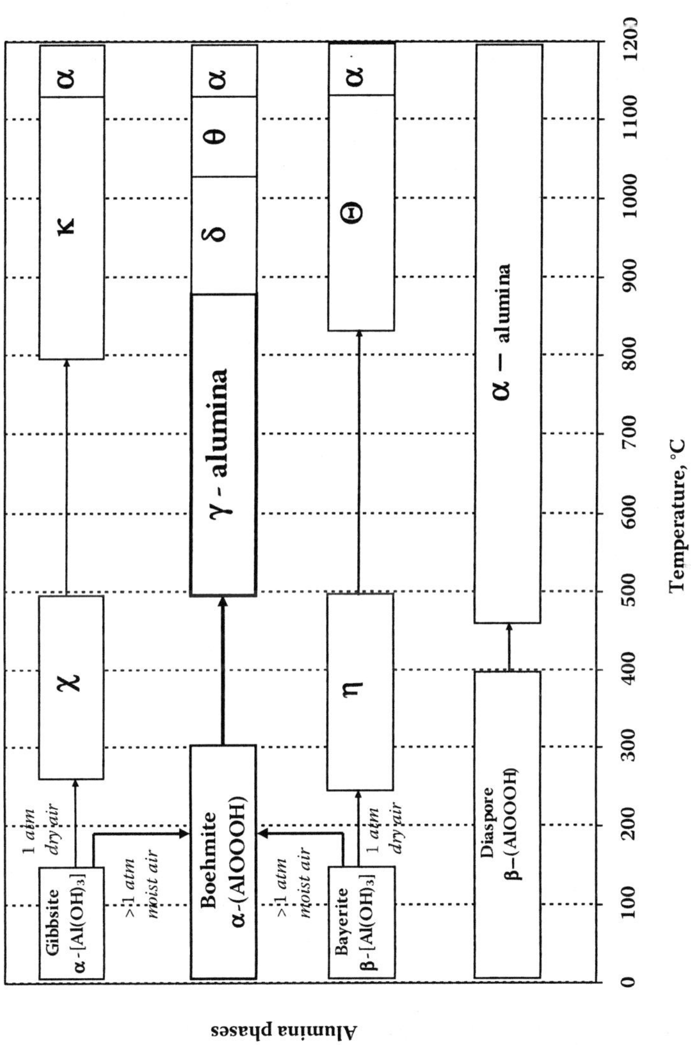

FIGURE 6.7 Transitional phases of alumina.

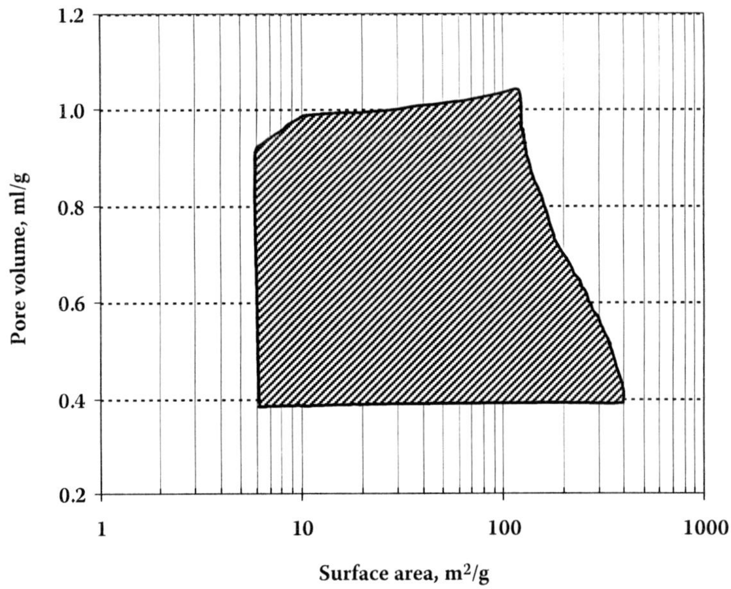

FIGURE 6.8 Typical pore volume vs. surface area patterns for commercial alumina supports. (Reprinted from Luck, F., *Bull. Soc. Chim. Belg.*, 100: 781–800, 1991. With permission.)

but due to the presence of the sulfate ion ($SO_4^=$), high acidity is generated on the surface, leading to coke formation and poor catalyst stability. In general, the below steps are followed for preparation of alumina in the laboratory:

1. Precipitation
2. Hydrothermal transformation (aging)
3. Centrifugation, filtration
4. Washing
5. Drying
6. Shaping of the support (crushing, grinding, extrudates, and so on)
7. Drying
8. Calcination

Each of the above steps has a particular effect on the properties of the final support, for example, the textural properties are affected by variations in the pH of the precipitating medium, as shown in Figure 6.9. The aging period and hydrothermal conditions also affect the properties of the final solid (Absi-Halabi et al., 1993; Moulijn et al., 1993).

The chemical composition and physical properties of the catalyst are crucial when high levels of HDS, hydrodenitrogenation (HDN), hyrdrodemetalization (HDM), and hydrocracking are targeted (Furimsky, 1998). For heavy feeds it is

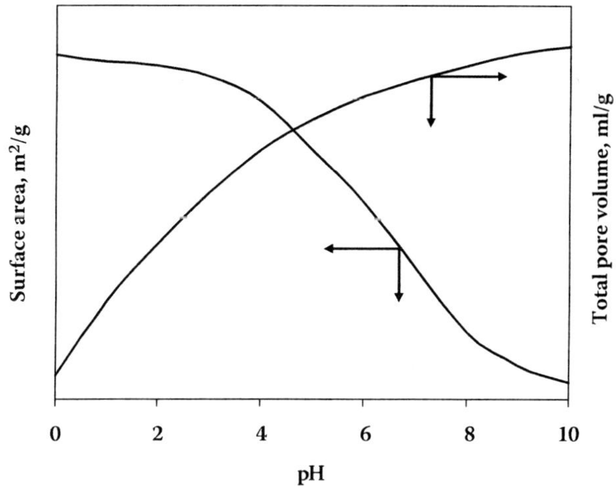

FIGURE 6.9 Effect of pH of precipitating medium on textural properties of alumina.

important to consider the presence of high concentrations of large molecules such as asphaltenes and metal (V and Ni) porphyrins, as well as S and N compounds. For catalyst design, in the hydroprocessing of conventional crudes the content of heteroatoms (S, N) is very important, whereas for heavy feeds the amount of asphaltenes and metals is of utmost importance. In the former case, the chemical composition of the catalytic surface and the specific surface area are determinant since poisoning and coke deposition are not crucial. In contrast, for heavy feeds porosity determines the activity and life of the catalyst. Besides, for heavy feeds the catalyst needs to have enough acidity to perform hydrocracking but not so much as to produce extensive coking. Since heavy feeds vary considerably, depending on their origin, it appears that there is not a universal catalyst for hydroprocessing such feeds. So, for light feeds, catalyst chemical composition and surface area are crucial, whereas chemical composition and porosity are more important in the case of heavy feeds. Although for both feeds the role of the support is important, in the latter case the support acidity and porosity have to be carefully designed to attain optimum catalyst performance.

6.3.2 Support Chemical Composition

In the past it was considered that the hydrotreating catalyst support was inert so far as the active component was concerned. However, we know now that this is not always the case, and that on many occasions the support interacts with the active component, leading to beneficial or detrimental interactions to catalytic activity (Imelik et al., 1982; Scott et al., 1987). Such interactions between support and precursor active phases depend on the solution chemistry of the precursor salts, method of deposition of the active phases, isoelectric point of the support, impregnation pH, calcination temperature, and so forth (Prins, 1992; Wachs et al., 1993; Kabe et al., 1999). The earlier view that the effect of support was just to stabilize the active component as small particles, thus increasing the dispersion without changing the specificity, has undergone enormous change as the results from different studies have shown that rate and selectivity of hydrotreating reactions over sulfide catalysts are affected by the support nature. Ramírez et al. (2004) demonstrated that increasing the amount of Ti in the Al_2O_3-TiO_2 support increases the hydrodesulfurization and hydrogenation rates of NiMo and NiW catalysts supported on Al_2O_3-TiO_2 mixed oxides due to an electronic promotion effect of the Ti present in the support over the Mo or W sulfided active phase. So, an electronic interaction between the support and the active component can alter the activity and selectivity of the catalysts. The support can also influence the active component by favoring the exposure of some crystallographic planes in preference to others. Morphological changes of the active phase induced by the support may also cause a change in the active metal–support interaction.

Although a great variety of supports have been proposed in the literature for hydrotreating catalysts, CoMo and NiMo catalysts supported on -Al_2O_3 continue to play, probably for economic reasons, the workhorse role, and have done so since the hydrotreating processes entered into petroleum refineries. Different amorphous and crystalline materials have been proposed as supports for hydrotreating catalysts. Among them, the use of mixed oxides has been reported to offer an interesting range of textural properties and different interactions with the active phases.

The support effect on the activity and selectivity of hydrotreating reactions has been the subject of several studies (Shimada et al., 1988; Luck, 1991; Breysse et al., 1991; Rana et al., 1999; Muralidhar et al., 2003). In general, the studies in this field present the relationship between different physical and chemical properties of the support with the activity of different hydrotreating reactions. Some studies concentrated on the acidic and basic nature of support and its effect on the nature of interaction with the active metal (Pratt et al., 1990; Qu et al., 2003; Muralidhar et al., 2003; Ji et al., 2004). Many of the studies carried out with model molecules have demonstrated that it is possible to obtain better catalytic performance by changing the nature of the support. Benefits have been reported with the use of TiO_2 (Ng and Guari, 1985; Nishijima et al., 1986; Okamoto et al., 1989; Ramírez et al., 1989, 1991, 2004; Maity et al., 2001a), ZrO_2 (Payen et al., 1991; Rao et al., 1992; Vrinat et al., 1994), MgO (Chary et al., 1991), carbon (Abotsi and Scaroni, 1989), CeO_2 (Muralidhar et al., 1984), SiO_2 (Laine et al., 1991), zeolites (Shimada et al., 1990), SiO_2-Al_2O_3 (Ramírez et al., 1989), SiO_2-MgO (Massoth et al., 1994), ZrO_2-Y_2O_3 (Vrinat et al., 1994), ZrO_2-Al_2O_3 (Lecrenary et al., 1998), ZrO_2-SiO_2 (Rana et al., 2004b), TiO_2-SiO_2 (Hanprasopwattana et al., 1998; Rana et al., 2003), Al_2O_3-B_2O_3 (Ramírez et al., 1995; Li et al., 1998; Usman et al., 2004), ZrO_2-TiO_2 (Nishijima et al., 1986; Weissman et al., 1993; Maity et al., 2001b), MCM-41 (Reddy et al., 1998), SBA-15 (Muralidhar et al., 2005; Sampieri et al., 2005), natural minerals clays (Maity et al., 1998), TiO_2-Al_2O_3 (Nishijima et al., 1986; Zhaobin et al., 1990; Wei et al., 1992;

Ramírez et al., 1993; Tanaka et al., 1996; Pophal et al., 1997; Olguin et al., 1997; Muralidhar et al., 2000), and so on. Although most of these studies were conducted with model molecules, some are considered good catalyst supports for gas oil and were found to be more active than γ-Al_2O_3-supported ones.

Particularly the use of Ti-containing supports has been studied in detail. Besides achieving higher dispersion of the active Mo or W sulfided phase, it was found that under reaction conditions some Ti^{4+} transforms into Ti^{3+}, which is capable of transferring electronic charge to the Mo or W sulfided phase, leading to an increase in the number of coordinatively unsaturated sites (CUSs) in MoS_2 and thus increasing the hydrodesulfurization and hydrogenation activities. In essence, Ti acts as an electronic promoter for the Mo sulfided phase. To produce this effect, it is necessary that the Ti oxide surface species are as polymeric (Ti-O-Ti-O) as those in TiO_2, rather than isolated (Al-O-Ti-O-Al) in the alumina matrix (Ramírez et al., 2004).

Although most of the work on the support effect has been carried out using model molecules, few different support systems have been tested in the hydroprocessing of crude oil and coal-derived liquids. Among them, TiO_2-Al_2O_3 (Nishijima et al., 1986; Rayo et al., 2004; Ramírez et al., 2005; Rana et al., 2005a; Maity et al., 2003, 2005), SiO_2-Al_2O_3 (Maity et al., 2004), MgO-Al_2O_3 (Caloch et al., 2004), and ZrO_2-Al_2O_3 (Rana et al., 2005b) have shown interesting features.

The use of mixed-oxide-supported catalysts in hydrotreating is not new; initially mixed oxides such as SiO_2-ZrO_2-TiO_2 (Hansford, 1964) and Al_2O_2-TiO_2 (Jaffe, 1968) were developed for hydrocracking applications due to their acidic nature. These supports were generally prepared by coprecipitation of appropriate Al and Ti salts, followed by washing, drying, and calcination. Hydrocracking was later dominated by the use of zeolitic supports due to their uncomplicated way of controlling or modifying the acidity (number and strength of acid sites). FAU, ZSM-5, erionite, and mordenite have been used. Among them, Y-zeolite is the most commonly used because it has the widest pores (van Veen, 2002), although ZSM-5, erionite, and mordenite have been used for shape-selective reactions controlled by pore geometry (Scherzer, 1990). The use of these microporous zeolites is, however, limited for heavy oil processing due to their small pore diameter. Nevertheless, the incorporation of a small amount of zeolite into alumina can have an important effect on the product selectivity and conversion.

Recently, silica-based mesoporous materials such as MCM-41 and SBA-15 have gained some attention in view of their interesting textural properties (surface area ~ 1000 m^2-g^{-1} and pore diameter ~ 3.0 to 10 nm), although due to their poor hydrothermal stability they are not likely to have a commercial impact as supports for catalysts dedicated to heavy oil hydroprocessing.

6.3.3 Shape and Size of Catalyst Particles

Shape and size of catalysts and therefore of supports are important parameters in the preparation of commercial hydroprocessing catalysts. To achieve good catalyst performance, it is important to match the size and shape of the catalyst with the properties of the feed, the process technology, and the type of reactor. In the case of heavy feeds, special attention has to be paid to the shape and size of the catalyst in view of the diffusion problems encountered during the hydroprocessing of the large molecules contained in these feeds. It is well known that in diffusion-controlled reactions the rate of diffusion is smaller than the rate of reaction. Therefore, a steep radial concentration profile of the reactants is established along the pellet radius, and the use of large catalyst particles can result in the central part of the catalyst particle not being used, because the reactants are exhausted by the chemical transformation before they reach the center of the catalyst particle. This problem can be avoided in some cases by decreasing the diffusion path of the molecules, in this way increasing the effectiveness factor of the catalyst. In practice, this is achieved by decreasing the particle diameter. However, there is a limit to the decrease of particle size after which particles disintegrate. Apparently, this limit is approached when the diameter of the particles comes within reach of 0.8 mm (1/32 in.). It must nonetheless be considered that such small catalyst particles will cause high-pressure drops in fixed-bed reactors.

The problem of a long diffusional path of reactants from the external surface to the center of the catalyst particle can be partially circumvented by the use of different shapes of catalyst. Some of these particle shapes, which may be suitable for hydroprocessing heavy feeds, are shown in Figure 6.10. However, to compare relative activities of different shapes of catalyst particles, it is necessary to establish a criterion to characterize particle size for different particle shapes.

For diffusion-limited processes, the catalyst effectiveness factor ($\eta = k/k_i = f(\phi)$) depends on the Thiele modulus (ϕ), which has been defined as

$$\phi = L_p(k_i C^{n-1}/D_{eff})^{0.5} \tag{6.1}$$

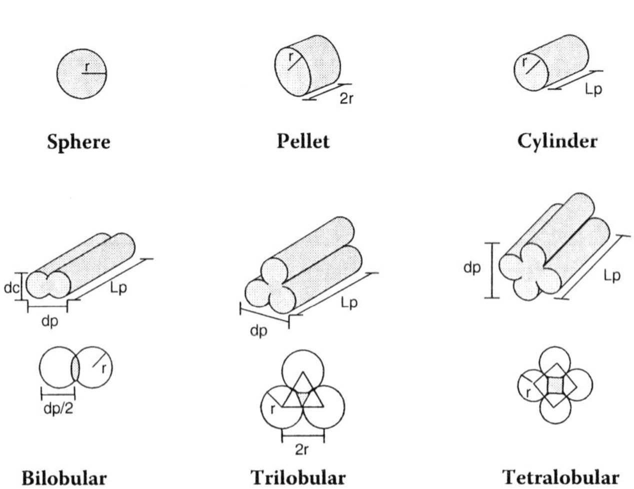

FIGURE 6.10 Particle shapes of industrial hydroprocessing catalysts. (From Macias, M.J. and Ancheyta, J., *Catal. Today*, 98: 243–250, 2004. With permission.)

where L_p is the particle size, k and k_i are the apparent and intrinsic rate constants, C is the reactant concentration, n is the reaction order, and D_{eff} is the effective diffusivity.

For spheres, cylinders, and flat plates, particle size can be defined as the ratio of particle volume (V_p) to surface area (S_p):

$$L_p = V_p/S_p \tag{6.2}$$

Suzuki and Uschida (1979) demonstrated that this definition of particle size can be extended to other noncylindrical particles. Recently, equations for calculating total geometric volume and external surface area of lobe-shaped catalytic particles as a function of easy-to-determine geometrical parameters were presented (Macias and Ancheyta, 2004; Ancheyta et al., 2005b).

The effects of the above parameters on catalyst performance and deactivation during hydroprocessing of heavy feeds have been reported in Furimsky and Massoth (1999).

The definition and evaluation of particle size is useful for comparing the relative activities of catalyst particles with different shapes. This is especially important when heavier feeds are processed. In this case, more severe process conditions are necessary, and many existing hydrotreaters are underdesigned with respect to catalyst inventory. So, it is important to obtain more catalytic activity in the same reactor volume. The introduction of more active catalysts appears as the solution. However, since hydroprocessing reactions in the case of heavy feeds are diffusion limited, improvement of the catalyst shape to obtain higher activity per volume of bed is necessary. For many refiners, the change from cylinder-shaped to multilobed catalysts has been the answer. Different techniques are used for shaping the catalyst into various forms; these techniques were reported in detail in the literature (Stiles, 1983; Richardson, 1989; Le Page et al., 1987; Moulijn et al., 1993). Usually, the use of multilobed catalyst shapes presenting big pores, small particle diameter, and large external area is preferred (Cooper et al., 1986; Bartholdy and Cooper, 1993).

Using the definitions given above, Cooper et al. (1986) compared the relative hydrodesulfurization activities of various catalyst shapes in the hydrotreatment of a heavy Arabian vacuum gas oil (VGO). They concluded that catalysts with the same volume-to-surface ratio exhibit the same effectiveness factors, and therefore the same activity per pound of catalyst, provided that the catalysts have the same physical and chemical properties. Concerning the relative volume activity, noncylindrical catalyst shapes exhibit higher void fraction, and therefore lower bed volume, activities than cylindrical catalysts with the same V_p/S_p ratio. However, pressure drop across the catalytic bed is also important in industrial applications. Since three-lobe catalysts have a higher void fraction and a larger hydrodynamic particle size, they present a lower pressure drop per foot of catalyst bed. In some cases it is relevant to compare the activities of particles giving the same pressure drop. By combining the plots of relative volume activity (RVA) vs. V_p/S_p with those of pressure drop vs. V_p/S_p, it is possible to compare different catalyst

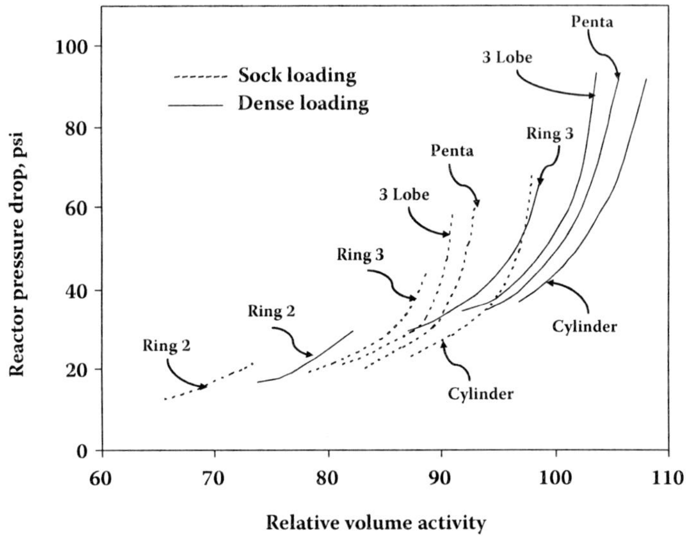

FIGURE 6.11 Effect of catalyst shape and loading on reactor pressure drop and activity. (From Cooper, B.H. et al., *Oil Gas J.*, 8: 39–44, 1986. With permission.)

shapes on the basis of equal pressure drop. Figure 6.11 shows the effect of catalyst shape on reactor pressure drop and activity.

From their results, presented in Table 6.6, Cooper et al. (1986) showed that when low-diffusion restrictions exist, a decrease in effective particle size has little effect on increasing activity. So, in this case, it is better to pack the reactor with as much catalyst as possible. This means that dense loading is preferable and that

TABLE 6.6
Effect of Particle Size and Shape on HDS Activity

Shape	Dimensions, mm	V_p/S_p, mm	Activity[a]
Cylinder	0.83 OD × 3.7 length	0.189	9.7
Cylinder	1.2 OD × 5.0 length	0.268	7.9
Cylinder	1.55 OD × 5.0 length	0.345	5.7
Ring	1.62 OD × 0.64 ID × 4.8 length	0.233	8.7
Ellipse	1.9 × 1.0 × 5.0	0.262	8.4
Three-lobe	Diameter = 1.0; length = 5.0	0.295	8.2
Crushed	0.25–0.45	~0.04	14.0

[a] HDS activity on catalyst weight basis (heavy gas oil, 622 K, 7 MPa).

Source: From Cooper, B.H. et al., *Oil Gas J.*, 8: 39–44, 1986. (With permission.)

cylinder-shaped catalysts with lower bed porosity are preferred. In contrast, when severe diffusional limitations are present, small particles with a high external surface will give the best performance for a given pressure drop. In this case, highest activities, significantly higher than those of cylinders, were obtained with sock-loaded-shaped catalysts. In general, shaped catalysts will give an improved performance with respect to cylinders when a large increase in effectiveness factor can be achieved with a relatively small decrease in particle size.

It is important to remember that under severe diffusion limitations performance can also be improved by an increase in catalyst pore size. To emphasize the importance of pore diameter for some of the reactions taking place during hydroprocessing, Figure 6.12 presents the activity of hydrodesulfurization (HDS), asphaltene conversion (HDAs), and vanadium removal (HDV) for a series of $CoMo/Al_2O_3$ catalysts as a function of average pore diameter. This figure shows that for HDS reactions involving small molecules, the optimum average pore diameter is about 10 nm. In contrast, for HDAs and HDV the optimum activity lies between 15 and 20 nm. However, it must be considered that the shape and position of the curves will change with the type of feed.

The relationship between catalyst activity and metal accumulation in the catalyst is presented in Figure 6.13 for catalysts with different surface areas and pore volumes. High surface areas and small pore volumes are best for model molecules and light feeds, while large pore volumes and consequently low surface areas work better for heavy feeds, where HDM and HDAs are important. Figure 6.14 shows an example of the pore size distribution required for HDS and HDM.

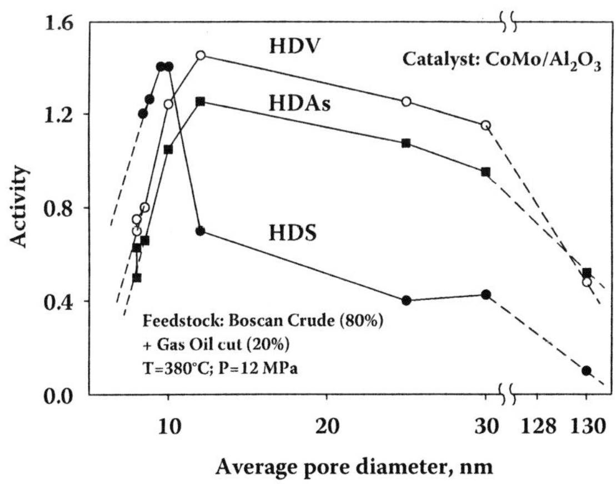

FIGURE 6.12 Sulfur, vanadium, and asphaltene removal activities of monomodal $CoMo/Al_2O_3$ catalysts of different average pore size (380°C, 1.2 MPa, gas oil-blended Boscan crude). (From Plumail, J.C. et al., in *Proceedings of the 4th International Conference on the Chemistry and Uses of Molybdenum*, Ann Arbor, MI, 1982, p. 389. With permission.)

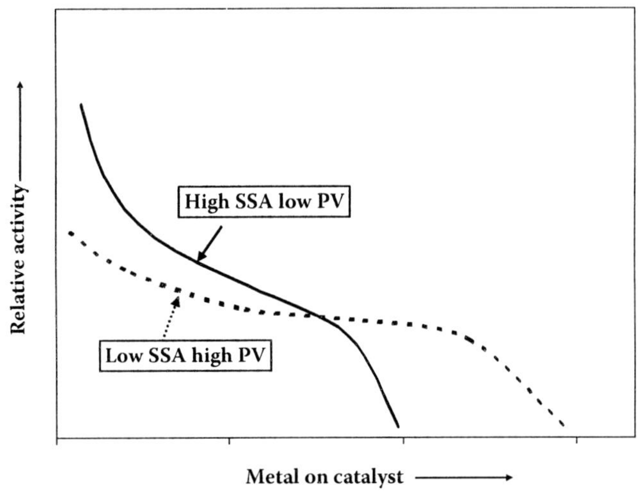

FIGURE 6.13 Relationship between catalyst activity and metal accumulation on the catalyst. SSA, specific surface area; PV, pore volume.

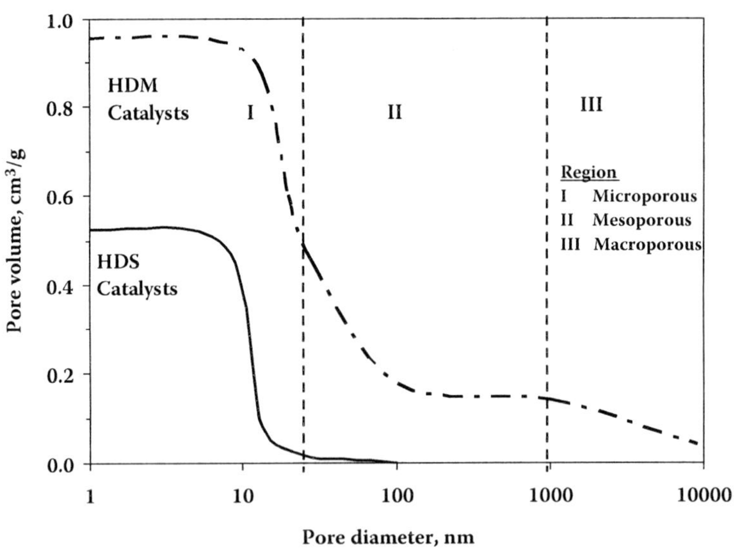

FIGURE 6.14 Pore size distribution for HDS and HDM catalysts. (From Peries, J.P. et al., Institut Fraçaise du Pétrole, Rapport IFP 34297, June 1986. With permission.)

It appears that no easy solution is available since small catalyst particles decrease diffusion paths but cause high bed pressure drops. On the other hand, catalysts with wide pores are less susceptible to diffusion limitations, but they necessarily have lower surface areas. So, to determine what option one must use in a particular hydroprocessing operation, an integrated approach that takes into account size, shape, pore size, and catalyst loading must be considered (Cooper et al., 1986). It must be remembered that for hydroprocessing heavy feeds, stability of the catalyst is at least as important as initial catalyst activity.

The problems of increased pressure drop (ΔP) and fouling of the catalytic bed can be partially solved with effective bed grading and the use of guard beds on the top of the reactor to trap particulate matter from dirty reactor feeds. In this way, it will be possible to minimize ΔP during operation and make more efficient use of the downstream catalyst bed. Porous solids like activated bauxite or alumina are used in the guard beds (Wolk and Rovesti, 1974; Wolk et al., 1975; Howell and Wilson, 1987; Hilbert et al., 1996). Other low-cost solids, such as magnesium silicate, have been also proposed (Shirota et al., 1979). Some other aspects of bed grading and reactor ΔP are given in Chapter 5 of this book.

6.3.4 Mechanical Properties

Mechanical properties of the catalyst are of industrial importance. Catalyst particles must have adequate mechanical strength to stand the weight of the catalyst bed itself to ensure proper operation. In this respect, catalysts for hydrotreatment of heavy feeds suffer more than those used for hydrotreatment of light feeds since the mechanical strength of a macroporous pellet is less than that of a microporous one. The breaking of the catalyst particles in a fixed-bed reactor can cause critical problems, such as higher pressure drop, ill distribution of flow, and shutdown of the operation due to collapse of the catalyst bed. Unsatisfactory resistance to attrition can cause similar problems in fixed beds and great loss of catalyst in ebullated beds. Plugging of the filters is also another consequence of low catalyst resistance to attrition. For ebullated beds, spherical particles have improved attrition resistance and offer lower pressure drops.

The mechanical and textural properties of the catalyst, mostly provided by the support, can be controlled during the preparation stage. Different factors like precipitate aging, pH, number of pH swing cycles, type of binder, and calcination temperature and time can affect the mechanical properties of the support (Ono et al., 1982; Snel, 1984a, 1984b, 1984c, 1987). The method of incorporating the active components to the support can also affect the mechanical properties of the final catalyst (Trimm and Stanislaus, 1986).

Hydrotreating catalysts are usually prepared by filling the support pores with a basic or acid solution containing the active metals and additive salt precursors. However, it is possible that although the chosen pH could be adequate to induce the formation of the desired active metal species in the impregnating solution, partial dissolution of the support pore walls can take place, reducing the mechanical strength of the final catalyst.

6.3.5 Catalyst Composition

The selection of the active catalyst components for the hydroprocessing of light and heavy feeds is quite different. For light feeds, which do not contain metals or asphaltenes, the chemical composition of the catalyst is a critical aspect and its choice must be based on the type and content of sulfur and nitrogen heteroatoms. In the case of heavy feeds containing high concentrations of heavy metals and asphaltenes, the physical properties of the catalyst (size, shape, porosity, and so on.) can be even more important than chemical composition, since in this case the presence of meso- and macropores is critical to achieve full utilization of the catalyst. Nevertheless, some modification of the chemical composition may be needed to improve the demetallization function of the catalyst.

The harmful effect of the presence of large concentrations of metals such as Ni and V in the feed is well known, and significant efforts have been made to develop catalysts that can withstand high concentrations of metals by providing high metal retention capacity without losing too much activity and life (Decroocq, 1997). To remove metals, the hydrodemetallization (HDM) catalyst must transform the resins and asphaltenes in which most of the metals are found. As mentioned before, the design of the catalyst pore system is critical for this task. For a microporous catalyst the metal deposits will occur close to the external surface of the catalyst pellet, leading to pore mouth poisoning. On the other hand, catalysts with a wide range of porosity or multimodal catalysts with wide pores avoid diffusion and plugging problems and are better suited for hydrodemetallization. Preparation of supports with "chestnut-bur"-like pores (see Figure 6.15), which avoid the problem of having internal pore walls, allows a homogeneous distribution of the metals on the catalyst surface with a 100 wt% metal retention capacity referred to the weight of the fresh

FIGURE 6.15 Scanning electron microscopic picture of a typical "chestnut bur" residue HDM catalyst. (From Kressmann, S. et al., *Catal. Today*, 43: 203–215, 1998. With permission.)

catalyst (Morel et al., 1997; Kressmann et al., 1998; Toulhoat et al., 1990). Catalysts with a large pore volume and appropriate pore size distribution will allow the diffusion of asphaltenes and resins into the pores and their adsorption on the catalyst sites. However, to obtain a high metal capacity in HDM catalysts, it is necessary to tune the activity provided by the NiMo active phase. If the activity is too high, the diffusion of resins and asphaltenes will be rate controlling and the metal deposit will be less homogeneous, with more metal deposit close to the external surface of the pellet. Nevertheless, sufficient hydrogenation activity is necessary to reduce the carbon deposits on the catalyst surface and improve demetallization activity. So, activity and porosity must be properly combined to achieve maximum utilization of the catalyst, acceptable activity, and life.

Concerning the formulation of the active phase, it is found that in general CoMo catalysts are more effective for HDS, whereas NiMo formulations have good activity for hydrogenation (HYD) and hydrodenitrogenation (HDN). Kellet et al. (1980) compared the performance of NiMo and CoMo catalysts and found that the HDS, HDN, and HYD trends did not change with space-velocity. However, this is not always the case. For cracked feeds or when the content of nitrogen is high, the NiMo catalyst would be preferred, since this formulation performs better the HDN function, in this way avoiding the poisoning of the active sites by nitrogen compounds, which interact strongly with the coordinatively unsaturated sites of the catalyst (Furimsky and Massoth, 1999, 2005). Plumail et al. (1982) compared the activity of a wide variety of catalytic formulations using a feed containing metals and asphaltenes and found higher HYD and HDS activities for the NiMo catalyst than for the CoMo formulation.

Usually, hydroprocessing of heavy feeds requires the use of mixed or multiple beds of catalysts. So, the catalyst formulation will depend on its place in the catalytic bed system. Furimsky and Massoth (1999) reviewed the literature on the use of mixed and layer beds and multiple-bed and multistage systems. Several works have been published on the use of a mixed bed consisting of two or more catalysts. A mixture of two catalyst particles in the same bed, one having high HDS and HDN and the other with high hydrocracking activity, offered good selectivity to liquid products and low deactivation (Habib et al., 1995, 1997). The incorporation of a bed of alumina particles impregnated with magnesium salts on top of the catalyst layer was claimed to give better performance than the bed without the alumina layer (Gardner and Kukes, 1989).

For multiple-bed systems, the first bed or catalyst layer is always designed to provide high HDM activity, the second is to provide some HDM but significant HDS, and the third is responsible for hydrocracking as well as HDS and HDN (Furimsky, 1998). The first bed or layer contains a large-pore catalyst, while in the second and third beds smaller pores and larger surface areas are required. Enough surface area must be supplied by mesopores greater than 120 Å to allow the heavy fractions to react, while most of the surface, comprised by pores between 80 and 120 Å, will be highly accessible to the lighter fractions of the feedstock, but much less so for residual asphaltenes (Decroocq, 1997). Clearly, the composition of each bed must take into consideration the composition of the feed and the different level of catalyst deactivation expected for each bed.

Several works have been published on the use of graded catalyst systems consisting of multiple beds with different composition. When the beds are in the same reactor, they operate at the same temperature and pressure. The patent literature on this subject has been well reviewed by Furimsky and Massoth (1999). For really difficult feeds, sometimes the use of three beds is necessary (Hensley and Quick, 1984; Gardner et al., 1987).

The use of multiple beds located in different reactors is more versatile but also more costly. In this case, the operating conditions (pressure and temperature) of each bed can be varied at will to suit the particular feed being processed. One of the advantages of this scheme is that part of the effluent from a reactor can be removed before introducing the remaining feed to the next reactor. For example, with this scheme it would be possible to remove light fractions from the effluent of the first stage, reducing the paraffinic character of the feed to the second reactor; this would reduce the possibility of precipitation of heavy components such as asphaltenes in the second reactor.

Sometimes heavy feeds are so problematic that they can not be processed in fixed-bed reactors. Several patents have been disclosed on the use of ebullated-bed reactors in series (Mounce, 1974), or combinations of ebullated-bed with fixed-bed reactors (Hammer and Clem, 1975; Kunesh, 1982). The use of multiple-reactor systems allows improvement of hydroprocessing of the feed by setting a processing scheme that allows efficient contact of the feed with hydrogen and catalyst. For example, in the multibed reactor described by Harrison et al. (1994), the makeup hydrogen is fed to the second bed while the H_2-containing gas exiting from the second bed is supplied to the first bed. The gas coming from the first bed is purged and introduced to the third bed.

6.4 CHARACTERIZATION OF CATALYSTS

Bearing in mind the great differences in feed properties, the selection of catalysts needs to be carefully considered to match the specifications of the feed. Consequently, a wide variety of catalyst formulations have been developed to perform the hydroprocessing of an equally large number of different feeds, processes, and reactors. To understand the performance of each catalyst formulation, deep characterization of the fresh and used catalyst is necessary. The scientific and patent literature on the preparation and characterization of catalysts is extensive and has been reviewed in the past by several authors (Grange, 1980; Trimm, 1980; Ratnasamy and Sivasanker, 1980; Massoth et al., 1982; Poncelet et al., 1983; Delannay, 1984; Anderson and Pratt, 1985; Le Page et al., 1987; Wachs et al., 1993; Topsøe et al., 1996; Thomas and Thomas, 1997; Delmon et al., 1998; Eartl, 1999). So, no attempt will be made here to review all the work in this field.

The catalysts used for heavy oil hydroprocessing are Mo and W sulfides promoted by Co or Ni supported mostly on γ-alumina, although some other supports, such as zeolites, silica, and silica-alumina, are also used. In some cases, modifiers such as Ti, P, B, or others are added to tune the surface properties.

The most important catalyst property is its ability to achieve the desired transformation of reactants at an acceptable industrial rate. However, it is also necessary that this transformation is carried out with the desired selectivity, maintaining stable operation for a long time and with the possibility to regenerate the catalyst at the end of the run cycle.

Activity: The catalytic activity is characterized by the magnitude of the overall rate constant or by the product of the intrinsic rate constant times the adsorption coefficient. However, frequently the rate equation is not well known. In this case, the activity can be evaluated as the variation in concentration, pressure, or number of moles or molecules per unit time at a given set of operating conditions for a specific type of reactor.

Selectivity: For consecutive and parallel reactions the selectivity can be evaluated from the ratio of activities or can be expressed as the yield at fixed conversion of the main reactant.

Stability: The life of a catalyst depends on its stability and regenerability. Catalyst stability is normally expressed as the time between two regenerations for a given catalyst load and feed input.

Regenerability: This is the number of regenerations the catalyst can undergo without losing unacceptable levels of activity, or before reaching too low operating times in the duration of the cycles between successive regenerations.

All the above factors are evaluated in different types of reactors, mainly batch, continuous stirred tanks, and continuous tubular reactors. The important aspects concerning the evaluation of activity, selectivity, mass transfer limitations, and so forth, have been reviewed by Wijngaarden et al. (1998).

For industrial applications it is of great importance to determine the morphological characteristics of the pellet grain in order to estimate the catalytic bed dimensions. Among the morphological characteristics of the catalyst pellet, shape, grain size distribution, skeletal density, and bed density must be evaluated. Additionally, it is also important to establish mechanical properties like resistance to crushing, abrasion, and attrition. These properties are easy to measure using different American Society for Testing and Materials (ASTM) methods in the laboratory; a detailed account of these techniques is given in Le Page et al. (1987).

To discover, design, or optimize a catalytic formulation, it is necessary to characterize the physicochemical properties of the catalyst. However, an effective approach to the chemical and physical properties of a catalyst requires the participation of a group of specialists and techniques that only a few laboratories can afford to have.

The review of all the characterization methods for supported sulfided catalysts is vast, and it goes beyond the scope of the present work. For more in-depth information on each of the techniques, refer to more specialized books in each field (Delannay, 1984; Anderson and Pratt, 1985; Le Page et al., 1987; Wachs, 1992; Eartl et al., 1999).

Textural properties: The study of the physicochemical characteristics of a catalyst requires knowledge about the texture of the support and catalyst (fresh and spent), the composition of the chemical elements, the nature and structure of the catalytic chemical species, and the state and dispersion of the active phases. Table 6.7 and Table 6.8 present a list of some of the techniques frequently used for catalyst characterization.

Heavy feedstocks contain several metals, such as Ni, V, Fe, and so on, which are present as porphyrins or chelating compounds. During

TABLE 6.7
Some Techniques Used for Catalyst Characterization

Properties	Methods
Chemical composition	Standard chemical analysis
	X-ray fluorescence
	Emission spectrometry
	Atomic absorption
	Flame spectrometry
	EDX
Textural properties: specific surface porosity, pore size distribution.	Nitrogen physisorption. BET methods
	Mercury porosimetry
Nature, state and structure of surface chemical species	X-ray diffraction
	Electron diffraction
	Nuclear Magnetic Resonance (NMR)
	Infrared and Raman spectroscopy
	Visible and ultraviolet spectroscopy
	Magnetic methods
	Thermogravimetric analysis (TGA)
	Differential Thermal Analysis (DTA)
	Mössbauer spectroscopy
	X-ray Photoelectron spectroscopy (XPS)
	Chemisorption
	Electron microscopy
	Electron microprobe analyzer
Electronic properties	EPR
	Visible and ultraviolet spectroscopy
	Conductivity, semiconductivity
	Electron extraction work functions

TABLE 6.8
Some Techniques Used for the Study of Textural Properties of Porous Solids

Techniques	Method	Property	Micropores	Mesopores	Macropores
N_2 adsorption (77 K)	BET, t-plot, α_s-plot	Surface	X	X	
	BJH	Surface area = f (pore size)		X	
	t-plot, α_s-plot, DRK, MP, Horvath–Karvazoe	Pore volume	X		
	Oliver–Conlin	Pore volume	X	X	
	BJH	Pore volume		X	
	Gurvitsch	Pore volume	X	X	
	Horvath–Karvazoe, Oliver–Conlin, BJH	Pore volume = f (pore size)	X	X	
Mercury porosimetry				X	X
Incipient wetness			X	X	X

hydrotreating these metals are transformed to sulfides (Ni_3S_2, V_3S_2, and V_3S_4) and deposit on the catalyst surface as crystallites of 2 to 30 nm in size. These deposited transition metal sulfides cause a decrease in the number of catalytic sites, hinder the transport of reacting molecules to the internal catalyst surface, and eventually cause the complete plugging of the catalyst pores. For these reasons, in the case of heavy oil hydroprocessing, characterization of the catalyst textural properties is of great importance given the large molecules to be processed. The textural properties of catalysts are evaluated by physical adsorption of inert gases, usually nitrogen at liquid nitrogen temperature. Total pore volume and pore size distribution are also calculated through the physical adsorption–desorption nitrogen isotherm. However, physical adsorption techniques are less likely to be used for the evaluation of textural properties of heavy oil catalysts because they contain a large fraction of macropores, which can not be deduced effectively from physisorption experiments, as shown in Figure 6.16. For these large-pore-diameter catalysts the pore size distribution must be estimated by mercury porosimetry. Nevertheless, by analyzing the data with suitable models, nitrogen adsorption at 77 K allows the estimation of the properties listed in Table 6.8. A thorough discussion of the theory of

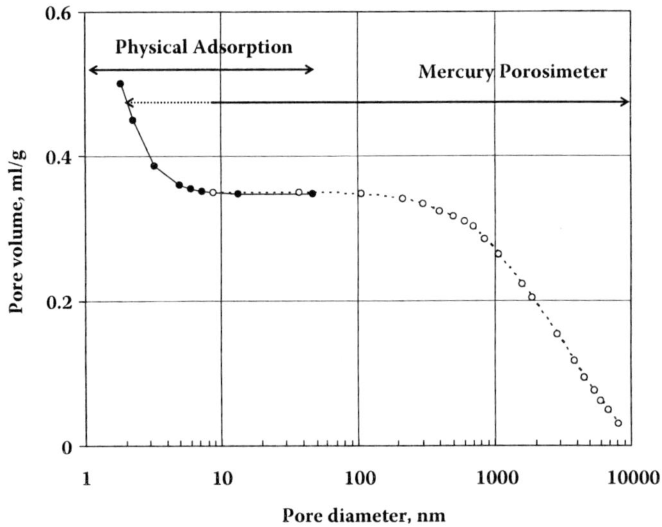

FIGURE 6.16 Comparing N_2 adsorption–desorption isotherms with mercury porosimeter (SiO_2-Al_2O_3; SSA, 430 m^2/g; TPV, 0.57 ml/g). (Adapted from Le Page, J.F. et al., *Applied Heterogeneous Catalysis. Design and Manufacture of Solid Catalysts*, Editions Technip, Paris, 1987. With permission.)

the different methods is given in Gregg and Sing (1982) and Leofanti et al. (1994).

Mercury porosimetry is utilized to obtain the pore size distribution of pores in the range of 7.5 to 1.5×10^4 nm (2000 atm). When working with this technique it is possible that some breakage of the pore walls occurs in some materials, like high porous silica, with a pore volume greater than 3.0 cm^3-g^{-1}. Details of the mercury porosimetry method can be found in Ritter and Drake (1945) and Lowell (1980).

The method of incipient wetness is simple and reliable to estimate the catalyst pore volume; the solid is impregnated with a volume of a nonsolvent liquid, usually water or hydrocarbons, enough to fill the pores. This method can be used for high porous silica, where other methods fail (McDaniel and Hottovy, 1980).

Surface properties: Although the chemical reaction occurs on the catalyst surface, not all of the surface is active, because only a fraction of the total surface area is occupied by the active phase. In fact, not even the complete surface of the active phases is active since normally the reaction occurs on specific centers of the active phase surface. In the case of acid catalysts, the acid centers not only occupy a small fraction of the surface, but also can differ in strength and nature (Lewis or Brønsted).

To understand the way a catalyst works, it is necessary to know how many active sites are present and what their nature is. In general, the methods that can render information on the characteristics of the catalyst surface at the atomic or molecular level can be divided into two groups:

1. Methods using probe molecules
2. Methods of direct study of the catalyst surface

Volumetric, gravimetric, static, and dynamic chemisorptions, adsorption calorimetry, spectroscopies of adsorbed molecules, and temperature-programmed adsorption (TPA) and desorption (TPD) methods belong to the first group. Table 6.9 gives the remainder of the information obtained with each technique and the possibility of quantitative analysis.

Some of the above techniques, like Fourier transform infrared (FTIR), can be used for *in situ* studies for direct observation of the surface under reaction conditions (Topsøe, 2006).

In general, the techniques of the second group are more expensive and specialized. These techniques consist of measuring the response of the solid when it is exposed to a radiation. The response may be the scattering or absorption of the radiation, or absorption followed by emission of another type of radiation. Table 6.10 presents the characteristics of some of these techniques.

Other techniques like low-energy diffraction (LEED), high-resolution electron energy loss spectroscopy (HREELS), and surface-extended x-ray absorption fine structure (SEXAFS) are also used. A classification and description of these

TABLE 6.9
Some Techniques for Surface Characterization Using Probe Molecules

Technique	Information	Quantitative Analysis
Volumetric adsorption	Amount adsorbed = f(P)	✓
Gravimetric adsorption	Amount adsorbed = f(weight)	✓
Calorimetry	ΔH_{ads}= f ()	✓
FT–IR spectroscopy	Support and adsorbate functional groups	Possible
Raman spectroscopy	Support and adsorbate functional groups	Possible
UV-vis spectroscopy	Electronic transitions of chemical species	Possible
Dynamic adsorption	Reversible and irreversible adsorption	✓
TPA, TPD	Amount of adsorbed (TPA) or desorbed (TPD) species = f(T)	✓

P = pressure, T = temperature, □ = surface coverage, ΔH_{ads}= heat of adsorption.

TABLE 6.10
Characteristics of Some Spectroscopic Techniques for Surface Characterization

Technique	Excitation	Response	Information	Monolayers Sampled	Quantitative Analysis	Gaseous Atmosphere
XPS	Photons	Electrons	Atomic composition. Nature of atoms, binding energies	2–20	Yes	Difficult
AES	Electrons	Electrons	Atomic composition	2–20	Possible	Difficult
SIMS	Ions	Ions	Atomic composition, short range order	1–3	Possible	No
ISS	Ions	Ions	Atomic composition	1	No	No

XPS: X-ray Photoelectron Spectroscopy; AES: Auger Electron Spectroscopy; SIMS: Secondary Ions Mass Spectrometry; ISS: Ion Scattering Spectroscopy.

techniques according to the incident radiation and the response has been given in Delannay (1984).

Among the surface spectroscopic techniques x-ray photoelectron spectroscopy (XPS) is widely used because it can give valuable information on the number of different atoms on the surface and their oxidation states.

The development of electron microscopy during the past years has enabled catalytic and other researchers to observe and analyze particle shapes, size distributions as well as structural features, and chemical information of the support and supported active phase species. The interaction of a high-energy electron beam with a solid surface gives rise to a variety of signals that can be used to study the nature of the solid specimen. Scanning electron microscopy (SEM), used in the analysis of bulk solids, utilizes an electron beam of low energy (typically 20 to 30 keV) and is used to obtain a high-magnification image of the sample. In transmission electron microscopy (TEM) a thin solid (<200 nm) is irradiated with an electron beam of sufficiently high energy to propagate through the specimen. Diffracted electrons are used to obtain a diffraction pattern of the specimen, and thus obtain information on the structure of the material. Transmitted electrons are used to form images that are analyzed to obtain information about the atomic structure of the solid and the defects present in the solid sample. Additional information on these and other surface techniques used for the analysis of sulfided and other types of catalysts can be obtained from reference textbooks (Delannay, 1984; Wachs, 1992; Topsøe et al., 1996).

Many of the above techniques have been used for the characterization of sulfided hydroprocessing catalysts; structural characterization of sulfided catalysts has been obtained using TEM (Delannay, 1985; Hensen et al., 2001; Eijsbouts et al., 2005), XPS (Prins et al., 1989), EXAFS (Boudart et al., 1983), and STEM (Lauritsen et al., 2004; Carlsson et al., 2004).

The dispersion and number of coordinatively unsaturated sites (CUSs) in sulfided catalysts has been evaluated using temperature-programmed reduction of sulfides (TPR-S) and the adsorption of probe molecules such as oxygen, NO, and CO (Massoth et al., 1982; Topsøe and Topsøe, 1983; Portela et al., 1995; Travert et al., 2001; Hédoire et al., 2003; Ramírez et al., 1998, 1999, 2004; Oyama et al., 2004; Choi et al., 2004). A correlation between the number of CUS associated with molybdenum or the promoters with HDS catalytic activity has been obtained in some cases. The chemical or spectroscopic characterization of fresh heavy oil hydroprocessing catalysts is quite similar to that of other hydrotreating catalysts; refer to the well-documented text by Topsøe et al. (1996).

Characterization of heavy oil hydrotreating spent catalysts has been performed to gain insight about the deposition of coke and metal sulfides on the catalyst surface. Deposits of metal sulfides have been confirmed by XRD and TPR for spent catalysts used in Maya crude hydrotreating (Rana et al., 2004a, 2005c). Deposited metal sulfides have been characterized by Smith and Wei (1991a) using TEM, EDX, STEM, and PS for the HDM of model molecules on a deactivated catalyst, and by Takuchi et al. (1985) by means of XRD, electron spin resonance (ESR), and SEM on catalysts used in HDM of heavy oil where the V_2S_3 phase was observed in acircular or rod-shaped crystallites about 100 nm in length. Toulhoat et al. (1987) reported that nickel is always associated with vanadium [$Ni(V_3S_4)$], and its crystallite grew perpendicular to the support surface. Apart from these metals, Ca, Mg, and Fe sulfides have also been reported to deposit on the exterior surface of the catalyst (Gosselink, 1998) and may play an important role in the plugging of catalyst pores.

The amount of deposits of vanadium and other metals also depends on the composition of feedstock. In a study using pure and diluted (50% Maya and 50% diesel) Maya crude, SEM-EDAX (Energy Dispersive X-ray) analysis indicated that vanadium deposition was almost five to six times higher for Maya crude than for the diluted feedstock, as shown in Figure 6.17. The typical U shape of the vanadium profile across the catalyst pellet diameter was observed by Galiasso and Caprioli (2005) for an ebullated-bed catalyst. Characterization of asphaltenes, which are precursors of coke formation and catalyst deactivation, has been reported in Absi-Halabi et al. (1991) and Furimsky and Massoth (1999).

The activity of spent catalysts has been characterized by using model molecule reactions over spent hydroprocessing catalysts in order to elucidate the role of deposited metals on the performance of active sites (Mitchell and Valero, 1982; Morales et al., 1982; Agrawal and Wei, 1984; Weitkamp et al., 1984; Ware and Wei, 1985; Mitchell et al., 1985; Dejonghe et al., 1990). Dejonghe et al. (1990) confirmed that part of the deposited vanadium is able to hydrogenate toluene by using vanadium deposited on pure alumina; XPS was used to estimate the dispersion and chemical states of the deposited elements. Ledoux et al. (1987) reported HDS and HDN activity over a vanadium-deposited NiMo commercial catalyst. In other work, it was confirmed by ESR and nuclear magnetic resonance (NMR) that a small amount of vanadium and nickel porphyrins have a drastic

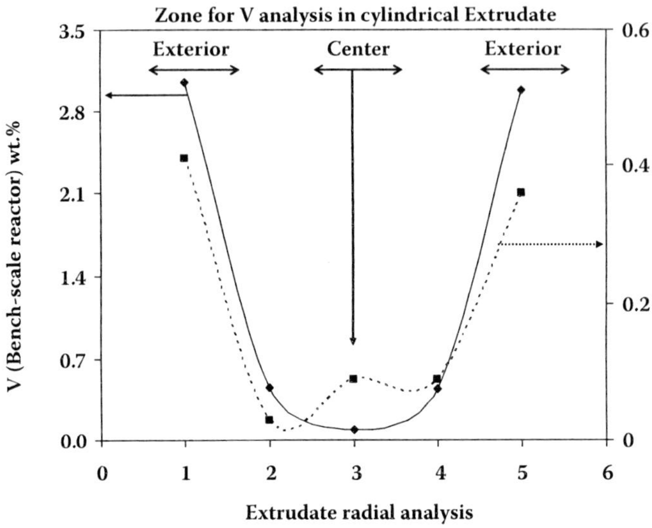

FIGURE 6.17 Deposited metal concentration profiles obtained by SEM-EDAX on the microflow and bench-scale reactors (radial distribution of vanadium). (From Rana, M. et al., *Catal. Today*, 109: 24–32, 2005. With permission.)

effect on the thiophene HDS activity of a conventional NiMo catalyst (Ledoux and Hantzer, 1990).

The characterization of a series of commercial alumina-supported nickel-molybdenum and cobalt-molybdenum hydrotreating catalysts exposed to different degrees of reaction length and severities using a variety of naphtha and gas oil feeds was reported by Weissman et al. (1996). Fresh and used catalysts were characterized in their HDS and HDN activities using gas oil as feedstock, chemical analysis of deposits, porosimetry, and high resolution transmission electron microscopy (HRTEM). Additionally, the deposits on the catalyst were characterized by C^{13}-NMR. It was found that deactivation occurred primarily by carbon deposits on the support. Changes in the structure of the MoS_2 phase as observed by TEM did not appear to contribute to activity changes. Two types of carbon were found, depending on the feed used; using gas oil as feedstock led to deposits of a compact carbon, resulting in important loss of HDN but only a slight loss in HDS activities. Processing with light feed (naphthas) led to the deposition of less dense carbon, which reduced nitrogen and sulfur removal proportional to the carbon loading. Some other recent studies have been made on the performance and characterization of catalysts for the hydroprocessing of heavy oil and other real feeds (Galarraga and Ramírez de Agudelo, 1992; Caloch et al., 2004; Maity et al., 2004; Rana et al., 2004a, 2005b, 2005c, 2005d; Ramírez et al., 2005).

6.5 CATALYST DEACTIVATION

Hydroprocessing catalysts are active for a number of important reactions: HDS, HYD, HDN, HDM, and, for coal-derived liquids, hydrodeoxygenation (HDO), all of which involve the hydrogenolysis of the heteroatom–carbon bond. For hydrocracking catalysts, breaking of C–C bonds also takes place. These catalysts consist of molybdenum supported on a carrier, typically alumina, and are promoted by nickel or cobalt and operate in the sulfided state. Hydroprocessing reactions take place on the active sites of the catalyst, which are believed to be located on the sulfur vacancies located on the edges of the MoS_2 crystallites. The vacancies associated with the Ni or Co promoters are considerably more active than those associated with Mo only. A detailed description of the more accepted model for the operation of hydrotreating catalysts has been proposed and refined by Topsøe during the past years (Topsøe et al., 2005 and references therein). It was shown that for molecules such as 4,6-dimethyldibenzothiophene, which mainly transform not through a direct desulfurization route but through hydrogenation followed by hydrodesulfurization, fully sulfided Mo sites (called Brim sites) participate in the first steps of the reaction mechanism. The final reaction step in the HYD-HDS mechanism occurs in a similar manner as that previously proposed for the direct desulfurization route, which involves the participation of coordinatively unsaturated sites (CUSs) and sulfur vacancies on the edge of the MoS_2 crystallites (Topsøe et al., 2005).

Clearly, since the sulfur vacancies have a Lewis acid character, poisoning of the hydrotreating reactions will occur when a strongly adsorbed molecule like a nitrogen compound, coke molecule, metal, or any other basic molecule occupies an active vacancy. Basically, poisoning of hydroprocessing catalysts occurs by single or multiple causes like active site blockage by strongly adsorbed species, active site coverage by deposits of coke or metals, pore mouth constriction/blockage, or sintering of the active phase (Dautzenberg et al., 1978; Furimsky and Massoth, 1999).

The degree and type of deactivation will depend on the feed characteristics and will be evidenced by an S-shaped curve of temperature vs. time-on-stream, as shown in Figure 6.18. Initially, coke deposition causes a rapid deactivation that in several hours attains a pseudo-steady state. Then deactivation by metal deposits is observed during a longer period until drastic deactivation caused by pore restriction and blockage is observed.

The deactivation of hydroprocessing catalysts through the above mechanisms has been thoroughly reviewed by Furimsky and Massoth (1999). Other reviews on the subject have been made for HDS reactions (Bartholomew, 1984), coal liquefaction (Thakur and Thomas, 1985), deactivation during hydroprocessing of heavy feeds and synthetic crude (Thakur and Thomas, 1984), deactivation during residue hydroprocessing (Bartholomew, 1994), and deactivation by coke (Gualda and Tulhoat, 1988; Menon, 1990; Absi-Halabi et al., 1991). Deactivation by metals was addressed in Dautzenberg and de Deken (1987).

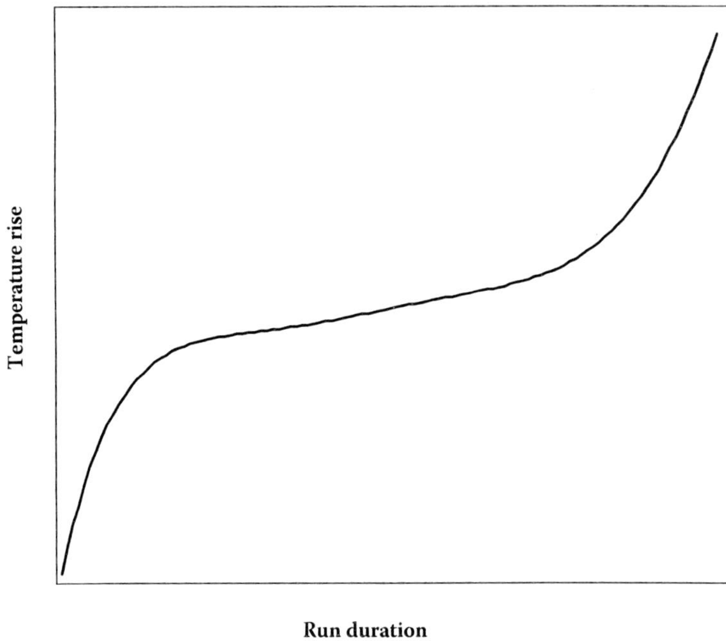

FIGURE 6.18 Typical S-shaped catalyst deactivation curve.

A poison is a substance that competes for adsorption on the active sites. The adsorption of the poison can be reversible or irreversible (Figure 6.19).

The effect of a poison (p) on the reaction rate (r_A) can be taken into account in the denominator term of a Langmuir–Hinshelwood mechanism:

$$r_A = \frac{k_A f(C_A)}{\left(1 + \sum K_i C_i + K_p C_p\right)^n} \tag{6.3}$$

where k_A represents the rate constant, C_A the concentration of reactant A and $f(C_A)$ its particular rate form, K_p and K_i the adsorption constants of the poison and the adsorbing species associated with the reaction, respectively, and n the power of the inhibition term (its value is associated with the number of active sites involved in the reaction mechanism). Equation 6.3 shows that the extent of inhibition will be related to the magnitude of the poison adsorption constant and concentration with respect to those of the species associated with the main reaction.

A typical example of a poisoning effect is that associated with the presence of nitrogen compounds, which due to their basic nature adsorb strongly on the coordinatively unsaturated sites of the MoS_2 structures and on the acidic sites of

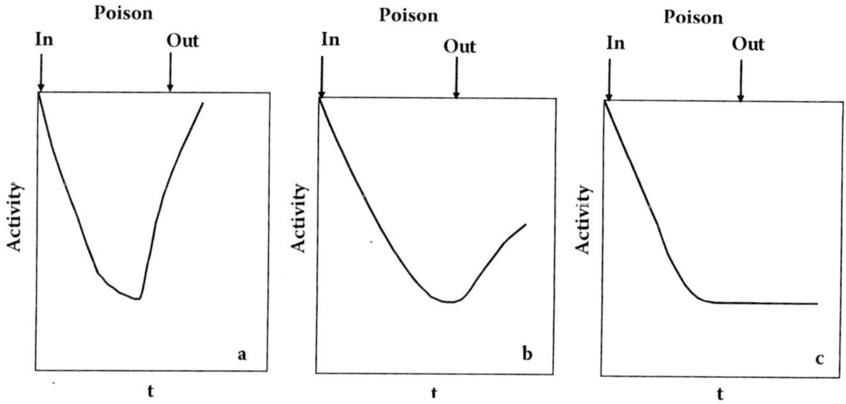

FIGURE 6.19 Activity-time (t) profiles for (a) reversible, (b) quasi-irreversible, and (c) irreversible poisoning. (From Furimsky, E. and Massoth, F.E., *Catal. Today*, 52, 381–495, 1999. With permission.)

the catalyst. A detailed discussion about the effects and nature of the poisoning by nitrogen compounds in hydroprocessing catalysts is given in Furimsky and Massoth (1999). Apart from nitrogen compounds, inhibition may be caused by other molecules, like H_2S, which is produced in the reaction (Girgis and Gates, 1991; Rana et al., 2004c; Furmisky and Massoth, 2005). However, adsorbed H_2S can enhance the catalyst acidity through sulfhydryl group formation (Topsøe et al., 1989; Rana et al., 2000; Breysse et al., 2002). Figure 6.20 shows the effect of

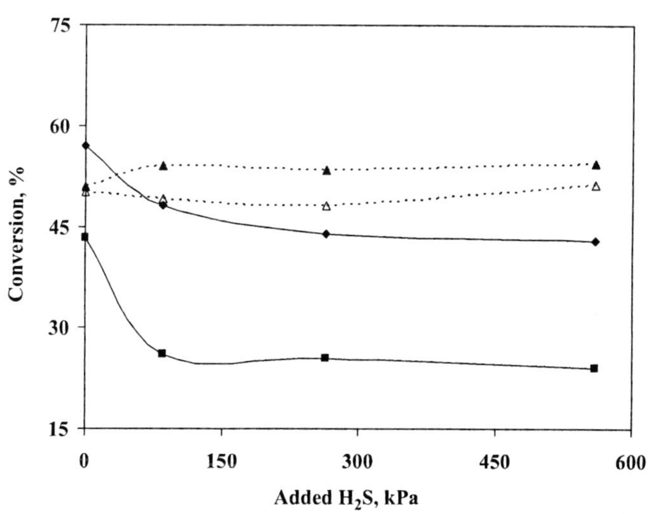

FIGURE 6.20 Effect of added H_2S partial pressure on steady-state conversion of Maya crude hydroprocessing: (▲) HDM, (Δ) HDAs, (◆) HDS, (■) HDN. (From Rana, M.S. et al., *Fuel*, 86(9) 1263–1269, 2007a. With permission.)

H_2S partial pressure on the steady-state conversion of Maya crude. According to the results of this study, increasing the amount of H_2S had no deleterious effect on HDM and HDAs, while HDS and HDN were moderately and strongly affected (Rana et al., 2007a).

Deactivation by coke and metals occurs to a greater extent in heavy feeds and follows a trend different from the one observed for light feeds. For heavy feeds, coke deposits rapidly during the earlier reaction stages and then reaches a steady state, while the deposition of metals increases continuously with time in a more or less linear fashion. Figure 6.21 shows the time evolution of coke and metals for catalysts at different locations in the catalytic bed.

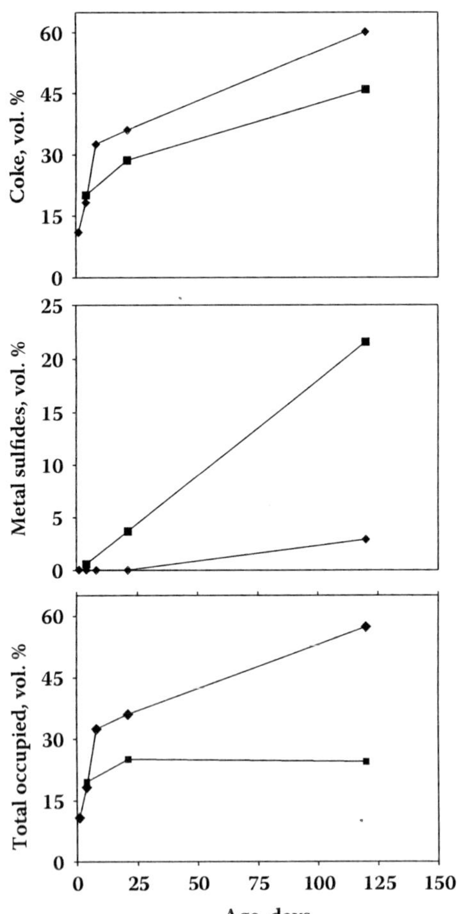

FIGURE 6.21 Evolution of (a) coke, (b) metals, and (c) total occupied volume as a function of time-on-stream: (◆) KO 012.bot, (■) KO 013.top. (From Fonseca, A. et al., *Fuel*, 75: 1363–1376, 1996. With permission.)

Coke deposition increases with the boiling range of the feed. However, for feeds with a similar boiling range, the one with the higher concentration of coke precursors, such as aromatics and heterocyclic compounds, will cause a more severe poisoning. This means that different catalysts will be needed for processing a naphtha fraction originated from coal liquefaction than for a conventional light feed.

Metals and coke affect to a different extent the various catalyst functionalities. Cable et al. (1981) compared the activities of fresh and regenerated catalysts and found that metals have a moderate effect on HDS and a weak effect on hydrogenation and cracking, while coke had a moderate effect on HDS but a strong and very strong effect on hydrogenation and cracking, respectively. These results are summarized in Table 6.11.

The composition of the catalytic surface is of major importance to control the extent of coke deposition. Coke precursors like alkenes, aromatics, and heterocyclics can transform into higher molecular weight species if sufficient active hydrogen is not available to prevent it (Furimsky, 1982; Wiwel et al., 1991; Furimsky and Massoth, 1999).

In the case of heavy feeds, the precipitation of asphaltenes from the feed is another important source of coke formation. Asphaltenes are stabilized by the presence of resins. Therefore, if during hydroprocessing resins are eliminated at a faster rate than asphaltenes, precipitation of the latter occurs on the surface of the catalyst, leading to the possibility of greater coke formation (Dautzenberg and de Deken, 1985). It is apparent from data in the literature that asphaltenes have a greater tendency than resins to coke formation (Banerjee et al., 1986).

Besides feed composition and catalyst properties, hydrogen partial pressure and operating temperature are important for controlling the extent of coke formation. The presence of a high concentration of active hydrogen on the surface can convert coke precursors to stable products. However, at high operating temperatures, the probability of this conversion decreases because of thermodynamical limitations. Figure 6.22 suggests the existence of an optimal combination of catalyst properties and process conditions, which ensures the best performance of the catalyst (Ternan et al., 1979).

TABLE 6.11
Relative Deactivation by Metals and Coke

	Relative Effect	
Catalyst Function	**Metals**	**Coke**
Hydrodesulfurization	Moderate	Moderate
Hydrogenation	Weak	Strong
Hydrocracking	Weak	Very strong

Source: Adapted from Cable, T.L. et al., *Fuel Process. Technol.*, 4, 265–275, 1981. (With permission.)

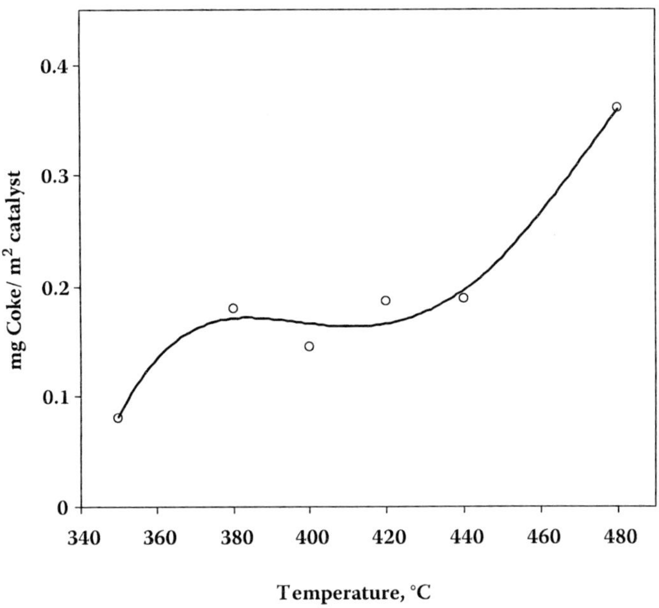

FIGURE 6.22 Effect of operating temperature on catalyst coke deposits. (From Ternan, M. et al., *Fuel Process. Technol.*, 2: 45–55, 1979. With permission.)

At low temperature, coke increases because the conversion of heavy species to lighter fractions is slow and the lifetime of heavy molecules on the surface is prolonged, increasing the possibility of polymerization reactions. Another reason for this observation could be that resins are converted faster than asphaltenes, causing the precipitation of the latter. Above 375°C, the transformation of asphaltenes may increase due to the higher temperature, but also because the hydrogenation of coke precursors competes succesfully with their polymerization. The drastic increase in coke formation above 440°C is possibly due to an increase in dehydrogenation, followed by polycondensation reactions. The shape of the curve in Figure 6.22 can be altered by changing the partial pressure of hydrogen. The results, shown in Figure 6.23, indicate that the steady-state level of carbon on the catalyst can be significantly altered by changing the hydrogen partial pressure. Different aspects of poisoning by coke in hydroprocessing catalysts have been addressed in Furimsky and Massoth (1999).

Pore volume and pore size distribution are important to maximize catalyst utilization when coke deposition is a problem. Figure 6.24 compares the nitrogen physisorption isotherms of fresh and spent catalysts with small- (6-nm) and large- (17-nm) diameter pores. The variations in the shape of the hysteresis loops were considered as representative of pore mouth coking (Rana et al., 2005c).

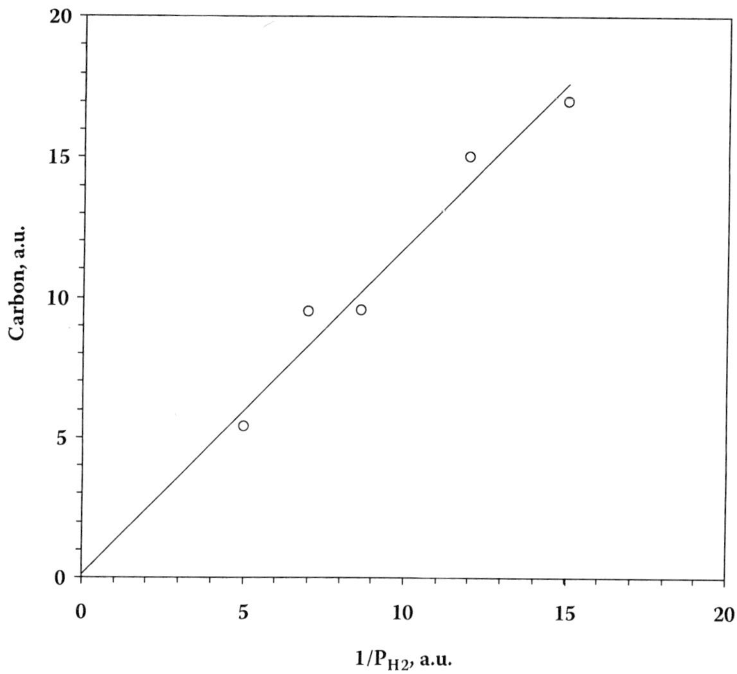

FIGURE 6.23 Steady-state level of carbon on catalyst as a function of H_2 pressure. (From Oelderick, J.M. et al., *Appl. Catal.*, 47: 1–24, 1989. With permission.)

As shown in Figure 6.4, catalyst life is greatly affected by the content of metals in the feed. For heavy feeds, which contain high concentrations of metals, the life of the catalyst can be several times lower than that for a light feed. Deactivation by metals occurs along with the deactivation by coke deposition. However, while coke reaches pseudo-steady-state equilibrium after an initial rapid buildup, the deposition of metals continuously increases with time-on-stream. One of the reasons for the continuous increase in catalyst metal content is that during hydroprocessing, metals, mostly V and Ni, are converted to sulfides, which autocatalyze the metal deposition reaction (Sie, 1980). Hence, the demetallization reaction can continue on the newly created sites after the original active sites of the catalyst have been covered by metal deposits. Permanent deactivation by metal deposition causes an important problem to refiners. Because of this, hydroprocessing schemes often use HDM catalysts to selectively remove metals and protect downstream catalysts.

The S-shaped catalyst deactivation profile observed for heavy feeds (see Figure 6.18) is then related to coke formation and metal deposits on the catalyst. The initial period of rapid deactivation is generally attributed to the buildup of an equilibrium coke load on the catalyst surface (Sie, 1980; Pazos et al., 1983)

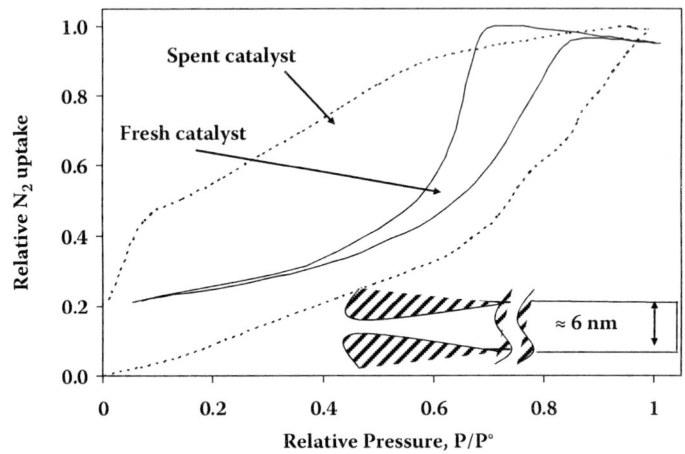

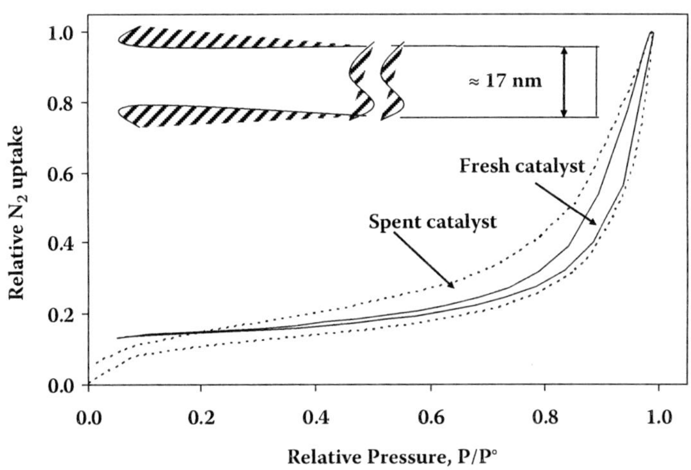

FIGURE 6.24 Fresh and spent catalysts adsorption–desorption isotherms. (a) CoMo/Al_2O_3 (6-nm average pore diameter) and (b) CoMo/Al_2O_3 (17-nm average pore diameter). (Adapted from Rana, M. et al., *Catal. Today*, 104: 86–93, 2005. With permission.)

but has also been ascribed in part to the buildup of monolayer or submonolayer metal deposits (Tamm et al., 1981; Johnson et al., 1986). The almost constant deactivation rate observed in the intermediate period is ascribed to increasing diffusional resistance caused by accumulation of metals on the pore walls. The final rapid deactivation that causes the end of catalyst life is due to pore plugging caused by metal sulfide deposits. However, studies performed by Smith and Wei (1991b) on deactivation of HDM catalysts with model molecules have shown evidence that metal deposits have a crystalline form that is inconsistent with the general idea of uniform metal deposits. It was also observed by the same authors that both a sulfided $CoMo/Al_2O_3$ catalyst and a low-promoter alumina carrier neither deactivated nor acquired catalytic activity as nickel sulfide deposits accumulated. These observations suggest that part of the active components of the catalyst remain active for HDM even in the presence of high levels of deposited nickel sulfide and are inconsistent with a contribution of the nickel deposits to the rapid initial deactivation of the HDM catalyst. In contrast, HDM experiments carried out with Vanadyl ethioporphyrin (VO-EP) in squalane were diffusion limited due to the low effective diffusion of VO-EP. The catalyst deactivation was consistent with the general idea that accumulation of metal sulfide deposits is responsible for the relatively slow steady deactivation that characterizes the intermediate period of the life of a hydroprocessing catalyst. In a study of deactivation of residue HDM catalysts it was shown that different results are obtained by carrying the HDM experiments in batch or continuous-flow reactors. The batch reactor yielded carbon deposits with almost no metals, whereas the continuous reactor provided used catalysts containing significant amounts of both carbon and metal. It was also found that a small amount of well-dispersed vanadium inside the catalyst grain was more deactivating than a large amount of carbon (Gualda and Kasztelan, 1996).

About one half of the Ni and V present in heavy feeds are present as porphyrins; the other part of metals are in less characterized forms, which include bonds with nitrogen, oxygen, and sulfur in the defect centers of asphaltene sheets (Mitchell, 1990; Beaton and Bertolacini, 1991). Due to the bulkiness of the porphyrins and asphaltene molecules, the demetallization reaction is affected by pore diffusion limitations. Consequently, metals are generally deposited on the outer zone of the catalyst particle. However, the characteristics of the catalyst can alter the radial profile of metal deposits, making it more shallow or broad. Figure 6.25 shows the radial profiles of vanadium obtained by electron microprobe analysis along the diameter of catalysts used in the hydroprocessing of residua. Profiles A, B, and C correspond to the catalysts labeled with the same letters in Table 6.12.

The pore size of the catalyst determines to a greater extent the type of metal penetration profile. For microporous catalysts the metal deposits are restricted to the outer shell of the catalyst particle, entailing pore mouth plugging. Catalysts with wide or intermediate pore diameter lead to deep or intermediate metal penetration and high and medium metal storage capacities, respectively. So, catalyst life will be determined to a great extent by the pore volume available for storing metal sulfide deposits. However, it must be considered that the increase in pore diameter and volume brings about a decrease in surface area, and therefore lower catalytic activity for other hydrotreatment reactions. Hence, porosity and relative activity in hydrotreatment must be balanced so as to obtain the best catalyst performance. The change in activity with pore diameter for the different reactions (HDS, HDM, HDAs), as shown in Figure 6.12, indicates that the optimum pore diameter is larger for hydrodemetallization and deasphaltenization than for hydrodesulfurization.

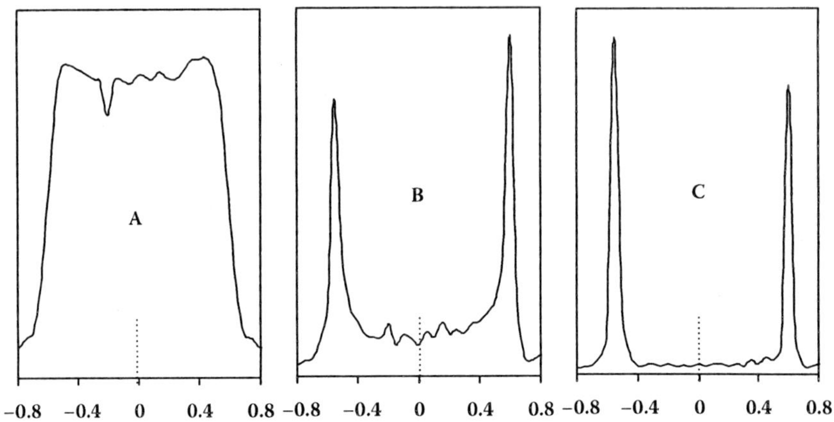

FIGURE 6.25 Radial concentration profiles of vanadium as determined by electron microprobe scans along the diameter of catalysts used in hydroprocessing of residues. Profiles A, B, and C are typical for the catalysts distinguished by the same letters in Table 6.12. (From Sie, S.T., *Appl. Catal. A Gen.*, 212: 129–151, 2001. With permission.)

The opposite effects of pore size variation on catalyst stability and activity are illustrated in Figure 6.26. It is clear that catalyst deactivation has important consequences for process design and operation. In general, the design of the catalyst has to reach a compromise between catalyst life and activity. So, neither of them must be maximum, but both must be adequate for the intended duty.

TABLE 6.12
Characteristics of Different Catalysts

Catalyst Type	A	B	C
Pore size	Wide	Intermediate	Narrow
Metals penetration	Deep	Intermediate	Shallow
Metal storage capacity	High	Medium	Low
Stability	Very good	Fair	Poor
HC/HDS activity	Low	Fair	High
Application	HDM catalyst	HC/HDS catalyst	HC/HDS catalyst
Location in reactor train	Front end	Middle	Tail end

Note: HC = hydroconversion.

Source: From Sie, S.T., *Appl. Catal. A Gen.*, 212, 129–151, 2001. (With permission.)

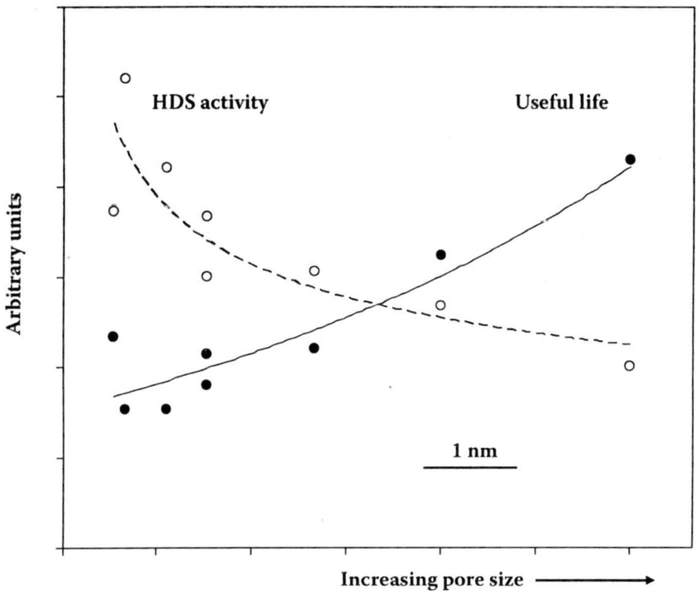

FIGURE 6.26 Effect of pore size on useful life and hydrodesulfurization activity for catalysts with narrow monomodal pore size distributions, tested under standard conditions. (From Sie, S.T., *Appl. Catal. A Gen.*, 212: 129–151, 2001. With permission.)

6.6 REGENERATION OF DEACTIVATED CATALYSTS

Deactivation results in poor catalyst stability or slow decrease in conversion with time. The normal action taken to maintain the overall activity is to increase the catalyst bed temperature gradually until a limit minimum level of activity is achieved. Ultimately, the catalyst will result in severe coke formation and undesirable product yield. Generally, coke is defined as soft and hard (Sahoo et al., 2004). Temperature-programmed oxidation (TPO) analysis confirms the two fractions of coke, one that can be removed easily and the other of refractory surface coke, which is strongly adsorbed on the catalyst support (Stanislaus et al., 1993; Marafi et al., 2005). Thus, the nature of support plays a crucial role to coke formation as well as on redispersion of the active metals after regeneration. So, deactivation and regeneration must be considered before selecting a support composition. The incorporation of small amounts (5 to 10 wt%) of a basic support component (MgO, ZrO_2, and so on.) as a mixed oxide with alumina was suggested to have better performance than an acidic component, although at the expense of the catalyst hydrogenation function (Rana et al., 2007b).

When the catalyst activity declines to a critical level, the first choice is the regeneration and reuse of catalyst; catalyst disposal is usually the last resort, especially in view of environmental considerations. Regeneration consists in the burning off of the carbonaceous deposits in the presence of oxygen and inert gas under controlled temperature. The complete regeneration of heavy oil spent catalysts is difficult due to the presence of metal sulfides, which can not be completely burned off and leave vanadium and nickel oxides on the surface of the catalyst. So, poisoning by metal deposition is typically irreversible (Gualda and Kasztelan, 1984). Metal deposits can be partially eliminated from the catalyst by mild chemical treatment (leaching) without destroying the support and avoiding excess removal of molybdenum (Stanislaus et al., 1993). Severe acid leaching is not advised because it will remove Ni and V deposits along with the promoter, Co (or Ni), and base catalytic metal, Mo (or W), leaving an almost inert support (Jocker, 1993; Inoue et al., 1993). Therefore, complete regeneration of the catalyst is practically impossible. Moreover, during the oxidative regeneration of the catalyst, sintering and redistribution of the active metallic phase may occur (George et al., 1988) with the consequent loss of active sites. A decrease in crushing strength and loss of active phase by attrition of the catalyst particle during regeneration might also occur. Thus, careful handling of the catalyst and control of regeneration temperature and gas composition (for example, oxygen/nitrogen) will be helpful to recover as much as possible of the original catalytic sites and activity (Stanford, 1986; Yui, 1991; Furimsky and Massoth, 1993).

6.7 HEAVY OIL HYDROPROCESSING REACTORS AND PROCESSES

Chapter 5 of this book is dedicated to reactors for hydroprocessing, while Chapter 9 is devoted to hydroprocesses. This section presents only a brief summary of these topics in order to provide a general idea of the relationship among catalysts, reactors, and processes.

A variety of hydroprocessing technologies have been developed and commercialized for heavy oil upgrading (Kressmann et al., 1998; Furimsky, 1998). For heavy feeds, catalyst deactivation by pore plugging has important consequences not only for catalyst design and selection but also for process design. Various processes have been developed to cope with catalyst deactivation. Some of the options available in the industry for residue processing are listed in Table 6.13. A compilation of residue process technologies throughout the world is given in Table 6.14, which indicates that for the feeds used today, fixed-bed reactor systems are more popular than ebullated beds or slurry systems.

The technological consequences associated with catalyst deactivation covering catalysts, reactors, and process configuration aspects were reviewed by Sie (2001). Fixed-bed reactors offer advantages due to their simplicity, and with the use of multiple beds of catalysts it is possible to obtain an acceptable catalyst life except

TABLE 6.13
Processes for Residue Hydroconversion

Concept	Provider
Continuous catalyst replacement	OCR/Chevron
Swing reactor	Hyval/IFP
Bunker-type reactor	Hycon/Shell
Ebullated-bed reactor	H-Oil/HRI
	LC-Finning/Lummus
Slurry system	Microcat/ExxonMobil
	Verba Combi-Cracking/Veba
	HDH/Intevep

for demanding feeds like residual oil processing. For heavy feeds where deactivation of the catalyst is faster, an alternative is the use of ebullated-bed reactors, in which the catalyst is maintained in a fluidized state by the upward liquid and gas stream. In this type of reactor the addition and taking out of catalyst can be performed during normal operation. So, with this reactor catalyst life is not an issue. However, because of the hydrodynamics inside the reactor, recycling of liquid is substantial, and therefore the reactor operates close to a continuous stirred tank reactor, which, according to theory, presents a lower rate of reaction than a fixed-bed reactor. Another disadvantage of this type of reactor is that because of the backmixing of solids, fresh and old catalysts are taken out each time a catalyst is withdrawn from the reactor. This problem was alleviated with the bunker flow reactor, which operates as a

TABLE 6.14
Residue Commercial Process Technologies in the World*

Process Technology	North and South America	Europe and Japan	Others	Total	% **
Hydroprocessing					
Fixed bed	12	21	27	60	9
Ebullated	7	4	1	12	2
Slurry	1	0	0	1	0
Total Hydroprocessing	20	25	28	73	12
Solvent Deasphalting	15	3	6	24	4
Thermal Processing	131	155	147	433	69
RFCC	19	26	57	102	16
Total **	185	209	238	632	100

* Units operating in March 2003; ** Based on total units.

moving-bed reactor with both liquid and solid moving close to a plug flow pattern (Sie, 2001). This type of reactor is particularly well suited for front-end demetallization. The expected axial deactivation profiles of fixed-bed, continuous stirred tank, and bunker flow reactors are depicted in Figure 6.27a. However, different profiles have been reported for end-run resid demetallation catalyst in an adiabatically operated fixed-bed reactor (Figure 6.27b). The explanation for the observed maximum vanadium and nickel concentration was associated with a consecutive mechanism of resid demetallation, where metal-bearing asphaltenes are first cracked to form resins, which subsequently demetallate; the top 10 to 20% of the catalyst bed functions as an asphaltenes cracking section. As new metal-bearing resins are generated, the demetallation rate increases until the supply becomes depleted and the exponential decline in metal deposition occurs. Nonisothermal behavior or unusual behaviors near the top of the bed were other possible causes (Fleisch et al., 1984). The increasing level of carbon down the bed was ascribed to the high temperatures at the end of the bed caused by the adiabatic operation of the reactor.

Usually, a fixed-bed reactor is preferred when the catalyst stability is more than a year, as is the case for middle distillate hydrotreating. For heavy oil processing where the expected stability is less than a year, the use of moving-bed or ebullated-bed reactor technology is preferred. Nevertheless, the process selection will depend on the feed composition and required conversion, as shown in Figure 6.28.

Ebullated-bed reactors are capable of converting the most problematic feeds, such as atmospheric residue, vacuum residue, and heavy oils (all of which have a high content of asphaltenes, metals, sulfur, and sediments) to lighter, more valuable products. Additionally, this type of reactor can perform both hydrotreating and hydrocracking functions; thus, ebullated-bed reactors are referred to as dual-purpose reactors. One of the problems for predicting the performance of these reactors is that in addition to data on feed composition and catalyst properties, knowledge of the hydrodynamic characteristics of the bed (incipient fluidization velocity, bed expansion, and bubble flow regime) at the severe operating temperature and pressure is necessary. Although these properties have been extensively studied at ambient conditions, little has been reported for systems at high temperature and pressure (Ruiz et al., 2004, 2005).

Ebullated-bed catalysts are made of pellets or grains that are less than 1 mm in size to facilitate suspension by the liquid phase in the reactor (Colyar and Wisdom, 1997; Courty et al., 1999). Moving- and ebullated-bed catalysts are chemically quite similar to those used in fixed-bed reactors, except that the mechanical strength, size, and shape of the catalyst particle must meet more demanding specifications.

Slurry reactors are another alternative for processing heavy feeds. However, the processes using this type of reactor require the separation of the catalyst from the products, which needs additional technology.

In general, the hydrotreating catalyst life is several years if the feedstock is light oil (middle distillate). However, for heavy oil processing, catalyst stability with time-on-stream is very low. The relationship between approximate

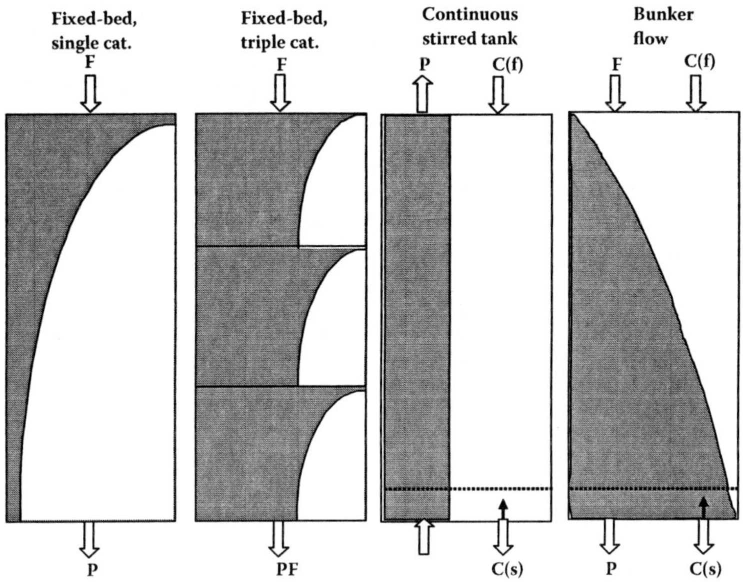

(a)

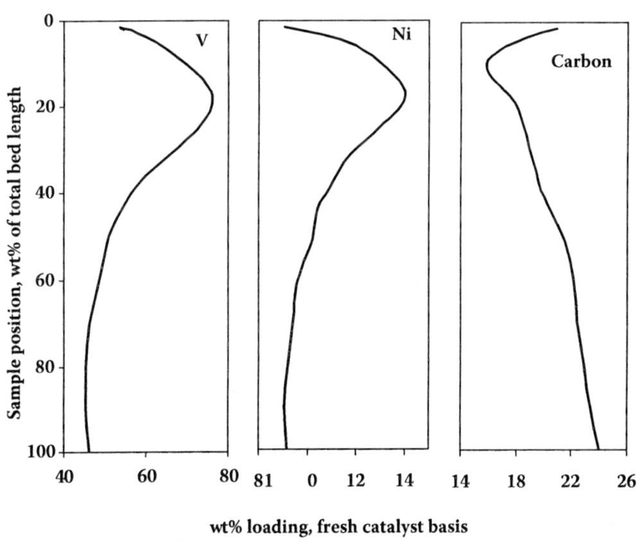

(b)

FIGURE 6.27 (a) Axial deactivation profiles at end-of-run conditions in fixed beds and during operation in a continuous stirred tank or bunker flow reactor. F, feed; P, product; C(f), fresh catalyst; C(s), spent catalyst. The hatched area denotes the deactivated part. (From Sie, S.T., *Appl. Catal. A Gen.*, 212: 129–151, 2001. With permission.) (b) V, Ni, and C concentration profile in weight percent (fresh catalyst basis at the end of demonstration run; 0 = top of bed, 100 = bottom of bed). (Adapted from Fleisch, T.H. et al., *J. Catal.*, 86: 147–157, 1984. With permission.)

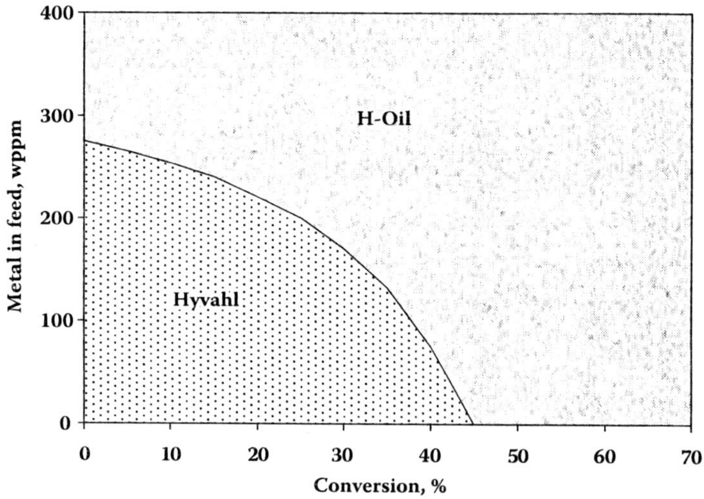

FIGURE 6.28 Process selection based on metal content in the feed. (From Plain, C. et al., Axens IFP Group Technologies, http://www.axens.net/upload/presentations/fichier/ options_ for _resid_conversion.pdf.)

deactivation timescale and type of process reactor is given in Table 6.15 (Moulijn et al., 2001).

Different technologies are in use for processing heavy feedstocks; thermal processes based on carbon withdrawal produce low-value products, while hydroprocessing processes, based on hydrogen addition, enhance liquid yield. A comparison of these processes is presented in Figure 6.29 for the hydroprocessing of

TABLE 6.15
Relationship between Approximate Deactivation Timescale and Type of Process Reactors

Timescale of Deactivation	Type of Reactor	Type of Regeneration	Feedstock
Years	Fixed bed	No regeneration	Middle distillate
Months	Fixed bed	Off-line	Heavy oil
Weeks	Fixed-bed swing mode Moving bed	Alternative	Heavy oil
Minute–days	Fluidized bed Slurry	Continuous	Heavy oil
Seconds	Entrained flow (riser)	Continuous	Heavy oil

Source: From Moulijn, J.A. et al., *Appl. Catal. A Gen.*, 212: 3–16, 2001. (With permission.)

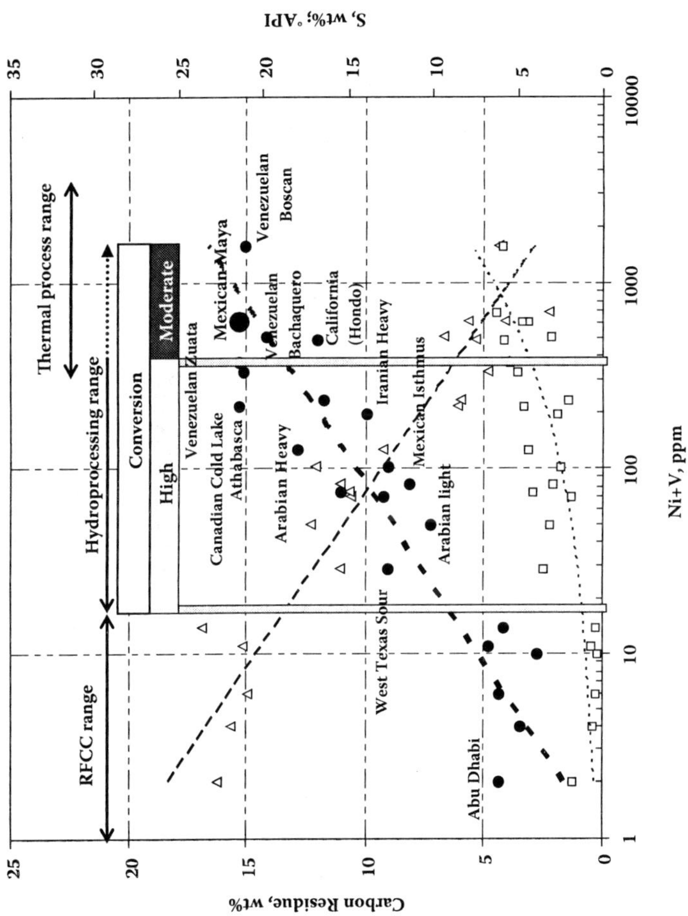

FIGURE 6.29 Operating range of different processes according to carbon, metals, and sulfur content of various atmospheric residues: (□) sulfur, (●) carbon residue, (Δ) API gravity. (From Rana, M.S. et al., *Fuel*, 86(9): 1216–1231, 2007c. With permission.)

atmospheric residue using parameters like metal, carbon, and sulfur contents. For heavier feeds, thermal processes are the option, although the range of hydroprocessing can be extended by the use of well-designed catalysts. Recently, the hydroprocessing application range has been extended to cover feeds with higher content of contaminants and lower API gravity than expected. This was possible by the use of a combination of catalysts with improved stability working under a moderate conversion regime to minimize sediment formation (Ancheyta et al., U.S. patent, filed 2006).

6.8 CATALYST BED PLUGGING

Catalyst bed fouling is present to some degree in all hydroprocessing units. When fouling significantly affects the hydrotreater cycle run length (pressure drop), some remedies are necessary. As mentioned earlier, a guard bed is the best choice to protect valuable downstream catalysts and maximize unit cycle length. The foulants that cause hydrotreater problems come from several different sources; primary foulants can be present already in the feed or can be generated in the equipment upstream from the hydrotreater. The principal causes, problems, and some of the solutions to bed plugging are shown in Table 6.16. Figure 6.30 illustrates the use of materials with different shapes and sizes to reduce ΔP problems due to particulate accumulation in the catalyst bed.

The use of high temperatures in the operation of residue hydroprocessing units results in shorter catalyst life and dry sludge (condensed carbonaceous material) formation in product oil. When residue conversion exceeds a certain limit, dry sludge precipitates because of the incompatibility of polymerized and

TABLE 6.16
Hydroprocessing Catalyst Bed Plugging and Its Principal Causes

Fouling	Principal Cause	Problem	Solutions
Particulates	Contamination of feedstock	Increased ΔP Void fraction of the catalyst bed is reduced Catalyst bed fully or partially plugged	Grading scheme Guard bed catalyst
Reactive molecules	Olefins and oxygenates	Carbon deposits Pore plugging	Increase in H_2 partial pressure Decrease in reaction temperature
Inorganic ions deposited	As, Ni, V, Fe, Ti, Si, and so on	Pore plugging due to metal deposits Buried catalytic sites	Guard bed catalyst

FIGURE 6.30 Example of shapes and sizes of supports and catalysts used in a typical catalytic bed. (Courtesy of Saint-Gobain NorPro.)

condensed asphaltenic components in the product. Once formed, dry sludge deposits in flash drums, fractionators, heat exchangers, and other downstream equipment. The maximum conversion attained is then determined by the limit of dry sludge formation rather than by the limit of catalyst activity. Analysis of dry sludge formation was performed by Takatsuka et al. (1989) with the aid of mathematical models. Figure 6.31 shows the ternary composition diagram obtained from the model calculations, demonstrating the dependence of asphaltene solubility on the components of product oil. Lines A and B in the ternary compatibility diagram mark the maximum conversion boundary regulated by asphaltene content and hydrogenation of product oil, respectively.

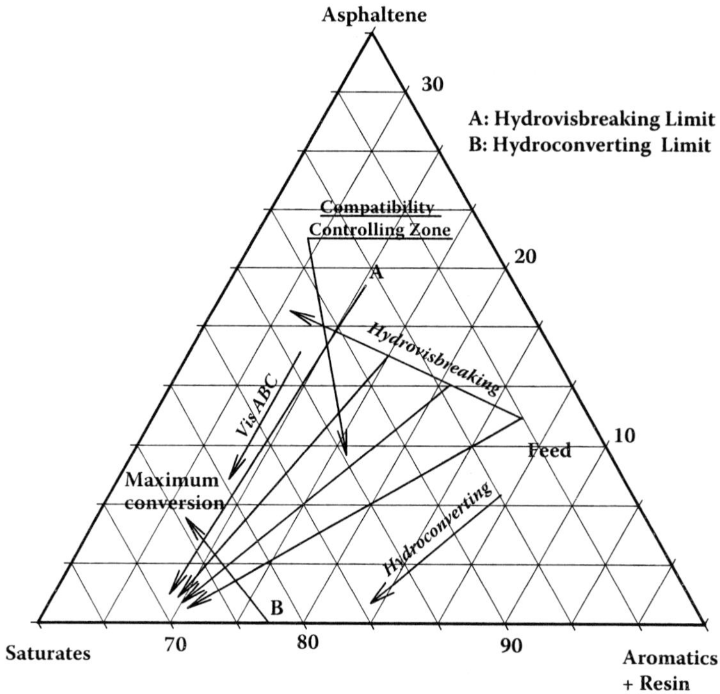

FIGURE 6.31 Compatibility control diagram for dry sludge. (From Takatsuka, T. et al., *J. Chem. Eng. Jpn.*, 22: 298–303, 1989. With permission.)

6.9 FINAL COMMENTS

Increased attention to the development of catalysts and processes for the hydroprocessing of heavy feeds will be necessary in view of the growing demand of distillates and the increasing need to process heavy feedstocks with high contents of sulfur, nitrogen, metals, aromatics, and asphaltenes.

The conversion of heavy feeds to valuable products must be maximized by a proper match of feed, catalyst properties, reactor type, and operating conditions. It appears necessary to develop better analytical methods for the characterization of heavy feeds to be able to better predict the behavior of catalysts and catalytic reactors when processing such feeds.

The design of improved catalyst supports is necessary. Supports with adequate pore size distribution, which minimize diffusional limitations of large molecules to the pore interior while providing enhanced metal retention capacity and as large a surface area as possible (without compromising the mechanical properties), are necessary. For the hydroprocessing of heavy feeds, catalyst life can be further prolonged by applying balanced textural properties (high pore volume and enough surface area). Besides, the chemical composition of the support must be capable of providing enough acidity to perform the hydrocracking of large molecules, but not so much as to lead to excessive coke formation. Studies of new catalyst formulations oriented to decrease catalyst deactivation and increase catalyst life and stability when processing heavy feeds are necessary.

The chemical composition of the active phase of the catalyst seems to depend on the processing scheme; the catalyst formulation will depend on the design of the catalytic bed system (single or multiple beds). For multiple-bed reactors, the first catalyst bed must provide high HDM activity with large metal retention capacity. The second bed provides high HDS and some HDM, and the third bed must have adequate hydrocracking, HDS, and HDN activities. The textural properties have to be adapted to the duty each catalyst bed performs. The shape and size of the catalyst particles have to be well designed to ensure smooth operation of each catalyst bed. Attention must be paid to the use of optimized graded beds of catalyst required to minimize the problem of increasing ΔP. Balanced catalyst formulations able to achieve maximum conversion with low formation of dry sludge for heavy feeds need to be developed. So, the chemical and physicochemical features of dry sludge formation and changes of asphaltene structure during hydroprocessing must be studied in depth in order to construct mathematical models that will allow the prediction of the conditions under which dry sludge is formed.

Knowledge obtained on the studies with model molecules and light feeds has to be integrated and applied to the processing of heavy feeds. However, any change or improvement in catalyst design or methodology of activation must take into consideration its feasibility for application in the refinery at the industrial scale. Increasing catalyst stability and life is of utmost importance. In fixed-bed units, catalyst deactivation during the run is compensated by a progressive increase in bed temperature, up to a certain value dictated by metallurgical constraints or product quality. Moving- or ebullated-bed reactors with catalyst takeout and replacement do not present the catalyst stability problem, but create safety and environmental problems associated with catalyst disposal. Nevertheless, it appears that moving-bed or ebullated-bed processes employing suitable catalysts with high metal retention capacity represent the most efficient way of handling petroleum bottoms and other heavy feeds. The catalyst particles used in these reactors must be small enough to decrease the diffusional path of the molecules to the catalyst interior without compromising the mechanical and other properties necessary for the proper handling of the catalyst in the reactor.

ACKNOWLEDGMENTS

We acknowledge P. Rayo and R. Contreras-Barbara for their help during the compilation of this work.

REFERENCES

Abotsi, G.M.K. and Scaroni, A.W. 1989. A review of carbon-supported hydrodesulfurization catalysts. *Fuel. Process. Technol.* 22:107–133.

Absi-Halabi, M., Stanislaus, A., and Al-Zaid, H. 1993. Effect of acidic and basic vapors on pore size distribution of alumina under hydrothermal conditions. *Appl. Catal. A Gen.* 101:117–128.

Absi-Halabi, M., Stanislaus, A., and Qabazard, H. 1997. Clean fuels technology/trends in catalysis research to meet future refining needs. *Hydrocarbon Process.* 76:5355.

Absi-Halabi, M., Stanislaus, A., and Trimm, D.L. 1991. Coke formation on catalysts during the hydroprocessing of heavy oils. *Appl. Catal. A Gen.* 72:19–215.

Agrawal, R. and Wei, J. 1984. Hydrodemetallization of nickel and vanadium porphyrins. 1. Intrinsic kinetics. *J. Ind. Eng. Chem. Process. Des. Dev.* 23:505–514.

Ancheyta, J., Betancourt, G., Marroquín, G., Centeno, G., Alonso, F., and Muñoz, J.A. Process for the Catalytic Hydrotreatment of Heavy Hydrocarbons of Petroleum. U.S. Patent, filed 2006.

Ancheyta, J., Muñoz, J.A.D., and Macias, M.J. 2005b. Experimental and theoretical determination of the particle size of hydrotreating catalysts of different shapes. *Catal. Today* 109:120–127.

Ancheyta, J., Rana, M.S., and Furimsky, E. 2005a. Hydroprocessing of heavy petroleum feeds: tutorial. *Catal. Today* 109:3–15.

Anderson, J.R. and Pratt, K.C. 1985. *Introduction to Characterization and Testing of Catalysts*. Academic Press, N.S.W., Australia.

Banerjee, D.K., Laidler, K.J., Nandi, B.N., and Patmore, D.J. 1986. Kinetic studies of coke formation in hydrocarbon fractions of heavy crudes. *Fuel* 65:480–484.

Bartholdy, J. and Cooper, B.H. 1993. Metal and coke deactivation of resid hydroprocessing catalysts. Preprint. *Am. Chem. Soc. Div. Petrol. Chem.* 38:386–390.

Bartholomew, C.H. 1984. Catalyst deactivation. *Chem Eng.* 91:96–112.

Bartholomew, C.H. 1994. In *Catalytic Hydroprocessing of Petroleum and Distillates*, Oballa, M.C. and Shih S.S. (Eds.). Marcel Dekker, New York.

Beaton, W.I. and Bertolacini, R.J. 1991. Resid hydroprocessing at Amoco. *Catal. Rev. Sci. Eng.* 33:281–317.

Bestougeff, M.A. and Byramjee, R.J. 1994. Chemical constitution of asphaltenes. *Dev. Petrol. Sci.* 40A:67–94.

Boudart, M., Arrieta, J.S., and Dalla, B. 1983. Correlation between thiophene hydrodesulfurization activity and the number of first sulfur neighbors as determined by EXAFS in sulfided CoMo/γ-Al_2O_3. *J. Am. Chem. Soc.* 105:6501–6502.

Breysse, M., Furimsky, E., Kasztelan, S., Lacroix, M., and Perot, G. 2002. Hydrogen activation by transition metal sulfides. *Catal. Rev. Sci. Eng.* 44:651–735.

Breysse, M., Portefaix, J.L., and Vrinat, M. 1991. Support effects on hydrotreating catalysts. *Catal. Today* 10:489–505.

Cable, T.L., Massoth, F.E., and Thomas, M.G. 1981. Studies on an aged H-coal catalyst. *Fuel Process. Technol.* 4:265–275.

Caloch, B., Rana, M.S., and Ancheyta, J. 2004. Improved hydrogenolysis (C-S, C-M) function with basic supported hydrodesulfurization catalysts. *Catal. Today* 98:91–98.

Carlsson, A., Brorson, M., and Topsøe, H. 2004. Morphology of WS_2 nanoclusters in WS_2/C hydrodesulfurization catalysts revealed by high-angle annular dark-field scanning transmission electron microscopy (HAADF-STEM) imaging. *J. Catal.* 227:530–536.

Chary, K.V.R., Ramkrishna, H., Rama Rao, K.S., Muralidhar, G., and Kanta Rao, P. 1991. Hydrodesulfurization on MoS_2/MgO. *Catal. Lett.* 10:27–34.

Choi, J., Maugé, F., Pichon, C., Olivier-Fourcade, J., Jumas, J., Petit-Clair, C., and Uzio, D. 2004. Alumina-supported cobalt-molybdenum sulfide modified by tin via surface organometallic chemistry: application to the simultaneous hydrodesulfurization of thiophenic compounds and the hydrogenation of olefins. *Appl. Catal. A Gen.* 267:203–216.

Colyar, J.J. and Wisdom, L.I. 1997. The H-Oil Process: A Worldwide Leader in Vacuum Residue Processing. Paper presented at Proceedings of the National Petroleum Refiners Association Annual Meeting, San Antonio, TX.

Cooper, B.H., Donnis, B.B.L., and Moyse, B.M. 1986. Technology: hydroprocessing conditions affect catalyst shape selection. *Oil Gas J.* 8:39–44.

Courty, Ph., Chamette, P., and Raimbault, C. 1999. Synthetic or reformulated fuels: a challenge for catalysis. *Oil Gas Sci. Technol. Rev. IFP* 54(3):357–363.

Dautzenberg, F.M. and de Deken, J.C. 1985. Modes of operation in hydrodemetallization. Preprint. *Am. Chem. Soc. Div. Petrol. Chem.* 30:8–20.

Dautzenberg, F.M. and de Deken, J.C. 1987. Modes of operation in hydrodemetallization (HDM). *ACS Symp. Ser.* 344:233–256.

Dautzenberg, F.M., van Klinken, J., Pronk, K.M.A., Sie, S.T., and Wijffels, J.B. 1978. Catalyst deactivation through pore mouth plugging during residue desulfurization. *ACS Symp. Ser.* 65:254–267.

Dawson, W.H., Chornet, E., Tiwari, P., and Heitz, M. 1989. Hydrogenation of individual components isolated from Athabasca bitumen vacuum resid. Preprint. *Div. Petrol. Chem. Am. Chem. Soc.* 34:384–394.

Decroocq, D. 1997. Major scientific challenges about development of new processes in refining and petrochemistry. *Rev. Inst. Franc. Petrole* 52:469–489.

Dejonghe, S., Hubaut, R., Grimblot, J., Bonnelle, J.P., Des Courieres, T., and Faure, D. 1990. Hydrodemetallization of a vanadylporphyrin over sulfided $NiMo/Al_2O_3$, Mo/Al_2O_3, and Al_2O_3 catalysts: effect of the vanadium deposit on the toluene hydrogenation. *Catal. Today* 7:569–585.

Dekkers, C. 1999. Solar-heating system studied for heavy-oil pipelines. *Oil Gas J.* 97:44–46.

Delannay, F. 1984. *Characterization of Heterogeneous Catalysts.* Marcel Dekker, New York.

Delannay, F. 1985. High resolution electron microscopy of hydrodesulfurization catalysts: a review. *Appl. Catal.* 16:i–ii.

Delmon, B., Jacobs, P.A., Maggi, R., Martens, J.A., Grange, P., and Poncelet, G. 1998. *Preparation of Catalysts VII.* Amsterdam, Elsevier.

Dickenson, R.L., Biasca, F.E., Schulman, B.L., and Johnson, H.E. 1997. Refiner options for converting and utilizing heavy fuel oil. *Hydrocarbon Process.* 76(2): 5–11.

Dickie, J.P. and Yen, T.F. 1967. Macrostructures of the asphaltic fractions by various instrumental methods. *Anal. Chem.* 39:1847–1852.

Eartl, G., Knozinger, H., and Weitkamp, J. (Eds.). 1999. *Preparation of Solid Catalysts.* Wyley-VCH Verlag GmbH, Weinheim, Germany.

Eijsbouts, S., van den Oetelaar, L.C.A., and van Puijenbroek, R.R. 2005. MoS_2 morphology and promoter segregation in commercial type 2 $Ni–Mo/Al_2O_3$ and $Co–Mo/Al_2O_3$ hydroprocessing catalysts. *J. Catal.* 229:352–364.

Fleisch, T.H., Meyers, B.L., Hall, J.B., and Ott, G.L. 1984. Multitechnique analysis of a deactivated resid demetallation catalyst. *J. Catal.* 86:147–157.

Fleischer, E.B. 1963. The structure of nickel etioporphyrin: I. *J. Am. Chem. Soc.* 85:146–148.

Fonseca, A., Zeuthen, P., and Nagy, J.B. 1996. ^{13}C n.m.r. quantitative analysis of catalyst carbon deposits. *Fuel* 75:1363–1376.

Furimsky, E. 1982. Characterization of deposits formed on catalysts surfaces during hydrotreatment of coal-derived liquids. *Fuel Process. Technol.* 6:1–8.

Furimsky, E. 1996. Spent refinery catalysts: environment, safety and utilization. *Catal. Today* 30:223–286.

Furimsky, E. 1998. Selection of catalysts and reactors for hydroprocessing. *Appl. Catal. A Gen.* 171:177–206.

Furimsky, E. and Massoth, F.E. 1993. Regeneration of hydroprocessing catalysts. Introduction. *Catal. Today* 17:537–659.

Furimsky, E. and Massoth, F.E. 1999. Deactivation of hydroprocessing catalysts. *Catal. Today* 52:381–495.

Furimsky, E. and Massoth, F.E. 2005. Hydrodenitrogenation of petroleum. *Catal. Rev. Sci. Eng.* 47:297–489.

Galarraga, C. and Ramírez de Agudelo, M.M. 1992. A stable catalyst for heavy oil processing. II. Preparation and characterization. *J. Cat.* 134:98–106.

Galiasso, R. and Caprioli, L. 2005. Catalyst pore plugging effects on hydrocracking reactions in an ebullated bed reactor operation. *Catal. Today* 109:185–194.

García-López, A.J., Cuevas, R., Ramírez, J., Ancheyta, J., Vargas-Tah, A.A., Nares, R., and Gutiérrez-Alejandre, A. 2005. Hydrodemetallation (HDM) kinetics of Ni-TPP over Mo/Al_2O_3-TiO_2 catalyst. *Catal. Today* 107/108:545–550.

Gardner, L.E., Hogan, R.J., Sughrue, E.L., and Myers, J.W. 1987. Hydrotreating Process Employing a Three Stage Catalyst System wherein a Titanium Compound Is Employed in the Second Stage. U.S. Patent 4,657,663, April 14.

Gardner, L.E. and Kukes, S.G. 1989. Hydrofining Employing a Support Material for Fixed Beds. U.S. Patent 4,828,683, May 9.

George, Z.M., Mohammed, P., and Tower, R. 1988. Regeneration of spent hydroprocessing catalyst. In *Proceedings of the 9th International Congress on Catalysis*, Calgary, 1, p. 230.

Girgis, M.J. and Gates, B.C. 1991. Reactivities, reaction networks, and kinetics in high-pressure catalytic hydroprocessing. *Ind. Eng. Chem. Res.* 30:2021–2058.

Gosselink, J.W. 1998. Sulfide catalysts in refineries. *CatTech* 2:127–144.

Goulon, J., Retournard, A., Friant, P., Ginet-Goulon, C., Berte, C., Muller, J.F., Poncet, J.L., Guilard, R., and Escalier, J.C. 1984. Structural characterization by x-ray absorption spectroscopy (EXAFS/XANES) of the vanadium chemical environment in Boscan asphaltenes. *J. Chem Soc. Dalton Trans.* 6:1095–1103.

Grange, P. 1980. Catalytic hydrodesulfurization. *Catal Rev. Sci. Eng.* 21:135–181.

Gray, M.R. 1994. *Upgrading Petroleum Residues and Heavy Oils*. Marcel Dekker, New York.

Gregg, S.J. and Sing, K.S.W. 1982. *Adsorption, Surface Area and Porosity*, 2nd ed. Academic Press, New York.

Gualda, G. and Kasztelan, S. 1984. Coke versus metal deactivation of residue hydrodemetallization catalysts. *Stud. Surf. Sci. Catal.* 88:145–154.

Gualda, G. and Kasztelan, S. 1996. Initial deactivation of residue hydrodemetallization catalysts. *J. Catal.* 161:319–337.

Gualda, G. and Tulhoat, H. 1988. Study of the deactivation of hydrotreating catalyst by coking. Bibliogric synthesis. *Rev. Inst. Franc. Petrole* 43:567–594.

Habib, M.M., Winslow, P.L., and Moore, R.O. 1995. Catalyst System for Combined Hydrotreating and Hydrocracking and a Process for Upgrading Hydrocarbonaceous Feedstocks. U.S. Patent 5439,860, August 8.

Habib, M.M., Winslow, P.L., and Moore, R.O. 1997. Catalyst System for Combined Hydrotreating and Hydrocracking and a Process for Upgrading Hydrocarbonaceous Feedstocks. U.S. Patent 5,593,570, January 14.

Hammer, G.P. and Clem, K.R. 1975. Ebullating Bed Process for Hydrotreatment of Heavy Crudes and Residua. U.S. Patent 3,887,455, June 3.

Hanprasopwattana, A., Sault, A.G., and Datye, A.K. 1998. TiO_2 promotion of hydrotreating catalysts. Preprint. *ACS Div. Petrol.* 43:90–93.

Hansford, R.C. 1964. Silica-Zirconia-Titania Hydrocracking Catalyst. U.S. Patent 3,159,588.

Harrison, G.E., McKinley, D.H., and Dennis, A.J. 1994. Multi-step Hydrodesulphurizations Process. U.S. Patent 5,292,428, March 8.

Hauser, A., Stanislaus, A., Marafi, A., and Al-Adwani, A. 2005. Initial coke deposition on hydrotreating catalysts. II. Structure elucidation of initial coke on hydrodematallation catalysts. *Fuel* 84:259–269.

Hédoire, C., Louis, C., Davidson, A., Breysse, M., Maugé, F., and Vrinat, M. 2003. Support effect in hydrotreating catalysts: hydrogenation properties of molybdenum sulfide supported β-zeolites of various acidities. *J. Catal.* 220:433–441.

Hensen, E.J.M., Kooyman, P.J., van der Meer, Y., van der Kraan, A.M., de Beer, H.J., van Veen, J.A.R., and van Santen, R.A. 2001. The relation between morphology and hydrotreating activity for supported MoS_2 particles. *J. Catal.* 199:224–235.

Hensley, A.L., Jr. and Quick, L.M. 1984. Three-Catalyst Process for the Hydrotreating of Heavy Hydrocarbon Streams. U.S. Patent 4,431,525.

Hilbert, T.T., Mazzone, N.D., and Sarli, M.S. 1996. Gasoline Upgrading Process. U.S. Patent 5,510,016, April 23.

Howell, J.A. and Wilson, M.E. 1987. Multi-stage Hydrofining Process. U.S. Patent 4,659,452.

Imelik, B., Naccache, C., Coudurier, G., Praliaud, H., Mariaudeau, P., Gallezot, P., Martin, G.A., and Vedrine, J.C. 1982. Metal-Support and Metal-Additive Effects in Catalysis. Paper presented at Proceedings of an International Symposium, Ecully, Italy, September 14–16.

Inoue, K., Zhang, P., and Tsyuyama, H. 1993. Recovery of Mo, V, Ni and Co from spent hydrodesulfurization catalysts. Preprint. *ACS Div. Petrol. Chem.* 38:77–80.

Jacobsen, A.C., Cooper, B.H., and Hannerup, P.N. 1987. Catalysts for hydrotreating of heavy oil fractions. *Proc. 12th Petrol. World Congr.* 4:97–107.

Jaffe, J. 1968. Coprecipitation Method for Making Multi-component Catalysts. U.S. Patent 3,401,125.

Ji, Y., Afanasiev, P., Vrinat, M., Li, W., and Li, C. 2004. Promoting effects in hydrogenation and hydrodesulfurization reactions on the zirconia and titania supported catalysts. *Appl. Catal. A Gen.* 257:157–164.

Jocker, S.J.M. 1993. The Metrex process, full recycling of spent hydroprocessing catalysts. Preprint. *ACS Div. Petrol. Chem.* 38:74–76.

Johnson, B.J., Massoth, F.E., and Bartholdy, J. 1986. Diffusion and catalytic activity studies on resid-deactivated HDS catalysts. *AIChE J.* 32:1980–1987.

Kabe, T., Ishihara, A., and Qian, W. 1999. *Hydrodesulfurization and Hydrodenitrogenation. Chemistry and Engineering*. Wiley-VCH, New York.

Kellet, T.F., Sartor, A.F., and Trevino, C.A. 1980. How to select hydrotreating: a comparison between Co/Mo and Ni/Mo catalyst shows applications for desulfurization, denitrogenation and hydrogen uptake. *Hydrocarbon Process.* 59: 139–142.

Kressmann, S., Morel, F., Harlé, V., and Kasztelan, S. 1998. Recent developments in fixed-bed catalytic residue upgrading. *Catal. Today* 43:203–215.

Kunesh, J.G. 1982. Hydrocracking and Hydrotreating Shale Oil in Multiple Catalytic Reactors. U.S. Patent 4,344,840, August 17.

Laine, J., Brito, J.L., and Severino, F. 1991. Structure and activity of Ni(Co)Mo/SiO_2 hydrodesulfurization catalysts. *J. Catal.* 131, 385–393.

Larson, O.A. and Beuther, H. 1966. Processing aspects of vanadium and nickel in crude oils. Preprint. *Am. Chem. Soc. Div. Petrol. Chem.* 11:B95–B103.

Lauritsen, J.V., Bollinger, M., Lægsgaard, V.E., Jacobsen, K.W., Nørskov, J.K., Clausen, B.S., Topsøe, H., and Besenbacher, F. 2004. Atomic-scale insight into structure and morphology changes of MoS_2 nanoclusters in hydrotreating catalysts. *J. Catal.* 221:510–522.

Lecrenary, E., Sakanishi, K., Mochida, I., and Suzuka, T. 1998. HDS activity of CoMo and NiMo catalysts supported on some acidic binary oxides. *Appl. Catal. A Gen.* 175:237–243.

Ledoux, M.J. and Hantzer, S. 1990. Hydrotreatment catalyst poisoning by vanadium and nickel porphyrin: ESR and NMR. *Catal. Today* 7:479–496.

Ledoux, M.J., Mchaux, O., Hantzer, P., Panissod, P., Patit, P., André, J.J., and Callot, H.J. 1987. Hydrodesulfurization (HDS) poisoning by vanadium compounds: EPR and metal solid NMR analysis. *J. Catal.* 106:525–537.

Leofanti, G., Tozzola, G., Padovan, M., Petrini, G., Bordiga, S., and Zecchina, A. 1994. Catalyst characterization. In *The Catalytic Process from Laboratory to the Industrial Plant*, Sanfilipo, D. (Ed.). Italian Chemical Society, Rimini.

Le Page, L.F., Chatila, S.G., and Davidson, M. 1992. *Resid and Heavy Oil Proceedings*, Editions Technip, Paris.

Le Page, J.F., Cosyns, J., Courty, P., Freund, E., Frank, J.P., Jacquin, Y., Juguin, B., Marcilly, C., Martino, G., Miquel, J., Montarnal, R., Sugier, A., and Van Landeghem, H. 1987. *Applied Heterogeneous Catalysis. Design and Manufacture of Solid Catalysts.* Editions Technip, Paris.

Li, D., Sato, T., Imamura, M., Shimada, H., and Nishijima, A. 1998. The effect of boron on HYD, HC and HDS activities of model compounds over Ni-Mo/-Al_2O_3-B_2O_3 catalysts. *Appl. Catal. B Environ.* 16:255–260.

Lowell, S. 1980. Continuous scan mercury porosimetry and the pore potential as a factor in porosimetry hysteresis. *Powder Technol.* 25:37–43.

Luck, F. 1991. A review of support effects on the activity and selectivity of hydrotreating catalysts. *Bull. Soc. Chim. Belg.* 100:781–800.

Macias, M.J. and Ancheyta, J. 2004. Simulation of an isothermal hydrodesulfurization small reactor with different catalyst particle shapes. *Catal. Today* 98:243–250.

MacKetta, J.J. 1992. *Petroleum Processing Handbook.* Marcel Dekker, New York.

Maity, S.K., Ancheyta, J., Alonso, F., and Rana, M.S. 2004. Preparation, characterization and evaluation of Maya crude hydroprocessing catalysts. *Catal. Today* 98:193–199.

Maity, S.K., Ancheyta, J., and Rana, M.S. 2005. Support effects on hydroprocessing of Maya heavy crude. *Energy Fuels* 19:343–347.

Maity, S.K., Ancheyta, J., Soberanis, L., Alonso, F., and Llanos, M.E. 2003. Alumina–titania binary mixed oxide used as support of catalysts for hydrotreating of Maya heavy crude. *Appl. Catal. A Gen.* 244:141–153.

Maity, S.K., Rana, M.S., Bej, S.K., Ancheyta-Juárez, J., Muralidhar, G., and Prasada Rao, T.S.R. 2001a. Studies on physico-chemical characterization and catalysis on high surface area titania supported molybdenum hydrotreating catalysts. *Appl. Catal. A Gen.* 205:215–225.

Maity, S.K., Rana, M.S., Bej, S.K., Ancheyta, J., Muralidhar, G., and Prasada Rao, T.S.R. 2001b. TiO_2-ZrO_2 mixed oxide as a support for hydrotreating catalyst. *Catal. Lett.* 72:115–119.

Maity, S.K., Srinivas, B.N., Prasad, V.V.D.N., Singh, A., Muralidahr, G., and Prasada Rao, T.S.R. 1998. Studies on sepiolite supported hydrotreating catalysts. *Stud. Surf. Sci. Catal.* 113:579–590.

Marafi, A., Stainslaus, A., Hauser, A., and Matsushita, K. 2005. Effect of diluents in controlling sediment formation during catalytic hydrocracking of Kuwait vacuum residue. *Petrol. Sci. Technol.* 23:385–390.

Massoth, F.E. and Muralidhar, G. 1982. Hydrodesulfurization catalysis. In *Proceedings of the 4th International Conference on Chemistry and Uses of Molybdenum*, Ann Arbor, MI, pp. 343–351.

Massoth, F.E., Muralidhar, G., and Shabtai, J. 1994. Catalytic functionality of supported sulfides. I. Effect of support and additive on the CoMo catalysts. *J. Catal.* 85:44–52.

McDaniel, M.P. and Hottovy, T.D. 1980. Total porosity of high pore-volume silicas by liquid adsorption. *J. Coll. Interface Sci.* 78:31–36.

Menon, P.G.J. 1990. Coke on catalysts: harmful, harmless, invisible and beneficial types. *J. Mol. Catal.* 59:207–220.

Miller, J.T., Fisher, R.B., Thiyagarajan, P., Winans, R.E., and Hunt, J.E., 1998. Subfraction and characterization of Mayan asphaltene. *Energy Fuels* 12:1290–1298.

Mitchell, P.C.H. 1990. Hydrodemetallisation of crude petroleum: fundamental studies. *Catal. Today* 7:439–445.

Mitchell, P.C.H., Scott, C.S., Bonnelle, J.P., and Grimblot, J. 1985. Binding and decomposition of oxovanadium (IV) phthalocyanine, tetraphenylprophyrion and etioporphyrin on hydrotreating catalysts studied by x-ray photoelectron and ultraviolet-visible spectroscopies. *J. Chem. Soc. Faraday Trans.* 81:1047–1056.

Mitchell, P.C.H. and Valero, J.A. 1982. Hydrodemetallization and promoter effect of vanadium compounds on a Co-Mo/Al_2O_3 hydrodesulfurization. *React. Kinet. Catal. Lett.* 20:219–225.

Mojelsky, T.W., Ignasiak, T.M., Frakman, Z., McIntyre, D.D., Lown, E.M., Montgomery, D.S., and Strausz, O.P. 1992. Structural features of Alberta oil sand bitumen and heavy oil asphaltenes. *Energy Fuels* 6:83–96.

Morales, A. and Galiasso, R. 1982. Adsorption mechanism of Boscan porphyrins on MoO_3, Co_3O_4 and CoMo/Al_2O_3. *Fuel* 61:13–17.

Morel, F., Kressmann, S., Harlè, V., and Kasztelan, S. 1997. Process and catalyst for hydrocracking of heavy oil and residue. *Stud. Surf. Sci. Catal.* 106:1–16.

Moulijn, J.A., van Diepen, A.E., and Kapteijn, F. 2001. Catalyst deactivation: is it predictable? What to do? *Appl. Catal. A Gen.* 212:3–16.

Moulijn, J.A., van Leeuwen, P.V.N.M., and van Santen, R.A. 1993. *An Integrated Approach to Homogeneous, Heterogeneous and Industrial Catalysis.* Elsevier Science, Amsterdam.

Mounce, W. 1974. Hydrocracking and Hydrodesulfurization Process. U.S. Patent 3,830,728, August 20.

Mullins, O.C. and Groenzin, H. 1999. Petroleum asphaltene molecular size and structure. *J. Phys. Chem. A* 103:11237–11245.

Muralidhar, G., Concha, E.B., Bartholmew, G.L., and Bortholomew, C.H., 1984. Characterization of reduced and sulfided supported molybdenum catalysts by O_2 chemisorption, XRD, and ESCA. *J. Catal.* 89:274–284.

Muralidhar, G., Kumaran, G.M., Kumar, M., Rawat, K.S., Sharma, L.D., Raju, B.D., and Rama Rao, K.S. 2005. Physico-chemical characterization and catalysis on SBA-15 supported molybdenum hydrotreating catalysts. *Catal. Today* 99:309–314.

Muralidhar, G., Rana, M.S., Maity, S.K., Srinivas, B.N., and Prasada Rao, T.S.R. 2000. In *Chemistry of Diesel Fuels*, Song, C., Hsu, S., and Mochida, I. (Eds.). Taylor & Francis, London, chap. 8.

Muralidhar, G., Srinivas, B.N., Rana, M.S., Kumar, M., and Maity, S.K. 2003. Mixed oxide supported hydrodesulfurization catalysts: a review. *Catal. Today* 86:45–60.

Ng, K.Y.S. and Guari, E. 1985. Molybdena on titania. I. Preparation and characterization by Raman FTIR. *J. Catal.* 92:340–354.

Nishijima, A., Shimada, H., Sato, T., Yoshimura, Y., and Hiraishi, J. 1986. Support effects on hydrocracking and hydrogenation activities of molybdenum catalysts used for upgrading coal-derived liquids. *Polyhedron* 5:243–247.

Oelderick, J.M., Sie, S.T., and Bode, D. 1989. Progress in the catalysis of the upgrading of petroleum residue: a review of 25 years of R&D on Shell's residue hydroconversion technology. *Appl. Catal.* 47:1–24.

Okamoto, Y., Maezawa, A., and Imanaka, T. 1989. Active sites of molybdenum sulfide catalysts supported on Al_2O_3 and TiO_2 for hydrodesulfurization and hydrogenation. *J. Catal.* 120:29–45.

Olguin, E., Vrinat, M., Cedeno, L., Ramírez, J., and Lopez-Agudo, A. 1997. The use of TiO_2-Al_2O_3 binary oxides as supports for Mo based catalysts in hydrodesulfurization of thiophene and dibenzothiophene. *Appl. Catal.* 165:1–13.

Ono, T., Ohguchi, Y., and Togari, O. 1982. Control of the pore structure of porous alumina. In *Preparation of Catalysts III*, Poncelet, G., Grange, P., and Jacobs, P.A. (Eds.). Elsevier Science Publishers B.V., Amsterdam, pp. 631–641.

Oyama, S.T., Wang, X., Lee, Y.K., and Chun, W.J. 2004. Active phase of Ni_2P/SiO_2 in hydroprocessing reactions. *J. Catal.* 221:263–273.

Payen, E., Gengembre, L., Mauge, F., Duchet, J.C., and Lavalley, J.C. 1991. Surface properties of zirconia catalysts carrier: interaction with oxomolybdates species. *Catal. Today* 10:521–539.

Pazos, J.M., Gonzalez, J.C., and Salazar-Guillen, A.J. 1983. Effect of catalyst properties and operating conditions on hydroprocessing high metals feeds. *Ind. Eng. Chem. Process Des. Dev.* 22:653–659.

Perego, C. and Villa, P. 1994. Catalyst preparation methods. In *The Catalytic Process from Laboratory to the Industrial Plant*, Sanfilipo, D. (Ed.). Italian Chemical Society, Rimini.

Peries, J.P., Billon, A., Hennico, A., Morrison, E., and Morel, F. 1986. Institut Française du Pétrole. Rapport IFP 34297, June.

Plain, C., Duddy, J., Kressmann, S., LeCoz, O., and Tasker, K. Axens IFP Group Technologies. http://www.axens.net/upload/presentations/fichier/options_for_ resid_ conversion.pdf.

Plumail, J.C., Jacquin, Y., and Toulhoat, H. 1982. In *Proceedings of the 4th International Conference on the Chemistry and Uses of Molybdenum*, Ann Arbor, MI, p. 389.

Poncelet, G., Grange, P., and Jacobs, P. (Eds.). 1983. *Preparation of Catalysts III*. Elsevier, Amsterdam.

Pophal, C., Kameda, F., Hoshino, K., Yoshinaka, S., and Segawa, K. 1997. Hydrodesulfurization of dibenzothiophene derivatives over TiO_2-Al_2O_3 supported sulfided molybdenum catalyst. *Catal. Today* 39:21–28.

Portela, L., Grange, P., and Delmon, B. 1995. The adsorption of nitric oxide on supported Co-Mo hydrodesulfurization catalysts: a review. *Catal. Rev. Sci. Eng.* 37:699–731.

Pratt, K.C., Sanders, J.V., and Chritov, V. 1990. Morphology and activity of MoS_2 on various supports: genesis of active phase. *J. Catal.* 124:416–432.

Prins, R. 1992. *Characterization of Catalytic Materials*, Wachs, I.E. and Fitzpatrick, L.E. (Eds.). Butterworth-Heinemann, Stonehan, MA, chap.6.

Prins, R., de Beer, V.H.J., and Somorjai, G.A. 1989. Structure and function of the catalyst and the promoter in Co-Mo hydrodesulfurization catalyst. *Catal. Rev. Sci. Eng.* 31:1–41.

Qu, L., Zhang, W., Kooyman, P.J., and Prins, R. 2003. MAS NMR, TPR, and TEM studies of the interaction of NiMo with alumina and silica–alumina supports. *J. Catal.* 215:7–13.

Ramírez, J., Castillo, P., Cedeno, L., Cuevas, R., Castillo, M., Palacios, J.M., and Agudo, L.A. 1995. Effect of boron addition on the activity and selectivity of hydrotreating CoMo/Al_2O_3 catalysts. *Appl. Catal. A Gen.* 132:317–334.

Ramírez, J., Cedeño, L., and Busca, G. 1999. The role of titania in Mo-based hydrodesulfurization catalysts. *J. Catal.* 184:59–67.

Ramírez, J., Cuevas, R., Gasque, L., Vrinat, M., and Breysse, M. 1991. Promoting effect of fluorine on cobalt-molybdenum titania hydrodesulfurization catalysts. *Appl. Catal.* 71:351–361.

Ramírez, J.F., Fuentes, S., Díaz, G., Vrinat, M., Lacroix, M., and Breysse, M. 1989. Hydrodesulphurization activity and characterization of sulfided molybdenum, and cobalt-molybdenum catalysts. Comparison of alumina, silica-alumina and titania supported catalysts. *Appl. Catal.* 52:211–224.

Ramírez, J. and Gutiérrez-Alejandre, A. 1998. Relationship between hydrodesulfurization activity and morphological and structural changes in NiW hydrotreating catalysts supported on Al_2O_3-TiO_2 mixed oxides. *Catal. Today* 43:123–133.

Ramírez, J., Macías, G., Cedeño, L., Gutiérrez-Alejandre, A., Cuevas, R., and Castillo, P. 2004. The role of titania in supported Mo, CoMo, NiMo and NiW hydrodesulfurization catalysts: analysis of past and new evidences. *Catal. Today* 98:1–30.

Ramírez, J., Rayo, P., Gutierrez-Alejandre, A., Ancheyta, J., and Rana, M. 2005. Analysis of the hydrotreatment of Maya heavy crude with NiMo catalysts supported on TiO_2-Al_2O_3 binary oxides: effect of the incorporation method of Ti. *Catal. Today* 109:54–60.

Ramírez, J., Ruíz, R.L., Cedeño, L., Harle, V., Vrinat, M., and Breysse, M. 1993. Titania-alumina mixed oxides as supports for molibdenum hydrotreating catalysts. *Appl. Catal. A Gen.* 93:163–180.

Rana, M.S., Ancheyta, J., Maity, S.K., and Rayo, P. 2005a. Maya crude hydrodemetallization and hydrodesulfurization catalysts: an effect of TiO_2 incorporation in Al_2O_3. *Catal. Today* 109:61–68.

Rana, M., Ancheyta, J., Maity, S.K., and Rayo, P. 2005c. Characteristics of Maya crude hydrodemetallization and hydrodesulfurization. *Catal. Today* 104:86–93.

Rana, M.S., Ancheyta, J., Maity, S.K., and Rayo, P. 2007b. Hydrotreating of Maya crude oil. I. Effect of support composition and its pore diameter on asphaltene conversion. *Petrol. Sci. Technol.*, 25(1–2): 187–200.

Rana, M., Ancheyta, J., and Rayo, P. 2005d. A comparative study for heavy oil hydroprocessing catalysts at micro-flow and bench-scale reactors, *Catal. Today* 109:24–32.

Rana, M.S., Ancheyta, J., Rayo, P., and Maity, S.K. 2007a. Heavy oil hydroprocessing over supported NiMo sulfided catalyst: an inhibition effect by added H_2S. *Fuel*, 86(9): 1263–1269.

Rana, M., Ancheyta, J., Rayo, P., and Maity, S.K. 2004a. Effect of alumina preparation on hydrodemetallization and hydrodesulfurization of Maya crude. *Catal. Today* 98:151–160.

Rana, M.S. and Furimsky, E. 2005d. Hydroprocessing of heavy oil fractions. *Catal. Today* 109:1–3.

Rana, M., Huidobro, M.L., Ancheyta, J., and Gomez, M.T. 2005b. Effect of support composition on hydrogenolysis of thiophene and Maya crude. *Catal. Today* 107/108:346–354.

Rana, M.S., Maity, S.K., Ancheyta, J., Muralidhar, G., and Prasada Rao, T.S.R. 2003. TiO_2-SiO_2 supported hydrotreating catalysts: physico-chemical characterization and activities. *Appl. Catal. A Gen.* 253:165–176.

Rana, M.S., Maity, S.K., Ancheyta, J., Muralidhar, G., and Prasada Rao, T.S.R. 2004b. MoCo(Ni)/ZrO_2-SiO_2 hydrotreating catalysts: physico-chemical characterization and activities studies. *Appl. Catal. A Gen.* 268:89–97.

Rana, M.S., Navarro, R., and Leglise, J. 2004c. Competitive effects of nitrogen and sulfur content on activity of hydrotreating CoMo/Al_2O_3 catalysts: a batch reactor study. *Catal. Today* 98:67–74.

Rana, M.S., Samano, V., Ancheyta, J., and Diaz, J.A.I. 2007c. A review of recent advances on process technologies for upgrading of heavy oils and residua. *Fuel*, 86(9): 1216–1213.

Rana, M.S., Srinivas, B.N., Maity, S.K., Muralidhar, G., and Prasada Rao, T.S.R. 1999. Catalytic functionalities of TiO_2 based SiO_2, Al_2O_3, ZrO_2 mixed oxide hydroprocessing catalysts. *Stud. Surf. Sci. Catal.* 127:397–400.

Rana, M.S., Srinivas, B.N., Maity, S.K., Muralidhar, G., and Prasada Rao, T.S.R. 2000. Origin of cracking functionality of sulfided (Ni)CoMo/SiO_2-ZrO_2 catalysts. *J. Catal.* 195:31–37.

Rao, K.S., Ramakrishna, H., and Muralidhar, G. 1992. Catalytic functionalities of WS_2/ZrO_2. *J. Catal.* 133:146–152.

Ratnasamy, P. and Sivasanker, S. 1980. Structural chemistry of Co-Mo-alumina catalyst. *Catal. Rev. Sci. Eng.* 22:401–429.

Rayo, P., Ancheyta, J., Ramírez, J., and G.-Alejandre, A. 2004. Hydrotreating of diluted Maya crude with NiMo/Al_2O_3-TiO_2 catalysts: effect of diluent composition. *Catal. Today* 98:171–179.

Reddy, K.M., Wei, B., and Song, C. 1998. Mesoporous molecular sieve MCM-41 supported CoMo catalyst for hydrodesulfurization of petroleum residues. *Catal. Today* 43:261–272.

Richardson, J.T. 1989. *Principles of Catalyst Development.* Plenum Press, New York.

Ritter, H.L. and Drake, L.C. 1945. Pore-size distribution in porous materials. Pressure porosimeter and determination of complete macropore-size distribution. *Ind. Eng. Chem. Anal. Ed.* 17:782–786.

Ruiz, R.S., Alonso, F., and Ancheyta, J. 2004. Minimum fluidization velocity and bed expansion characteristics of hydrotreating catalysts in ebullated-bed systems. *Energy Fuels* 18:1149–1155.

Ruiz, R.S., Alonso, F., and Ancheyta, J. 2005. Pressure and temperature effects on the hydrodynamic characteristics of ebullated-bed systems. *Catal. Today* 109:205–213.

Sahoo, S.K., Ray, S.S., and Singh, I.D. 2004. Structural characterization of coke on spent hydroprocessing catalysts used for processing of vacuum gas oils. *Appl. Catal. A Gen.* 278:83–91.

Sakanishi, K., Yamashita, N., Witehurst, D.D., and Mochida, I. 1997. Depolymerization and demetallization treatments of asphaltene in vacuum residue. Preprint. *Am. Chem. Soc. Div. Petrol. Chem.* 42:373–377.

Sampieri, A., Pronier, S., Blanchard, J., Breysse, M., Brunet, S., Fajerwerg, K., Louis, C., and Pérot, G. 2005. Hydrodesulfurization of dibenzothiophene on MoS_2/MCM-41 and MoS_2/SBA-15 catalyst prepared by thermal spreading of MoO_3. *Catal. Today* 107/108:537–544.

Sanfilipo, D. (Ed.). 1994. *The Catalytic Process from Laboratory to the Industrial Plant*. Italian Chemical Society, Rimini.

Savage, P.E. and Klein, M.T. 1989. Asphaltene reaction pathways. 5. Chemical and mathematical modelling. *Chem. Engi. Sci.* 44:393–404.

Schabron, J.F. and Speight, J.G. 1997. An evaluation of the delayed cooking product yield of heavy feedstocks using asphaltene content and residue. *Rev. Inst. Franc. Petrole* 52:73–85.

Scherzer, J. 1990. *Octane-Enhancing Zeolite FCC Catalysts, Scientific and Technical Aspects*. Marcel Dekker, New York.

Scherzer, J. and Gruia, A.J. 1996. *Hydrocracking Science and Technology*. Marcel Dekker, New York.

Scheuerman, G.L., Jonson, D.L., Reynolds, B.E., Bachtel, R.W., and Threlkel, R.S. 1993. Advances in Chevron RDS technology for heavy oil upgrading flexibility. *Fuel Process. Technol.* 35:39–54.

Schuetze, B. and Hofmann, H. 1997. How to upgrade heavy feeds. *Hydrocarbon Proc. Int. Ed.* 63:75–82.

Scott, A., Stevenson-Dumesic, J.A., and Baker, R.T.K. 1987. In *Metal Support Interaction in Catalysis, Sintering and Redispersion*, Ruckenstein, E. (Ed.). Van Nostrand Reinhold Company, New York.

Shimada, H., Sato, T., Yoshimura, Y., Hinata, A., Yoshitomi, S., Mares, A.C., and Nishijima, A. 1990. Application of zeolite-based catalyst to hydrocracking of coal-derived liquids. *Fuel Process. Technol.* 25:153–165.

Shimada, H., Sato, T., Yoshimura, Y., Hiraishi, J., and Nishijima, A. 1988. Support effect on the catalytic activity and properties of sulfided molybdenum catalysts. *J. Catal.* 110:275–284.

Shirokoff, J.W., Siddiqui, M.N., and Ali, M.F. 1997. Characterization of the structure of Saudi crude asphaltenes by x-ray diffraction. *Energy Fuels* 11:561–565.

Shirota, Y., Fukui, Y., Ando, M., and Homma, Y. 1979. Two-Step Hydrodesulfurization of Heavy Hydrocarbon Oil. U.S. Patent 4,166,026, August 28.

Sie, S.T. 1980. Catalyst deactivation by pore plugging in petroleum processing. In *Catalyst Deactivation*, Delmon, B. and Froment, G.F. (Eds.). Elsevier, Amsterdam, pp. 545–569.

Sie, S.T. 2001. Consequences of catalyst deactivation for process design and operation. *Appl. Catal. A Gen.* 212:129–151.

Smith, B.J. and Wei, J. 1991a. Deactivation in catalytic hydrodemetallation. II. Catalyst characterization. *J. Catal.* 132:21–40.

Smith, B.J. and Wei, J. 1991b. Deactivation in catalytic hydrodemetallation. III. Random-spheres catalyst models. *J. Catal.* 132:41–57.

Snel, R. 1984a. Control of the porous structure of amorphous silica-alumina. I. The effects of sodium ions and syneresis. *Appl. Catal.* 11:271–280.

Snel, R. 1984b. Control of the porous structure of amorphous silica-alumina. II. The effects of pH and reactant concentration. *Appl. Catal.* 12:189–200.

Snel, R. 1984c. Control of the porous structure of amorphous silica-alumina. III. The influence of pore-regulating agents. *Appl. Catal.* 12:347–357.

Snel, R. 1987. Control of the porous structure of amourphous silica-alumina. IV. Nitrogen bases as pore-regulating agents. *Appl. Catal.* 33:281–294.

Speight, J.G. 1981. *The Desulfurization of Heavy Oils and Residua.* Marcel Dekker, New York.

Speight, J.G. 1991. *Chemistry and Technology of Petroleum*, 2nd ed. Marcel Dekker, New York.

Speight, J.G. 1994. Chemical and physical studies of petroleum asphaltenes. In *Asphaltenes and Asphalts*, Vol. 1, *Developments in Petroleum Science*, Yen, T.F. and Chilingarian, G.V. (Eds.). Elsevier Science, Amsterdam, chap. 2.

Speight, J.G. 1998. Asphaltenes and the structure of petroleum. In *Petroleum Chemistry and Refining*, Speight, J.G. (Ed.). Taylor & Francis, Washington, DC, p. 103.

Speight, J.G. 1999. *The Chemistry and Technology of Petroleum.* Marcel Dekker, New York.

Speight, J.G. 2004. New approaches to hydroprocessing. *Catal. Today* 98:55–60.

Speight, J.G. and Moschopedis, S.E. 1979. Some observations on the molecular nature of petroleum asphaltenes. Preprint. *ACS Div. Fuel Chem.* 24:910–923.

Speight, J.G. and Moschopedis, S.E. 1981. *Chemistry of Asphaltenes*, Advances in Chemistry Series, Comstock, M.J. (Ed.), chap. 1.

Stanford, E. 1986. Regeneration of catalysts from hydrotreating bitumen derived coker gas oil: correlation of catalyst activity with regeneration condition and measure catalyst properties. In *Preprints of the 10th Canadian Symposium on Catalysis.* Kingston, Ontario, pp. 589–598.

Stanislaus, A., Marafi, M., and Absi-Halabi, M. 1993. Studies on the rejuvenation of spent catalysts: effectiveness and selectivity in the removal of foulant metals from spent hydroprocessing catalysts in coked and decoked forms. *Appl. Catal. A Gen.* 105:195–203.

Stiles, A.B. 1983. *Catalyst Manufacturer.* Marcel Dekker, New York.

Suzuki, T., Itoh, W., Takegami, W., and Watanabe, T. 1982. Chemical structure of tar-sand bitumens by ^{13}C and ^{1}H NMR spectroscopic methods. *Fuel* 61:402–410.

Suzuki, T. and Uschida, T. 1979. Numerical analysis of the effectiveness factor for non-cylindrical extruded catalyst. *J. Chem. Eng. Jpn.* 12:425–429.

Takatsuka, T., Wada, Y., Hirohama, S., and Fukui, Y. 1989. A prediction model for dry sludge formation in residue hydroconversion. *J. Chem. Eng. Jpn.* 22:298–303.

Takuchi, C., Asaoka, S., Nakata, S., and Shiroto, Y. 1985. Characteristics of residue hydrodemetallization catalysts. Preprint. *Am. Chem. Soc. Div. Petrol. Chem.* 30:96–107.

Tamm, P.W., Harnsberger, H.F., and Bridge, A.G. 1981. Effects of feed metals on catalyst aging in hydroprocessing residuum. *Ind. Eng. Chem. Process Des. Dev.* 20:262.

Tanaka, H., Boulinguiez, M., and Vrinat, M. 1996. HDS of thiophene, dibenzothiophene and gas oil on various Co-Mo/TiO_2-Al_2O_3 catalysts. *Catal. Today* 29:209–213.

Ternan, M., Furimsky, E., and Parsons, B.I. 1979. Coke formation on hydrodesulphurization catalysts. *Fuel Process. Technol.* 2:45–55.

Thakur, D.S. and Thomas, M.G. 1984. Catalyst deactivation during coal liquefaction: a review. *Ind. Eng. Chem. Process Res. Dev.* 23:349–360.

Thakur, D.S. and Thomas, M.G. 1985. Catalyst deactivation in heavy petroleum and synthetic crude processing: a review. *Appl. Catal.* 15:197–225.

Thomas, J.M. and Thomas, W.J. 1997. *Principles and Practice of Heterogeneous Catalysis.* VCH, New York

Topsøe, H., Clausen, B.S., and Massoth, F.E. 1996. In *Hydrotreating Catalysis Science and Technology*, Anderson, J.R. and Boudart, M. (Eds.). Springer-Verlag, New York, p. 11.

Topsøe, H., Hinnemann, B., Nørskov, J.K., Lauritsen, J.V., Besenbacher, F., Hansen, P.L., Hytoft, G., Egeberg, R.G., and Knudsen, K.G. 2005. The role of reaction pathways and support interactions in the development of high activity hydrotreating catalysts. *Catal. Today* 107/108:12–22.

Topsøe, N.-Y. 2006. *In situ* FTIR: a versatile tool for the study of industrial catalysts. *Catal. Today* 113:58–64.

Topsøe, N.-Y. and Topsøe, H. 1983. Characterization of the structures and active sites in sulfided CoMo/Al_2O_3 and NiMo/Al_2O_3 catalysts by NO chemisorption. *J. Catal.* 84:386–401.

Topsøe, N.Y., Topsøe, H., and Masooth, F.E. 1989. Evidence of Brønsted acidity on sulfided promoted and unpromoted Mo/Al_2O_3 catalysts. *J. Catal.* 119:252–255.

Travert, A., Dujardin, C., Maugé, F., Cristol, S., Paul, J.F., Payen, E., and Bougeard, D. 2001. Parallel between infrared characterisation and *ab initio* calculations of CO adsorption on sulphided Mo catalysts. *Catal. Today* 70:255–269.

Trimm, D.L. 1980. *Design of Industrial Catalysts.* Elsevier, Amsterdam.

Trimm, D.L. and Stanislaus, A. 1986. The control of pore size in alumina catalyst supports. *Appl. Catal.* 21:215–238.

Toulhoat, H., Plumail, J.C., Houpert, C., Szymanski, R., Bourseau, P., and Muratet, G. 1987. Advances in residue upgrading. *Am. Chem. Soc. Div. Petrol. Chem.* 32:463–463.

Toulhoat, H., Szymansky, R., and Plumail, J.C. 1990. Interrelations between initial pore structure, morphology and distribution of accumulated deposits, and lifetimes of hydrodemetallisation catalysts. *Catal. Today* 7:531–568.

Tynan, E.C. and Yen, T.F. 1970. General purpose computer program for exact ESR spectrum calculations with applications to vanadium chelates. *J. Magn. Reson.* 3:327–335.

Usman, Kubota, T., Araki, Y., Ishida, K., and Okamoto, Y. 2004. The effect of boron addition on the hydrodesulfurization activity of MoS_2/Al_2O_3 and Co-MoS_2/Al_2O_3 catalysts. *J. Catal.* 227:523–529.

van Kessel, M.M., van Dongen, R.H., and Chevalier, G.M.A. 1987. Catalysts have large effect on refinery process economics. *Oil Gas J.* 85:55–59.

van Veen, J.A.R. 2002. Hydrocracking. In *Zeolites for Cleaner Technologies*, Guisnet, M. and Gilson, J.-P. (Eds.). Imperial College Press, London.

Venuto, P.B. and Habib, E.T., Jr. 1978. Catalyst-feedstock-engineering interactions in fluid catalytic cracking (FCC). *Catal. Rev. Sci. Eng.* 18:1–150.

Vrinat, M., Breysse, M., Geantet, C., Ramírez, J., and Massot, F. 1994. Effect of MoS_2 morphology on the HDS activity of hydrotreating catalysts. *Catal. Lett.* 26:25–35.

Wachs, I.E. (Ed.). 1992. *Characterization of Catalytic Materials*. Butterworth-Heinemann, Stonehan, MA.

Wachs, I.E., Deo, G., Kim, D.S., Vuurman, M.A., and Hu, H. 1993. Molecular design of supported metal oxide catalysts. In *New Frontiers in Catalysis. Proceedings of the 10th International Congress on Catalysis*, Budapest, p. 543.

Ward, J.W. 1983. Design and preparation of hydrocracking catalysts. *Stud. Surf. Sci. Catal.* 16:587–618.

Ware, R.A. and Wei, J. 1985. Catalytic hydrodemetallation of nickel porphyrins. III. Acid-base modification of selectivity. *J. Catal.* 93:131–151.

Wei, Z., Xin, Q., and Xiong, G. 1992. Investigation of the sulfidation of Mo/TiO_2:Al_2O_3 catalysts by TPS and LRS. *Catal. Lett.* 15:255–267.

Weiss, H. and Schmalfeld, J. 1999. Low Cost Process for Refinery Residue Conversion. *Refining PTQ*, Summer, p. 83.

Weissman, J.G. and Edwards, J.C. 1996. Characterization and aging of hydrotreating catalysts exposed to industrial processing conditions. *Appl. Catal. A Gen.* 142:289.

Weissman, J.G., Ko, E.I., and Kaytal, S. 1993. TiO_2-ZrO_2 mixed oxide aerogels as supports for hydrotreating catalysts. *Appl. Catal. A Gen.* 94:45–59.

Weitkamp, J., Gerhardt, W., and Scholl, D. 1984. Hydrodemetallation of nickel porphyrins over sulfided and reduced CoO-MoO_3/γ-Al_2O_3. In *8th International Congress of Catalysis*, Berlin, pp. II269–II280.

Wiehe, K.S. and Liang, K.S. 1996. Asphaltenes, resins, and other petroleum macromolecules. *Fluid Phase Equilib.* 117:201–210.

Wijngaarden, R.J., Kronberg, A., and Westerterp, K.R. 1998. *Industrial Catalysis*. Wiley-VCH, New York.

Wiwel, P., Zeuthen, P., and Jacobsen, A.C. 1991. In *Catalyst Deactivation*, Bartholomew, C.H. and Butt, J.B. (Eds.). Elsevier, Amsterdam, 1991, p. 257.

Wolk, R.H., Nogbri, G., and Rovesti, W.C. 1975. Multi-zone Method for Demetallization and Desulfurizing Crude Oil or Atmospheric Residual Oil. U.S. Patent 3,901,792, August 26.

Wolk, R.H. and Rovesti, W.C. 1974. Low Sulfur Fuel Oil from High Metals Containing Petroleum Residuum. U.S. Patent 3,819,509, June 25.

Yui, S.M. 1991. Using Fresh and/or Regenerated NiMo Catalysts on Coker Gas Oil Hydrotreating. Paper AM-91-60, presented at the NPRA Annual Meeting, San Antonio, TX, March 17–19.

Zhaobin, W., Qin, X., Xiexian, G., Sham, E.L., Grange, P., and Delmon, B. 1990. TiO_2 modified HDS catalysts. I. Effect of preparation techniques on morphology and properties of TiO_2:Al_2O_3 carrier. *Appl. Catal.* 63:305–317.

7 Maya Heavy Crude Oil Hydroprocessing Catalysts

Mohan S. Rana, S.K. Maity, and Jorge Ancheyta

CONTENTS

7.1 INTRODUCTION

Crude oil is a naturally occurring wide-boiling-range mixture of hydrocarbons acquired under the earth and sea that is often classified as light, medium, heavy, or extra-heavy, referring to its gravity as measured on the American Petroleum Institute (API) scale. In general, crude oils are commercially defined as follows: light crude oil has an API gravity higher than 31.1°, medium oil has an API gravity between 31.1° and 22.3°, heavy oil possesses an API gravity between 22.3° and 10°, and extra-heavy oil or bitumen has an API gravity less than 10°, which cannot flow without being heated or diluted (Riazi, 2005). Crude oil produced by different petroleum fields differs importantly in viscosity and sulfur content. The more viscous and high-sulfur-content (>2.0 wt%) crudes are called heavier or sour, whereas low-sulfur-content (<0.5 wt%) crudes are called sweet (Gary and Handwerk, 1994). Most of the refineries are not capable of processing heavy or sour crudes due to their complex nature, which compels refiners to process the crude oil in some of the selective refineries that are able to upgrade them to more valuable products.

On the other hand, stricter environment legislation on transportation fuel quality is being implemented or planned throughout the world. These increasingly stringent standards reduce the yield of gasoline and diesel per barrel of crude even if the quality of the crude inputs is not declining. Since the quality of petroleum feed, configuration, and complexity of refinery processes can vary substantially, each refinery has its own strategy to meet the new goals of fuel specification. Each country develops new product specifications, but the time of introduction of them varies depending on the region and origin of fuel oils. Apart from the several impurities, the pressure on crude oil and oil products prices is linked to a deficit of upstream (oil production) and downstream (oil refining) capacities to fulfill the demand over an extended period. The major oil fields in the word are in the Arab countries and North (Canada, Mexico, the United States) and South America (Venezuela, Argentina, Brazil, and Bolivia). However, the largest downstream capacity is located in North America (mainly in the United States). In addition, the role of the United States as the world's largest petroleum importer has been influential to Canada and Mexico (exporters), including Latin American countries.

In Mexico, the first petroleum well was discovered in 1869, but commercial production of crude oil began in 1901 near the central Gulf Coast of Mexico. Mexico started commercial export of crude oil in 1911 (PEMEX, 2007). Presently, Mexico produces three different commercial crudes: Olmeca, Isthmus, and Maya. The main properties of these three crudes are given in Table 7.1 and Table 7.2 along with their atmospheric and vacuum residua. These crudes are classified as extra-light or sweet Olmeca 39 °API, light Isthmus 33 °API, and heavy Maya 21 °API. The production of Olmeca crude oil is now almost negligible, while Isthmus production shows a decreasing trend (Ancheyta et al., 2005a). Thus, the production of light crudes has been dwindling over 9% per year, while heavy crude oil (Maya) output has grown at about 4.5% per year. Moreover, the Mexican

TABLE 7.1
Physical Properties of Isthmus and Olmeca Crude Oils and Their Residua

	Isthmus			Olmeca		
Properties	**Base**	**AR, 343°C+**	**VR, 538°C+**	**Base**	**AR, 343°C+**	**VR, 538°C+**
API gravity	33.3	15.64	8.36	39.4	17.18	10.45
S, wt%	1.8	2.38	3.14	0.93	2	2.78
N, wt%	0.14	0.26	0.48	0.51	0.18	0.36
Ni + V, wppm	99.7	121.89	299.68	10.44	27.8	84.54
Asphaltene, wt% (ins. nC_7)	3.06	5.91	14.2	0.41	1.11	3.33
CCR, wt%	4.13	8	19.5	2.22	5.9	17.9

TABLE 7.2
Physical Properties of Maya Crude Oil and Its Residua

Properties	**Base Maya Crude**	**Maya AR, 345°C+**	**Maya VR, 543°C+**
API gravity	21.31	7.14	1.43
Density, 20/4°C	0.9232	1.0177	1.0615
S, wt%	3.52	4.60	5.7
N, wt%	0.32	5086	6591
Ni, wt%	49.5	86.61	127
V, wt%	273.0	488.96	684
Asphaltene, wt% (ins. nC_7)	12.7	17.74	26.3
Ramscarbon, wt%	10.87	17.66	26.21
Distillation, °C	TBP	ASTM D-1160	
IBP/5, vol%	19	340	—
10	131	410	—
20	201	454	
30	273	497	
40	352	—	
50	430	—	
60	509	—	
70	586	—	

TABLE 7.3
Effect of Feed Contamination on Hydrocracking (HCR) and Fluid Catalytic Cracking (FCC)

Contaminants	Effect on HCR or FCC Catalyst	Process Remedies
Metals	Structural damage of zeolite (US-Y)	HDM
	Vanadium passivation	
	Metal deactivation on catalytic sites	
Nitrogen	Deactivation on acid sites	HDN
	Competitive adsorption on catalytic sites	
Sulfur	SOx formation (flue gas)	HDS
Carbon (low H/C ratio)	Formation of Conradson carbon residue	HYD
	Carbon deposition on catalytic sites	(high H_2 pressure)
	Pore plugging (diffusion limitations)	

reserve position indicates that the production of heavy or extra-heavy oils will continue in coming years. At present, the major production in Mexico is of Maya crude oil, and most of the research work is oriented to the problems presented by this crude. On average, PEMEX exports ca. 87% heavy crude oil (that is, Maya), with the rest being light and extra-light crude oil (Isthmus and Olmeca). Seventy-nine percent of the total crude oil is exported to the United States, while the remaining 21% is exported to Europe (10%), the Far East (2%), and the rest of the world (9%). With respect to the clean fuel demand, Mexico's oil company PEMEX, the world's ninth largest oil company and third largest producer of crude oil (*Petroleum Intelligence Weekly*, 2003), is in the process of an upgrading program to reduce the fuel sulfur level in all its refineries. PEMEX's major petroleum research and process technology work is carried out at the Instituto Mexicano del Petroleo (IMP), especially for development of hydroprocessing catalysts (Ancheyta et al., 2005b; Rana et al., 2005a). The aim of upgrading these raw crude oils is to enhance the gasoline production, which can be obtained by using two routes of crude oil processing; the primary or most common way is the distillation of heavy crude oil, then residue hydrotreating before hydrocracking or fluid catalytic cracking. The second option is the preparation of *synthetic crude oil* (SCO) by upgrading heavy or extra-heavy crudes (bitumen) to reduce metals (Ni and V primarily), asphaltenes, Conradson carbon residue (CCR), sulfur, and nitrogen levels. Generally, these contaminants have individual effects (Table 7.3) on hydrocracking and fluid catalytic cracking (FCC) processes, the subsequent processes used in refineries.

7.2 COMPOSITION OF MAYA HEAVY CRUDE OIL

Petroleum is a complex mixture of hydrocarbons, which extends from the simplest alkanes to highly complex structures with molecular weights in the thousands. The composition of crude oil classically depends on the origin, for example,

Boscan (10.1° API), Kern River (13.6° API), Maya (21° API), offshore California, and so forth. These crudes are paraffinic, naphthenic, or aliphatic in nature. Boscan is aliphatic, Kern River is naphthenic, whereas Maya is intermediate in nature (Quann et al., 1988). Constituents of Maya crude, which are the principals concerned when upgrading, are sulfur, nitrogen, metals, and asphaltene. Detailed properties of Maya crude and its atmospheric residue (AR) and vacuum residue (VR) are presented in Table 7.2.

The foremost problem for upgrading of crude oil is asphaltene due to its complex nature (Absi-Halabi et al., 1991; Furimsky and Massoth, 1999). In general, asphaltene molecules are difficult to process because they are very large, polar, and noncrystalline, have multiple stacked structures, and also contain high amounts of hetero-atoms, which can easily deactivate the hydroprocessing catalysts during the operation (Gosselink, 1998). Furthermore, the chemistry of asphaltene is not totally understood and its molecules are precursors to the most problematic organic deposits and sediment formation. The organic deposits or solid precipitation occurs due to asphaltene flocculation and formation of random aggregates (Mansoori, 1997), which is likely to form steric colloids in the presence of resins and paraffins (Branco et al., 2001). These colloids and floc aggregates can rarely convert into some other products at hydrotreating conditions. Moreover, these aggregates deposit over the catalytic sites as well as near the pore mouths, leading to diffusion limitations. Asphaltene molecules carry a core of stacked, flat sheets of condensed aromatic rings linked at their edges by aliphatic or naphthenic chains. The condensed aromatic rings exist in the form of a nonhomogeneous flat sheet, as shown in Figure 7.1, along with the organometallic chelates. The asphaltene elemental analysis and other properties of Maya and Isthmus crudes are shown in Table 7.4. Due to the biogenic origin of petroleum, most of the metals are associated with a metal–porphyrin structure of chlorophyll type, and the stability of these complexes vary in this order: V > Ni > Mg (Falk, 1964).

During hydroprocessing asphaltene molecules are allowed to diffuse into the catalyst pore and their metal atoms (V and Ni) are removed from the ring structures. The asphaltene molecules may be partially hydrogenated or cracked over the catalyst surface, depending on the reaction conditions, and produce relatively small-sized molecules, called maltenes, but it is also likely that these molecules may thermally decompose and polycondense to coke precursors. This leads to instability and finally to dry sludge, which is the dead point for any kind of catalyst stability. The structure of the asphaltenes has been the subject of several investigations and is now believed to consist of polycyclic aromatic clusters, substituted with varying alkyl side chains. The composition becomes more complex in the bottom of distillation, such as atmospheric and vacuum residues that contain large percentages of heteroatoms (O, S, N) and organometallic constituents (Ni, V, Fe). The residue oils generally have a relatively lower percentage of paraffin and a higher percentage of metals, sulfur, nitrogen, and asphaltene than virgin Maya crude oil.

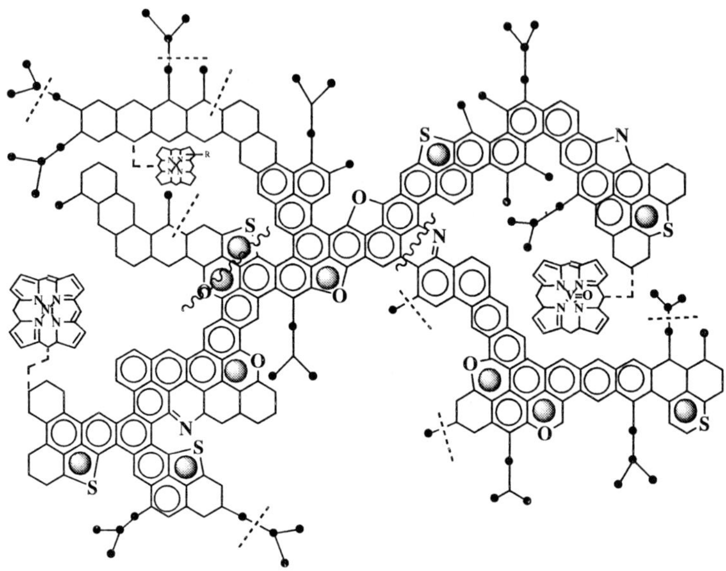

FIGURE 7.1 Hypothetical asphaltene molecule in Maya crude and its structural cleavage during the hydroprocessing reaction: (----) primary cleavage of weak bonds, (∿∿∿) secondary cleavage. (From Rana, M.S. et al., *Petrol. Sci. Technol.*, 25(1–2): 201–214. With permission.)

TABLE 7.4
Properties of Asphaltene in Different Crudes

Elements, wt%	Maya	Isthmus
Carbon	81.62	83.99
Hydrogen	7.26	7.77
Nitrogen	1.46	1.35
Sulfur	8.42	6.48
Oxygen	1.02	0.79
H/C	1.067	1.043
Ni	0.032	0.018
V	0.151	0.074
n	6.8	5.0
fa	0.52	0.57
AMW	5190	3375

Note: AMW = average molecular weight; fa = aromaticity factor; n = average number of carbon per alkyl side chains.

Source: From Trejo, F. et al., *Fuel*, 83, 2169–2175, 2004. (With permission.)

7.3 SELECTION OF SUPPORT AND CATALYST FOR MAYA CRUDE HYDROPROCESSING

The principal target of heavy oil processing is the selection of a proper catalyst, which consists of a support, active metal, and promoter. Based on the heavy oil feedstock properties, the composition of the catalyst is selected, for example, low metal loading and high pore diameter. However, the selection of the support nature (acidic or basic) is based on the selectivity and stability of a catalyst with time-on-stream (TOS). In addition to the fundamental catalyst properties, external morphology (shape and size) and mechanical properties are also very important. Apart from these properties, one must perceive the technical and economic aspects behind the industrial scale (Armor, 2005; Mehrotra, 2006). Thus, the development of industrial catalysts is an evolutionary process that requires significant pieces of information, which cannot be obtained with just a few formulations prepared in the lab (Farrauto and Bartholomew, 1997). There are many papers dealing with fundamental research regularly published in different journals, but most of them do not explain the commercial applicability of these catalysts, that is; most likely because either these catalysts are not at all applicable to the industrial process or the cost of the catalyst is too high. Therefore, the preparation of the industrial catalyst remains *an art*. A large amount of basic information of a catalyst formulation is hidden in support preparation, active metal impregnation, and its drying and calcinations. Most of the catalyst properties are not independent; thus, if one of them is varied with a view to improvement, the others may also be modified, not necessarily having the same effect. In hydroprocessing, the stability of the catalyst is less important if the feedstock is in the middle distillates range, but using heavy crude like Maya, the catalyst stability becomes crucial due to carbon and metal deposition during the operation. Therefore, the support plays an important role since it provides the textural properties of the catalyst and surface for dispersion of active metals.

7.3.1 Preparation of Supports

The catalytic performance of a catalyst depends on its different components, such as support and active metal. Usually, the behavior of the active phase is influenced by the type of support material; thus, it becomes necessary to select an optimum support material according to the criteria of interaction with active phases, acid-base properties (inertness), textural properties, stability, extrusion properties, and cost at commercial scale. These properties become critical when the catalyst is used for heavy oil processing. Therefore, taking all parameters into consideration, the most used support for hydroprocessing Maya crude has been alumina (Rana et al., 2004a, 2005a). Alumina combined with small amounts (5 to 10 wt%) of other oxides such as TiO_2 (Maity et al., 2003a, 2004, 2005a, 2006; Rayo et al., 2004; Rana et al., 2005b; Ramírez et al., 2005), ZrO_2 (Rana et al., 2005c), SiO_2 (Maity et al., 2003b), and MgO (Caloch et al., 2004) has also been used as a catalyst support. The major component of these supports remains γ-alumina due to its outstanding mechanical properties, nature of interaction with active phases (Mo, W, Co, Ni), and cost. Contrary to middle distillate or model molecules hydrotreating catalysts, heavy oil hydroprocessing catalytic activity depends not only on the active phases, but also on the textural properties of the solid (Trimm, 1980).

Catalyst particle size, pore diameter, and pore size distribution are main issues in the development of a suitable catalyst for hydroprocessing of heavy crudes. These characteristics of a catalyst are endowed by the support materials. The desired textural properties, pore volume, and surface area of a catalyst can be controlled during the preparation of the alumina support through the mixing of the aluminum source (commonly aluminum nitrate) and precipitating agents (Huang et al., 1989). The control of textural properties is also possible by using additives (Trimm and Stanislaus, 1986), aging (Johnson and Mooi, 1968; Ono et al., 1983; Chuah et al., 2000), washing, drying, and calcinations (White et al., 1989; Huang et al., 1989). The effects of seeding (Papayannakos et al., 1993), pH swing (Maity et al., 2003a, 2004, 2006), and different precipitating agents (Rana et al., 2004a, 2005b) are also reported for such a purpose. In general, alumina supports are precipitated at around pH Ca 8 to obtain pseudoboehmite with moderate pore volume and unimodal pore diameter, while precipitation at pH > 10 forms barite, having a bimodal pore size distribution (Huang et al., 1989). The bimodal type of support can also be prepared by using additives or combustible fibers during the extrusion of boehmite (Tischer, 1981; Tischer et al., 1985; López-Salinas et al., 2005), while the effect of steaming agents on the texture of the alumina extrudate has been studied (Absi-Halabi et al., 1993; Walendziewski and Trawczynski, 1993, 1994). These support preparation methods were observed at laboratory scale, whereas very few have been reported at commercial scale using an integrated minipilot plant (Kaloidas et al., 2000). The impurities trapped in the precursor may influence the support pore structure as well as some of the chemical properties, such as isoelectric point (IEP), which are responsible for active metal dispersion, and thus the catalytic activity. The source of the alumina precipitating agent also requires attention, for example, traces of Na and SO_4 ions must be removed during the washing.

Bimodal alumina is prepared by variation of pH from acid to basic or vis-à-vis (Maity et al., 2004). In this method, aluminum sulfate and sodium aluminum oxide salts are used to prepare acid and basic solutions, respectively. A basic salt solution is added to acid salt solution to change the pH of the mixture from the acid to the basic side, and then the mixture is aged several minutes. In this way, the pH of the mixture is changed several times, and finally the precipitate is aged in basic media. The common pore size distribution of the calcined support obtained by this method is given in Figure 7.2. However, when urea is added into the above mixture at the beginning of the precipitation, a wide pore range of alumina is observed instead of bimodal alumina (Maity et al., 2005a).

The effect of hydrolyzing agents such as ammonia, ammonium bicarbonate, ammonium carbonate, and urea on alumina preparation have been studied (Rana et al., 2004a). Changing the hydrolyzing agent in the alumina preparation caused changes in the pore structure of alumina-supported CoMo catalysts (Figure 7.3), indicating that alumina prepared by ammonium carbonate has ample pore structure

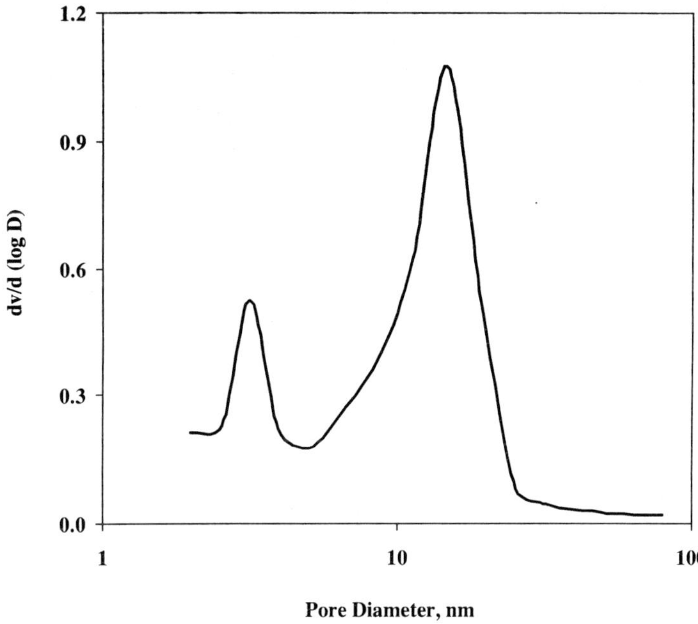

FIGURE 7.2 Bimodal pore size distribution of $CoMo/Al_2O_3$-SiO_2 using the pH swing method for support preparation. (From Maity, S.K. et al., *Catal. Today*, 98: 193–199, 2004. With permission.)

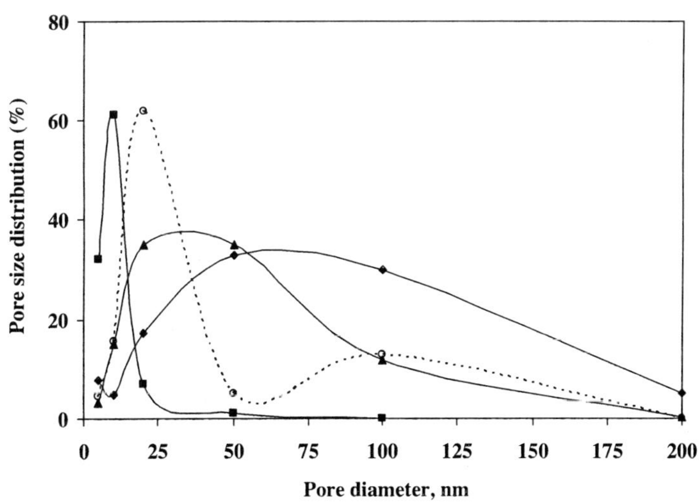

FIGURE 7.3 Effect of different hydrolyzing agents on the pore size distribution of CoMo-supported catalysts: (♦) Al_2O_3-u, (▲) Al_2O_3-acs, (■) Al_2O_3-am, (O) reference catalyst. (From Rana, M.S. et al., *Catal. Today*, 98: 151–160, 2004a. With permission.)

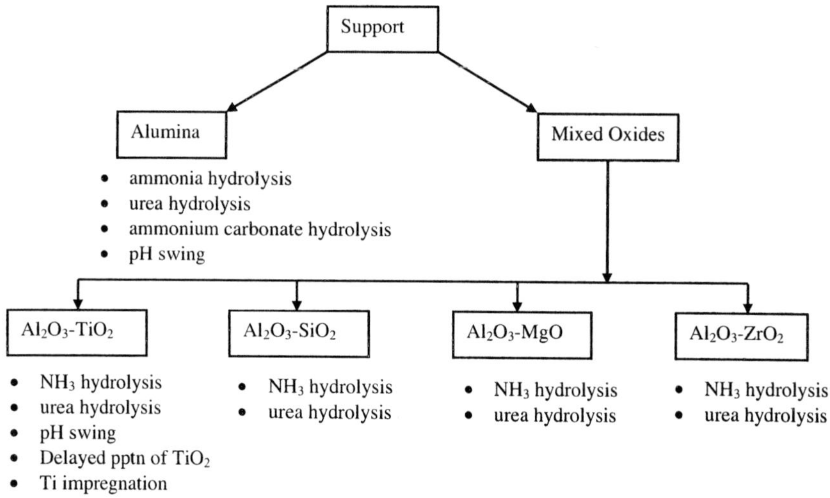

SCHEME 7.1 Different methods used for preparation of mixed-oxide supports.

compared with other hydrolyzing agents. The resulting wide range of pore diameters was explained by the presence of CO_3^- ions, which are coming from ammonium carbonate or urea, trapped in between aluminum oxyhydroxide (Rana et al., 2004a). These trapped carbonate ions lead to wide-pore alumina during calcination.

Among the mixed oxides, alumina–titania support is widely used for the preparation of hydrotreating catalysts due to its favorable acid–base properties. The different procedures employed to prepare this binary oxide have been discussed in the literature (Rana et al., 2005b; Maity et al., 2006). In these methods Al_2O_3-TiO_2 mixed-oxide supports are prepared by the coprecipitation method using different precipitating agents, urea, and ammonia. It is reported that Al_2O_3-TiO_2 supports prepared by sodium aluminate and titanium chloride salts have higher-pore-volume and larger-pore-diameter solids than the same support prepared by other methods (Maity et al., 2006). A simplified flow diagram of the different supports and their preparation methods is presented in Scheme 7.1.

The second step of support preparation is shaping and sizing, which also affect the mechanical and pore texture of the carrier. Thus, at the kneading stage a balanced use of precursor, additive, peptizing agent (HNO_3), and its concentration (Kaloidas et al., 2000) is required to produce a homogeneous paste to form suitable extrudates (Le Page et al., 1987; Benbow and Bridgewater, 1987; Benbow et al., 1987). Moreover, extrudate drying and calcination play important roles to generate suitable textural and mechanical properties (for example, crushing strength and attrition), which can be affected during the removal of water, additive, and peptizing agent.

7.3.2 Preparation of Catalysts

A hydroprocessing catalyst is composed of a support and active metals such as cobalt (nickel) and molybdenum (tungsten) in sulfided state. Various parameters such as impregnation procedure, sequence of impregnation, and presence of additives during the preparation strongly affect the morphology or surface structure of the catalyst (Topsøe et al., 1996). Thus, the catalytic performance of the working sulfided catalyst (CoMoS or NiMoS) is strappingly dependent on the catalyst preparation and its activation conditions. Molybdenum is introduced as ammonium hepta molybdate, while cobalt and nickel nitrates are the usual precursor salts for the promoters. Usually, CoMo catalyst is more active for hydrodesulfurization (HDS), while NiMo is for hydrodenitrogenation (HDN). The more expensive NiW catalyst is used in instances where high saturation and moderate cracking of low-sulfur feedstock are desired (McCulloch, 1983; Bartholomew, 1994). A number of other transition and noble metals have been tested in laboratory scale for hydrotreating (Ru, V, Fe, Mn, Cr, Pt, Rh, and Pd), but do not appear to be used commercially. The effects of the amounts of these or other active precursors with their optimum concentration (monolayer formation) over different supports are given elsewhere (Pecoraro and Chianelli, 1981; Topsøe et al., 1996). The choice of active metal (CoMo or NiMo) and support selection may depend on the feed composition as well as the reactor location in the refinery. The method of metal loading is one of the most important parameters with respect to the metal dispersion to obtain high catalytic activities (Kasztelan et al., 1986). Usually, impregnation of the active metal is carried out by wetting the support by either an excess or incipient wetness method. Typically, catalyst is prepared by introducing Mo(W) first and then Co or Ni promoter atoms. However, a coimpregnation method can be applied when an additive such as P, urea, chelating agents, and so on, is used. A small quantity of additives is frequently used in hydroprocessing catalyst to stabilize the alumina support and avoid nickel and cobalt diffusion into the support (Rana et al., 2007d). To prepare hydroprocessing catalysts for Maya crude, both subsequent and coimpregnation methods are used. Typical metal loadings of these catalysts and textural properties are given in Table 7.5 and Table 7.6.

7.3.3 Characterization of Fresh Catalysts

Generally, hydrotreating catalysts are characterized using different techniques; most of the characterization is physicochemical, including surface characterization of active sites. Apart from this, heavy oil catalysts must have superior mechanical as well as morphological properties due to the complex nature of the feedstock or pressure drop in the reactor. Abrasion and attrition problems are complicated and can be eliminated by using the required characteristics of the catalyst, as shown in Table 7.7. These properties are calculated using a standard industrial method and are part of the *know-how* of commercial catalyst formulation. More details about the effect of these properties on reactor pressure drop are discussed in Chapters 5 and 6. Usually the above-mentioned properties are related, for example, the larger the pore diameter, the greater the attrition loss and low crushing strength.

TABLE 7.5
Catalysts Textural Characterization and Composition

Support Preparation		Textural Properties			Composition, wt%	
Support	Preparation Method	SSA (m^2/g)	PV (ml/g)	APD, nm	Mo	Co (Ni)
Al_2O_3-am (6.4 nm)	Ammonia	164	0.27	6.4	7.8	(2.4)
Al_2O_3-am (6.5 nm)	Ammonia	169	0.27	6.5	7.5	2.5
Al_2O_3-u (12.9 nm)	Urea	136	0.39	12.9	7.5	2.5
Al_2O_3-ac (6.0 nm)	Ammonium carbonate	183	0.37	6.0	7.3	2.5
Al_2O_3-acs (17.3 nm)	Ammonium carbonate-pH swing	160	0.47	17.3	6.8	2.5
Al_2O_3-22.2	Catapal-200	60.0	0.48	22.2	7.2	2.3
Al_2O_3-u-ac	Urea-ammonium carbonate	184	0.48	10.0	5.93	2.1
Ti-Al_2O_3	Ti/Catapal-1	176	0.40		6.56	(4.1)
Al_2O_3-MgO	Ammonia	169	0.31	7.4	7.9	2.3
Al_2O_3-ZrO_2	Ammonia	193	0.33	6.2	7.2	3.6
Al_2O_3-TiO_2	Ammonia	163	0.320	6.3	7.5	2.5
Al_2O_3-B_2O_3	Ammonia	48.0	0.34	24.7	7.1	2.2

Note: APD = average pore diameter; SSA = specific surface area; PV = pore volume.

Source: Adapted from Rana, M.S. et al., *Catal. Today*, 98: 151–160, 2004; Rana, M.S. et al., *Catal. Today*, 109: 24–32, 2005; Rana, M.S. et al., *Catal. Today*, 107/108: 346–354, 2005. (With permission.)

The aim of different alumina support (Table 7.5 and Table 7.6) preparations is to illustrate a concept of support preparation method on hydrotreating (HDT) catalysts' performance. Different γ-Al_2O_3 support preparation methods have been employed to vary the porosity and pore size distribution, which play important roles in stability as well as in the metal retention capacity of the catalyst. Recently, Institut Français du Pétrole (IFP) has reported a microcrystalline that forms "chesnut-bur"-like pores and nonacidic material. This catalyst has a large pore volume and an appropriate pore size distribution, which allow the diffusion of resins and asphaltenes into the catalyst pellet and their adsorption on the active sites. It is reported that this catalyst has better metal retention capacity and asphaltene elimination than traditional HDS catalysts. The catalyst has high dispersion of metal over acicular alumina platelets, large pore volume, and an appropriate pore size distribution, which prevents plugging of the pore network and a high metal retention up to 100% in relation to its weight as a fresh catalyst (Kressmann et al., 1998).

The characterization of the catalysts before and after reaction provides important information about the role of textural properties on metal deposition.

TABLE 7.6
Effect of Support and Catalyst Preparation on Textural Properties

Support Preparation and Its Composition		Catalyst Textural Properties			Catalyst Composition, wt%			
Support	Preparation Method	SSA, m²/g	TPV, ml/g	APD, nm	Mo (W)	Co (Ni)	P	Catalyst Name
TiO_2-Al_2O_3	AlSul-TiCl-NH_3	236	0.24	3.6	6.7	2.36		A
(5/95, w/w)	AlSul-Ti-Iso-NH_3	216	0.25	4.4	6.7	2.36		B
	AlSul-Ti-Iso-urea	258	0.62	9.6	6.7	2.36		C
	AlSul-Ti-Iso-urea	250	0.67	10.4	6.7	2.36	0.8	D
SiO_2-Al_2O_3	AlSul-NaSi-urea	216	0.30	5.4	6.7	2.36		E
(5/95, w/w)	AlSul-NaSi-urea	154	0.31	7.9	6.7	2.36	0.8	F
	AlSul-NaSi-urea	167	0.31	7.4	6.7	(3.36)	0.8	G
Al_2O_3	$AlNO_3$-NaAl-pH	244	0.78	12.2	10	3.44		
		240	0.78	12.3	10	(3.44)		
		213	0.70	12.8	(15.8)	3.14		
		225	0.73	12.5	(15.8)	(3.14)		
		243	0.82	12.9	6.7	2.36	0.8	
Al_2O_3	Commercial Al_2O_3	223	0.64	11.4	6.7	2.36		
	$AlNO_3$-NaAl-pH	262	0.49	7.3	6.7	2.36		
	$AlNO_3$-NaAl-pH	257	0.49	7.4	6.7	(2.36)		
Al_2O_3		241	0.75	12.0	6.7	2.36		
Al_2O_3-SiO_2		218	0.30	5.6	6.7	2.36		
Al_2O_3-TiO_2		257	0.62	9.6	6.7	2.36		

Note: AlSul = aluminum sulfate; TiCl = titanium chloride; Ti-Iso = titanium isobutoxide; NaSi = sodium silicate; $AlNO_3$ = aluminum nitrate; NaAl = sodium aluminate.

Source: Adapted from Maity, S.K. et al., *Appl. Catal. A Gen.*, 244: 141–153, 2003; Maity, S.K. et al., *Appl. Catal. A Gen.*, 250: 231–238, 2003; Maity, S.K. et al., *Appl. Catal. A Gen.*, 253: 125–134, 2003. (With permission.)

Moreover, the hydrodemetallization (HDM) catalysts' behavior can be observed in two ways, which depend on the textural properties of the catalyst: (1) microporous and low-range mesoporous catalyst decreases activity rapidly due to pore mouth blockage and diffusion limitations of complex molecules into the pores, and (2) the catalysts containing macropores, in which deposited metal sulfides (V and Ni) behave as catalytic sites, showed comparatively more stability with TOS. Apart from the textural properties, the nature and quantity of the supported species, their structure and size, the chemical state of the elements, their dispersion on the support, and the interaction between the molybdenum entities with the support or promoter can be characterized by using several spectroscopic techniques (Topsøe et al., 1996).

TABLE 7.7
Morphological and Mechanical Properties of Laboratory-Prepared $CoMo/Al_2O_3$ Catalysts

Properties	
Morphological Characteristics	
Shape of extrudate	Cylindrical
Length of extrudate	3–3.5 mm
Diameter of extrudate	1.8–2 mm
Density of extrudate	0.85 g/ml
Mechanical Properties	
Attrition loss (rotating tube)	<1.5%
Grain-to-grain crushing strength	5–7 kg

The use of a mixed-oxide-supported catalyst in heavy oil processing modifies the metal support interaction and facilitates the sulfidation of the active phases, and consequently the number of catalytic sites (that is, coordinated unsaturated sites (CUS) or anion vacancies), which are proportional to the increased catalytic activity (Topsøe et al., 1984). To analyze the state of dispersion of the sulfided phase in the different catalysts used for hydroconversion of Maya crude, NO adsorption is on the sulfided samples. It is well known that NO titrates the sulfur vacancy sites or coordinated unsaturated site, which is related to the hydrotreating activity. Figure 7.4 shows the results of NO adsorption over various NiMo-supported catalysts normalized per square meter of catalyst surface (Ramírez et al., 2005). Not surprisingly, this trend is the same as the one observed for catalytic activity. It appears the incorporation of Ti into the catalyst support increases the number of sulfur vacancies associated with the active sites. This can also be an indication of increased dispersion of the sulfided phase. It was reported (Agudo et al., 1984; Redey et al., 1988) that the NO chemisorption correlates better with catalytic activity than O_2 adsorption (Zmierczak et al., 1982) due to the fact that NO chemisorption is more selective than O_2. Moreover, the mixed-oxide-supported catalyst has higher acidity than alumina. In Figure 7.4, the differences between the two supported catalysts (Al_2O_3 and Al_2O_3-TiO_2) are relatively small, but they appear to be sufficient to explain the behavior of the catalysts. Nevertheless, the sulfided catalysts showed a negligible amount of Brönsted acidity at 1541 cm^{-1} after 200°C pyridine desorption, while an adequate number of Lewis acid sites remains up to 400°C. The infrared (IR) bands at 1598 and 1445 cm^{-1} are specific to pyridine interaction with the Lewis acid sites, and quantification of these bands is shown in Figure 7.5 (Rana et al., 2005a).

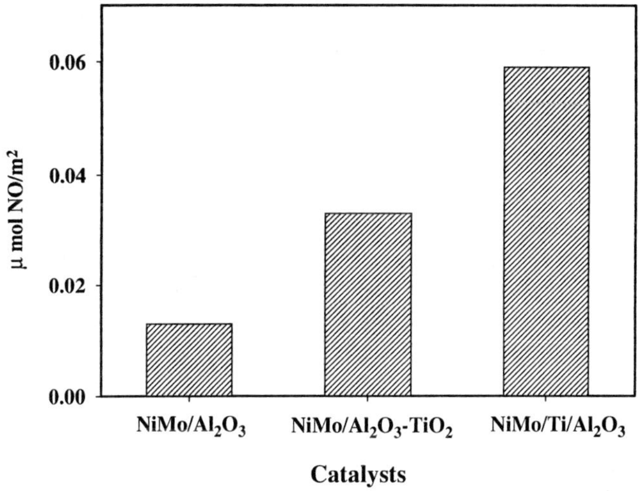

FIGURE 7.4 Dynamic NO chemisorption over sulfided NiMo catalysts as an effect of Ti incorporation method. (From Ramirez, J. et al., *Catal. Today*, 109: 54–60, 2005. With permission.)

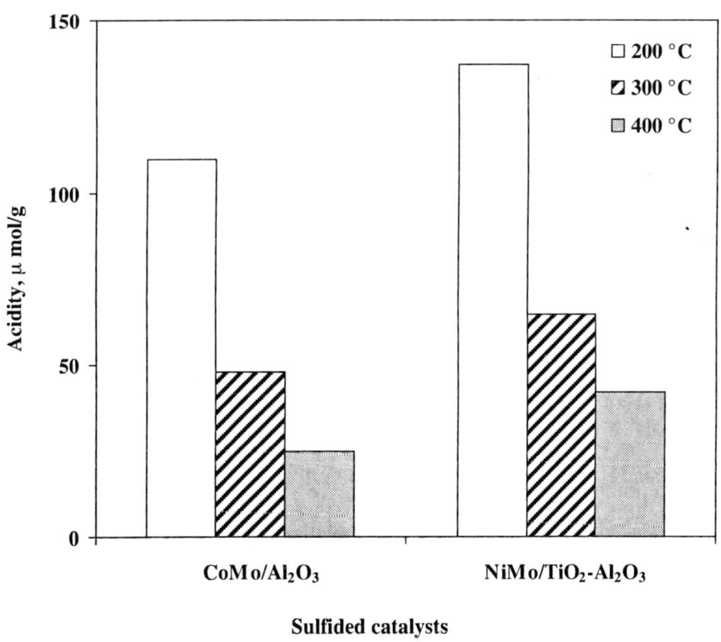

FIGURE 7.5 Lewis acidity of sulfided catalysts with different support composition. (From Rana, M.S. et al., *Catal. Today*, 109: 24–32, 2005. With permission.)

7.4 CATALYTIC ACTIVITIES FOR MAYA HEAVY CRUDE OIL HYDROPROCESSING

7.4.1 Catalyst Loading and Pretreatment

Typically, CoMo and NiMo catalysts, which are sulfided *in situ* or *ex situ* before the activity test, are loaded to the reactor in the oxide state. In our experiments (Maya crude oil processing), oxide catalyst was loaded with an equal volume of diluent (SiC), dried at 120°C and atmospheric pressure, and then allowed to soak with gas oil. The sulfiding agent, a mixture of dimethyldisulfide (DMDS) with gas oil having 2.7 wt% total sulfur, was introduced after soaking. An increase of temperature during sulfidation was 1.5°C/min; the first sulfiding step was performed at 260°C for 3 h and subsequently at 320°C for 5 h. Both low- and high-temperature sulfidations were performed at 28 kg/cm^2 pressure. The detailed catalyst activation procedure for testing catalyst activity during Maya heavy oil hydroprocessing was reported elsewhere (Marroquín et al., 2004). Usually, the better or higher degree of sulfidation corresponds to higher catalytic activity. After sulfidation, the flow was switched to the corresponding feedstock (Table 7.8) and the operating conditions were adjusted as shown in Table 7.9.

7.4.2 Maya Crude Oil Hydroprocessing Feedstock and Analysis

Different catalyst formulations were tested close to the industrial conditions to find out the optimum catalyst using diluted as well as pure Maya crude feed composition, as shown in Table 7.8. The dilution of Maya heavy oil was used to minimize the experimental problems occurring in microreactor processing due to the high viscosity of Maya crude. The analysis of feed and product composition was carried out by different American Society for Testing and Materials (ASTM) techniques: metals (Ni, V) were analyzed by flame atomic absorption spectrometry (ASTM D-5863-00a), S content was determined with HORIBA model SLFA-2100/2800 equipment using energy-dispersive x-ray fluorescent (XRF) x-ray beam, total nitrogen was measured by oxidative combustion and chemiluminescence (ASTM D-4629-02), and asphaltene was defined as the insoluble fraction in n-heptane.

7.4.3 Active Sites and Catalytic Activity

Comprehensive catalytic studies of sulfided catalysts have been carried out with Maya heavy crude oil using different diluents as well as pure crude oil to determine the hydrodesulfurization (HDS), hydrodenitrogenation (HDN), hydrodemetallization (HDM), and hydrodeasphaltenization (HDAs) activities. Generally, these reactions take place on the sulfided catalyst, a sulfur ion vacancy that has a deficiency of electrons, which adsorb S, N organic molecules through unpaired electrons, for example, porphyrin of metal chelates in the case of hydrodemetallization. On the other hand, the adsorption of organic molecules competes with

TABLE 7.8
Physicochemical Characterization of Feed for Different Reactors

	Bench-Scale Reactor Feed		Microflow Reactor Feed		
Properties	**Maya Crude**	**Maya HDT**	**HDM[a]**	**HDS[b]**	**HDS[c]**
	Elemental Analysis				
C, wt%	86.9	85.5	84.2	83.2	
H, wt%	5.3	7.2	8.8	9.5	
N, wt%	0.32	0.1852	0.184	0.118	0.0873
S, wt%	3.52	1.217	2.21	0.648	0.437
	Metal, wppm				
Ni	49.5	36.76	26.21	18.9	14.93
V	273.0	107.98	124.78	81.66	61.59
(Ni + V)	322.5	144.74	150.99	100.56	76.52
Ca	11.26	—	5.0	—	
Mg	2.04	—	1.01	—	
Na	44.83	—	21.2	—	
K	20.25	—	10.2	—	
Fe	2.16	—	1.02	—	
Asphaltene, wt% (ins. n-C_7)	12.7	6.87	8.43	—	4.35
	Physical Properties				
Density, 20/4°C	0.925	0.877	0.88	0.865	0.8401
Pour point, °C	−30		−15	−24	
Ramscarbon, wt%	10.87	8.0	5.45	5.54	
API gravity	21.31	31.14	29.29	32.10	
	Viscosity, g/cm sec				
At 50°C			3.08	2.63	
At 100°C			9.45	8.29	

Note: Maya HDT = Maya hydrotreated partially.

[a] HDM feed: Maya + diesel, 50/50, wt/wt.
[b] HDS feed: Maya HDT + diesel, 50/50, wt/wt.
[c] HDS feed: Maya HDT + naphtha, 50/50, wt/wt.

Source: From Rana, M.S. et al., *Catal. Today*, 109: 24–32, 2005. (With permission.)

the presence of H_2S or H_2 dissociation species (HS^-, H^+, H^-) for the same site. The adsorption of dissociated species converts CUS into the saturated sites (that is, the sulfhydryl group), which is a Brönsted acid site center (Topsøe and Topsøe, 1993; Rana et al., 2000, 2004b, 2004c; Breysse et al., 2002). Since these saturated sites depend on the partial pressure of H_2S (P_{H2S}), thus the optimum concentration of H_2S is a critical parameter for maintaining the number CUS. The role of

TABLE 7.9
Reaction Conditions for Fixed-Bed Integral Reactors

Conditions	Microflow Reactor	Bench-Scale Reactor
Temperature, °C	380	400
Pressure, MPa	5.4	7.0
Hydrogen flow, l/h	4.6	90
Flow of Maya crude, ml/h	10	100
LHSV, h^{-1}	1.0	1.0
Hydrogen/oil ratio, m^3/m^3	356	891.0
Mode of operation	Upflow	Downflow
Time-on-stream, h	120	200
Catalyst volume, ml (g)	10 (≈8.5)	100 (≈85.0)
Catalyst shape	Cylindrical extrudate	Cylindrical extrudate
Catalyst size, in.	1/16	1/16
Feed composition tested		
Pure Maya crude	—	√
Maya HDT	—	√
HDM feed	√	—
HDS feed	√	—

sulfhydryl (–SH) groups for heavy oil hydroprocessing is confirmed, and it is a source of hydrogen for hydroprocessing reactions (Breysse et al., 2002), which are based on the transfer of hydrogen to reactant. Since heavy oil feedstock is also rich in nitrogen content, poisoning of the active sites also takes place by competitive adsorption. On the other hand, it is also reported that H_2S enhances HDM (Rankel, 1981; Ware and Wei, 1985a, 1985b; Bonné et al., 2001; Rana et al., 2007b). According to Bonné et al. (2001), H_2S coordinates to the central transition metal atom (Ni or V) and as a result weakens the metal–nitrogen bond, while Rana et al. (2007b) proposed that sulfhydryl groups are responsible for the destabilization of the metal–nitrogen bond. It is also expected that the increased presence of –SH groups enhances the cracking of the asphaltene molecule or HDAs (Rana et al., 2007b). Thus, the formation of coke and metal deposits near CUS is expected. The loss of catalyst activity during hydroprocessing of heavy feeds will be caused mainly by coke and metals (Furimsky and Massoth, 1999).

Apart from the heterometal atoms, a major problem in Maya heavy crude oil hydroprocessing is the high content of asphaltenes, which affect the conversion and make more difficult the characterization due to its complex nature (Ancheyta et al., 2001, 2002a, 2002b, 2003a, 2003b, 2004, 2005c; Trejo et al., 2004, 2005b; Trejo and Ancheyta, 2005a). Recently, small-angle neutron scattering characterization of Maya crude asphaltene indicated that a wide range of molecular changes occur with temperature and its fractal network, as shown in Figure 7.6 (Tanaka et al., 2003). Thus, it is a theme of further studies that asphaltene conversion follows catalytic conversion or thermal cracking mechanisms. Comparison of

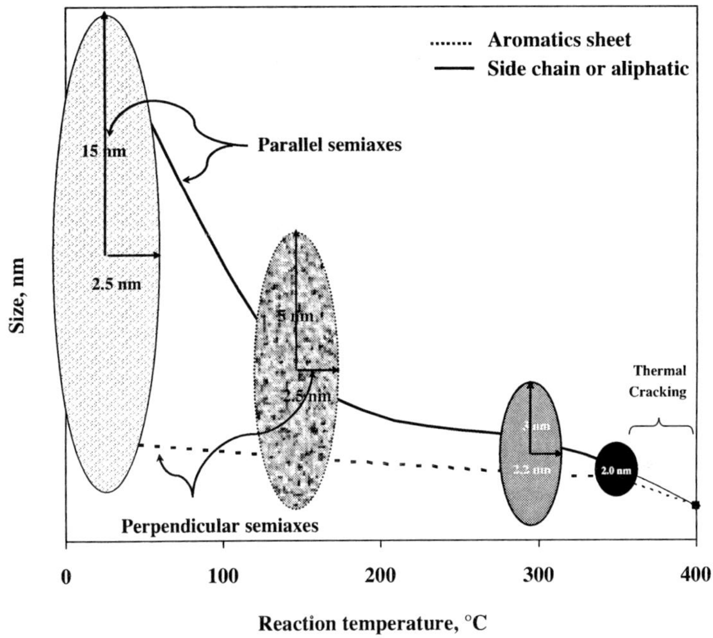

FIGURE 7.6 Temperature dependency of the shape and size of Maya asphaltene. (Data adapted from Tanaka, R. et al., *Energy Fuels*, 17: 127–134, 2003. With permission.)

Maya with other crude asphaltenes such as Iranian light oil and Arabian heavy oil (Khafji) showed that Maya asphaltene is more *refractory* in nature. The decaline precipitated Maya asphaltene showed a fractal network even at 350°C. Thus, thermal cracking starts at around these temperatures, which further indicates the high coke-making tendency of Maya asphaltene. The size of asphaltene in Maya crude is about 2 nm in radius. Moreover, metals (Ni + V) in the asphaltene fraction are believed to be present as organometallic compounds associated in the form of micelles (Figure 7.1). The metal chelate molecules in crude oil, which have been extensively characterized, mostly exist in the form of etioporphyrine; their diameter is approximately 1.6 nm (Fleischer, 1963). Therefore, the required catalyst pore diameter is necessarily higher than 2 nm, to allow for diffusion of these molecules to the catalytic sites.

7.4.4 Naphtha-Diluted Feedstock

Maya crude diluted with desulfurized naphtha was used as feedstock to evaluate catalyst activity by Maity et al. (2003a, 2003b, 2003c) and Rayo et al. (2004). The HDS, HDN, HDM, and HDAs activities of four different catalysts are compared in Figure 7.7. The physicochemical properties of these catalysts are presented in Table 7.6. It was observed that the HDM and HDAs activities are higher for the

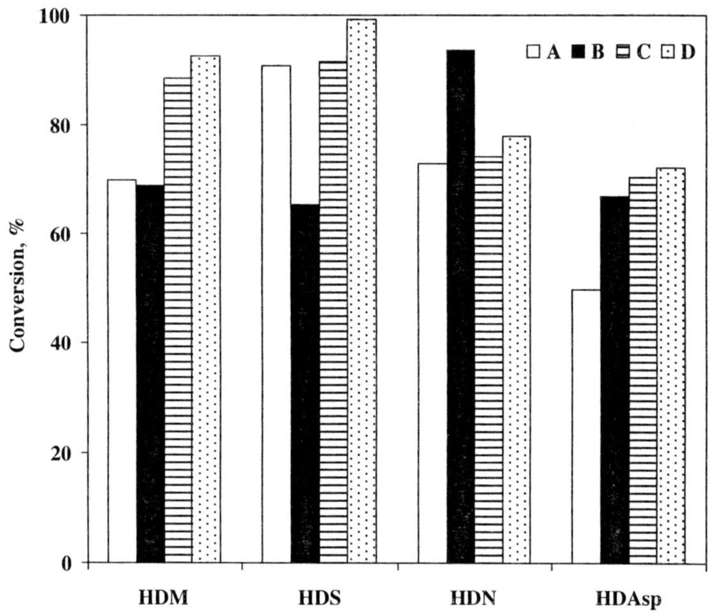

FIGURE 7.7 Effect of catalyst composition on Maya crude/hydrodesulfurized naphtha hydroprocessing: CoMo/AT-1 (A), CoMo/AT-2 (B), CoMo (C), PCoMo (D) supported on Al_2O_3-TiO_2. (Adapted from Maity, S.K. et al., *Appl. Catal. A Gen.*, 250: 231–238, 2003. With permission.)

catalyst having a higher average pore diameter. It was stated that the internal active surface of the catalyst having a bigger pore diameter is accessible to the metal and asphaltene molecules. On the other hand, HDS and HDN activities do not directly correlate with pore diameter. It was suggested that the lack of correlation of HDS and HDN activities with pore diameter was due to the presence of sulfur and nitrogen compounds of significantly different size. One type of sulfur compounds is bigger in size, and these are generally attached to an asphaltene structure. The other sulfur compounds are smaller, like benzo-thiophene and dibenzo-thiophene (Gorbaty et al., 1990; Kelemen et al., 1990; Mullins, 1995). Therefore, the conversion of the sulfur compounds may depend not only on pore diameter but also on the active site of the catalyst.

The effect of phosphorous on Maya crude hydrotreating catalysts was studied (Maity et al., 2003a). Catalysts C and D (Table 7.6) are both supported on the same Al_2O_3-TiO_2 support, but the latter has 0.8 wt% P. Though the P-containing catalyst exhibited high initial activity, a rapid decrease was observed with TOS. Similar results were also found for catalysts supported on alumina (Maity et al., 2003c). It was assumed that P increases acidity of the catalyst, and hence its initial activity. The increase in catalyst acidity also increases coke formation, and therefore catalyst activity decreases rapidly with TOS. Li alkali metal was introduced to the catalyst to reduce acidity. However, the addition of Li did not improve catalyst stability. The synergetic effect of phosphorous, however, is not always the same, and it may depend on the support and preparation method. Though P increased significantly the activities of Al_2O_3- and Al_2O_3-TiO_2-supported catalysts (Maity et al., 2003a, 2003b), it did not show any synergetic effect on the Al_2O_3-SiO_2-supported catalyst. For this catalyst, the effect of promoters (Co or Ni) on hydrotreating activities of Maya crude are compared in Figure 7.8. Catalyst F, promoted by Co, shows higher HDM and HDS activities than catalyst G, promoted by Ni.

The variation of catalyst activity with time-on-stream during hydroprocessing of heavy crude indicated that severe catalyst deactivation takes place, and most probably near the pore mouth. These results were observed by the analysis of metal and coke deposition in the spent catalyst (Maity et al., 2003a, 2003b, 2003c). It is reported in the literature (Stanislaus et al., 1988) that phosphorus decreases the number of strong acid sites, which impede coke formation on the catalyst. Thus, this may be the reason for inhibition of coke formation in P catalysts. Kushiyama et al. (1990) stated that phosphorus may interact with vanadium in heavy feed, preventing deactivation of the catalyst.

The above activity results for Maya crude hydroprocessing over mixed-oxide-supported catalysts indicated high initial activity but showed rapid deactivation

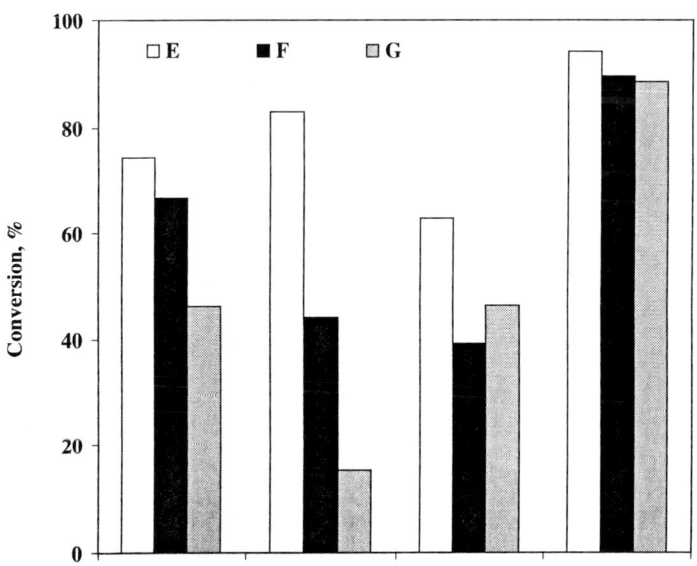

FIGURE 7.8 Effect of Al_2O_3-SiO_2 support preparation and catalyst composition on Maya crude activities. (Adapted from Maity, S.K. et al., *Appl. Catal. A Gen.*, 244: 141–153, 2003. With permission.)

with time. The reason behind the fast deactivation could be explained in two ways. The first is related to the preparation method of the support, which contains high acidity, particularly supports prepared with aluminum sulfate. When sulfate ions remain on the solid, they can produce superactive acid sites during or after the sulfidation, which can have very high conversion, especially asphaltenes at the initial stage, but with time acid sites deactivate, and consequently a drop in conversion is observed. Another reason for the fast catalyst deactivation could be the diluent used (that is, naphtha), which contains a significant amount of paraffins, which induces the precipitation of asphaltenes over the catalyst surface, leading to a continuous drop in catalyst activity. In this regard, a detailed study (Rayo et al., 2004) of the diluent effect (HDS naphtha and HDS diesel) on Maya crude hydrotreating indicated that diluent has an important role in HDS and HDAs conversions, as shown in Figure 7.9 and Figure 7.10, respectively, over NiMo/TiO_2-Al_2O_3-supported catalyst. The authors conclude from their results that naphtha diluent leads to an increase in the poisoning of the catalyst surface by carbon deposition originated by the insolubility of asphaltene in naphtha. These results indicate that careful attention must be paid to the type of diluent used when processing heavy Maya crude. Dilution of Maya crude with hydrodesulfurized diesel avoids this insolubility problem. Thus, this dilemma leads the research group to switch to a Maya crude dilution with hydrodesulfurized diesel.

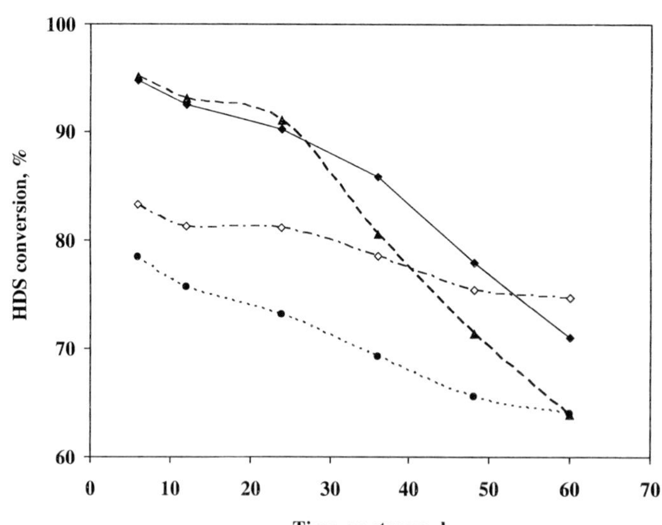

FIGURE 7.9 Effect of diluent (HDS naphtha or HDS diesel) or feed composition on the HDS conversion of Maya crude oil: ▲, NiMo/Al_2O_3 (Maya HDT + naphtha); ◆, NiMo/Al_2O_3-TiO_2 (Maya HDT + naphtha); ◇, NiMo/Al_2O_3-TiO_2 (Maya HDT + diesel);●, NiMo/Al_2O_3-TiO_2 (Maya + diesel). (From Rayo, P. et al., *Catal. Today*, 98: 171–179, 2004. With permission.)

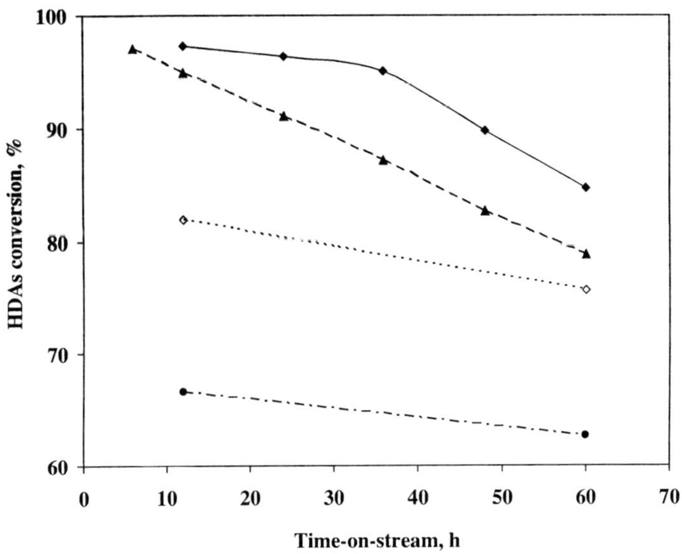

FIGURE 7.10 Effect of diluent on the HDAs conversion of Maya crude oil: ▲, $NiMo/Al_2O_3$ (Maya HDT + naphtha); ◆, $NiMo/Al_2O_3$-TiO_2 (Maya HDT + naphtha); ◇, $NiMo/Al_2O_3$-TiO_2 (Maya HDT + diesel);●, $NiMo/Al_2O_3$-TiO_2 (Maya + diesel). (From Rayo, P. et al., *Catal. Today*, 98: 171–179, 2004. With permission.)

7.4.5 Diesel-Diluted Feedstock

7.4.5.1 Effect of Support Composition

The benefits of using mixed oxides as supports in hydrotreating of light molecules are well recognized. However, their use is not common for heavy oil hydroprocessing due to catalyst stability problems and higher cost than alumina. It has been shown that the support properties are crucial for defining the desirable porosity of catalysts. The advantage of using mixed-oxide supports for hydrotreating is their acid–base properties. This in turn improves hydrocracking activity, which is required for achieving a desirable conversion of asphaltenes and resins to distillate fractions. At the same time, more acidic supports have an adverse effect on other catalyst functionalities. Therefore, the activity of catalysts for hydroprocessing of heavy feeds must be optimized to achieve a desirable level of hydrocracking with considerable activities of HDM, HDS, and HDN. In this regard, literature reports showed that the catalyst functionalities can be modified by incorporating a small amount of TiO_2 into Al_2O_3, using different incorporation methods (Rayo et al., 2004; Rana et al., 2005b; Maity et al., 2005b, 2006; Ramírez et al., 2005). In the case of neutral supports like carbon, a CoMo catalyst was shown to be less sensitive to poisoning by N bases and better able to restrict coke or sediment formation than alumina (Fukuyama et al., 2004); however, their

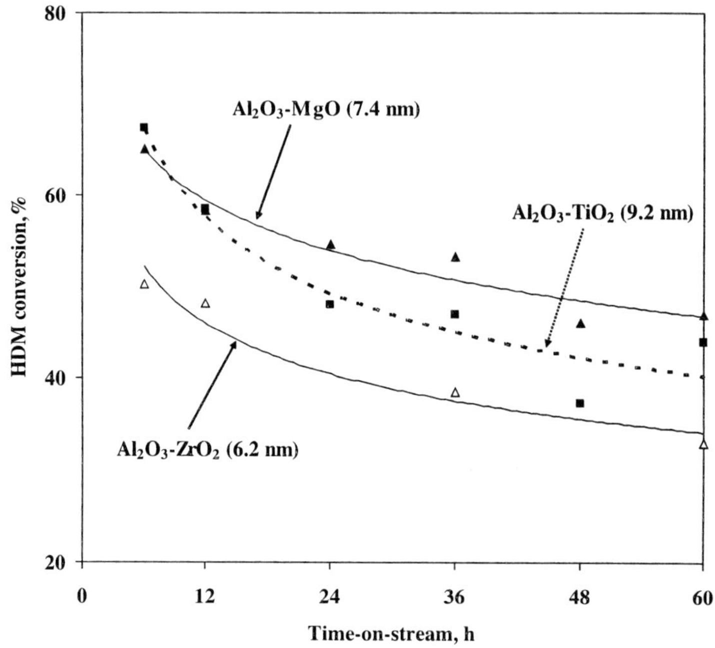

FIGURE 7.11 Effect of the nature of catalyst support on HDM conversion. (From Caloch, B. et al., *Catal. Today*, 98, 91–98, 2004; Rana, M.S. et al., *Catal. Today*, 107/108: 346–354, 2005c. With permission.)

hydrocracking activity is low, which is subject to the unique modification of carbon.

Catalyst activity with Maya crude oil using different-natured supports (acid or base) is reported in Figure 7.11. Initially, the TiO_2-Al_2O_3-supported catalyst was more active due to its higher acidity. However, CoMo catalysts on basic supports (for example, MgO-Al_2O_3, ZrO_2-Al_2O_3) were slightly more stable and more selective for HDM than TiO_2-Al_2O_3-supported catalyst (Caloch et al., 2004; Rana et al., 2005c). The variation in activity was attributed to the small amounts of MgO, ZrO_2, and TiO_2 (7.2, 9.2, and 9.0 wt%, respectively) incorporated into γ-Al_2O_3. On the other hand, the use of catalysts supported on basic supports in hydroprocessing of heavy feeds may be limited because their hydrogenation and hydrocracking activities are low. A comparison of HDS and HDM conversions (120 h TOS) over different TiO_2-Al_2O_3- and alumina-supported CoMo catalysts is shown in Figure 7.12 (Rana et al., 2005b). In all cases, the initial conversion (not shown) was higher for TiO_2-Al_2O_3-supported catalysts than for the alumina-supported catalyst, but the results of initial conversion are not justifiable for heavy oil processing. The TiO_2-containing catalyst showed enhanced HDS activity (120 h TOS), due to the structural promotional effect of TiO_2, while HDM conversion either remained the same as in alumina or became slightly less, which seems to

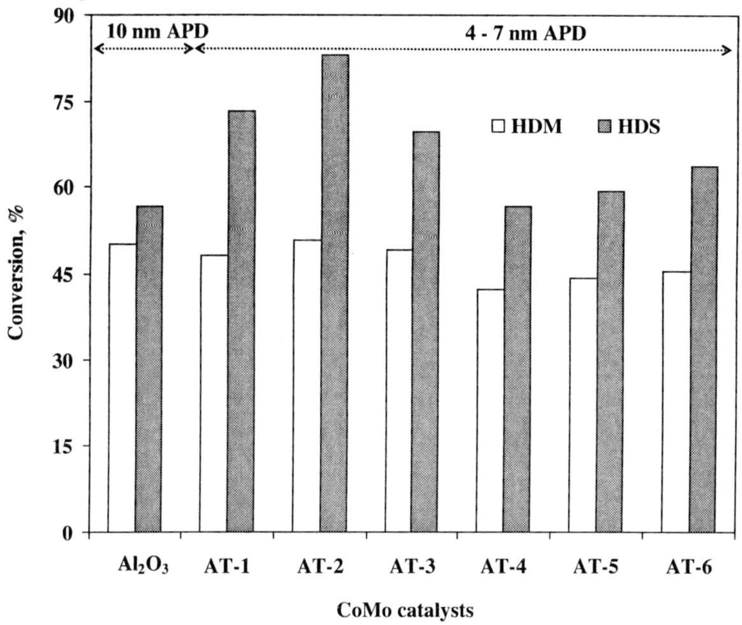

FIGURE 7.12 Comparison between HDS and HDM conversions (120 h TOS) over $CoMo/Al_2O_3$ and TiO_2-Al_2O_3 (AT) catalysts as a function of the TiO_2 (10 wt%) precursor ($TiCl_4$ = AT-1, AT-2, and Ti iso-propoxide = AT-3, AT-4, AT-5, AT-6) and its incorporation method in alumina. (From Rana, M.S. et al., *Catal. Today*, 109: 61–68, 2005. With permission.)

be an effect of average pore diameter (APD) or the stability of the catalysts. The TiO_2-Al_2O_3-supported catalysts have an APD in the range of 4 to 7 nm, while the alumina-supported catalyst has an APD of 10 nm, which might have a corresponding effect on the stability of the catalyst.

7.4.5.2 Effect of Textural Properties

To study the effect of pore diameter on different catalysts for Maya crude hydroprocessing, various aluminas were prepared using different preparation methods to vary the pore diameter. The composition and textural properties of these catalysts along with the support preparation method are reported in Table 7.5 (Rana et al., 2004a). Preparation methods of these supports are also reported in the table. A steady-state conversion comparison for these catalysts is shown in Figure 7.13. The results with a $CoMo/$-Al_2O_3-22.2 catalyst having a pore diameter of 22.2 nm show that pore diameter is not the only factor because activities of this catalyst are comparatively lower (except HDAs) than those of the other catalysts having smaller pores, probably due to the lack of catalytic sites on the low-surface-area (Ca.60 m^2/g) catalyst. Since the composition of all catalysts is

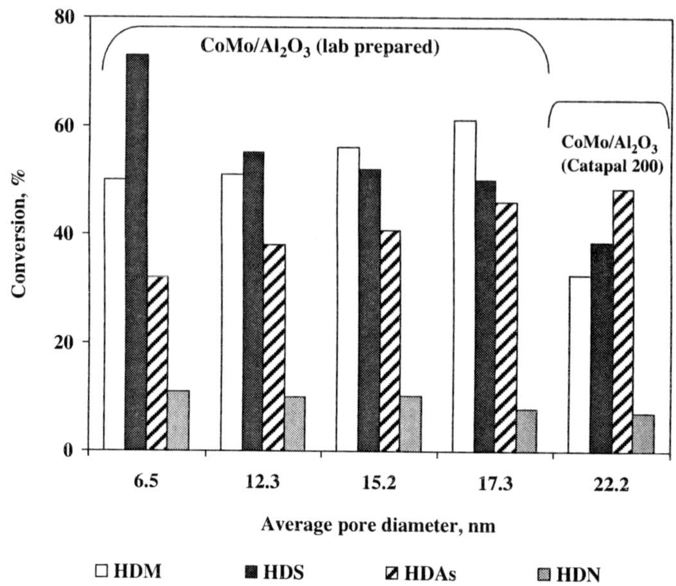

FIGURE 7.13 Effect of average pore diameter (APD) on Maya crude conversion after 60 h TOS (feedstock: Maya + HDS diesel, microreactor). (Adapted from Rana, M.S. et al., *Catal. Today*, 98: 151–160, 2004; Rana, M.S. et al., *Petrol. Sci. Technol.*, 25(1–2): 187–200, 2007c. With permission.)

similar, these results reveal that asphaltene conversion is less affected by the metallic function (CoMo active sites); rather, it is more affected by the pore diameter of the catalyst. The comparison between HDM and HDS shows opposite trends, which reveals that the HDM catalyst should be essentially macroporous in nature (Toulhoat et al., 1990). The Maya crude contains a significant amount of asphaltenes, which are responsible for catalyst deactivation in hydroprocessing along with metal deposition. It is observed that HDM and HDAs significantly depend on the catalyst pore structure, while HDS may depend on the dispersion of the active metal (Eijsbouts et al., 1993). The result in Figure 7.13 indicates that HDS activity distinctively differs from that of HDM and HDAs. Most probably, HDS activity appreciably depends on the surface area and the number of active sites. The $CoMo/Al_2O_3$-6.5 (6.5 nm) catalyst shows the lowest activity for HDM and HDAs, while higher activity can be seen for HDS. Thus, HDM and HDAs conversions are limited due to the diffusional limitations of complex metalloid and asphaltene molecules for the catalysts having pore diameters smaller than 10 nm. Thus, the performance of the heavy oil HDT process with respect to different functionalities, such as HDM, HDS, and HDAs, is clearly linked to the catalyst porosity and nature of the heavy crude oil. Hence, the effect of support preparation on the pore size distribution and average pore diameter apparently controls the catalytic activities along with the metal dispersion of active phases.

7.4.6 Pure Maya Crude Oil

In previous sections catalyst activities were studied with diluted (naphtha or diesel) Maya crude oil and the primary focus was on the screening of catalysts. The results indicated that mixed-oxide-supported catalysts are of great interest for heavy oil hydroprocessing. This might be due to the textural properties and modified acid–base sites. The next step is to study catalyst stability with time-on-stream for prolonged periods using pure Maya crude; to this end a bench-scale reactor (100-ml catalyst) was used. Additionally, to avoid dry sludge formation with prolonged temperature and time-on-stream, or compatibility of asphaltene and maltenes (Takatsuka et al., 1988; Inoue et al., 1998), that is, controlled hydrogenation of asphaltene, it is a good idea to use a two-stage fixed-bed reactor system, where the combination of the two can be effectively used to handle the real feedstock (Ancheyta et al., 2006).

In this respect, an alumina support was prepared by using urea hydrolysis for the first reactor (CoMo, HDM catalyst), while Catapal C1 and TiO_2 (5 wt%, impregnated) were prepared and considered the second reactor catalyst (NiMo, HDS catalyst) (Rayo et al., 2005). The composition and textural properties of these catalysts are given in Table 7.5. Thus, the aim is to compare the behavior of catalysts for Maya crude hydrotreating with adequate stability with time-on-stream at bench-scale reactors (pilot plant). The comparison is made considering a CoMo catalyst with high porosity (HDM) and a NiMo catalyst that is more prone to the deep HDS, and hydrogenating using Al_2O_3 and TiO_2/Al_2O_3 as supports, respectively, as shown in Figure 7.14. Thus, the HDM catalyst (first reactor) is designed with high porosity and low surface area, while the HDS catalyst (second reactor) is characterized by higher surface area and moderate pore size distribution (Gosselink, 1998; Toulhoat et al., 2005).

The results for different activities against TOS are shown in Figure 7.15 and Figure 7.16 for CoMo and NiMo catalysts, respectively. Figure 7.15 shows that all activities marginally decrease with TOS. As expected, the CoMo catalyst

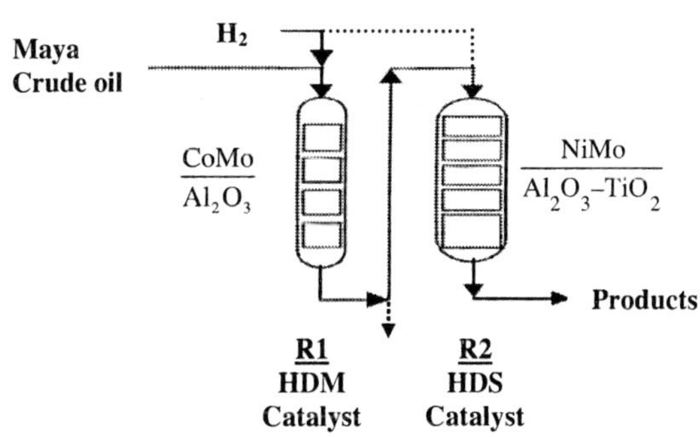

FIGURE 7.14 Schematic diagram of the two-stage reactor for hydrotreating of Maya crude. (From Ancheyta, J. et al., Process for the Catalytic Hydrotreatment of Heavy Hydrocarbons of Petroleum, U.S. Patent, pending, 2006.)

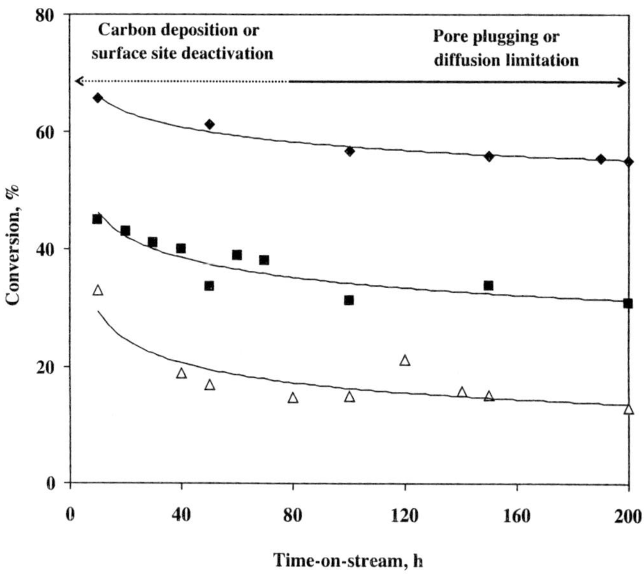

FIGURE 7.15 Catalytic activities of a $CoMo/Al_2O_3$ catalyst with time-on-stream during hydroprocessing of Maya crude with a bench-scale reactor: (◆) HDS, (■) HDM, (Δ) HDAs. (Adapted from Rana, M.S. et al., *Catal. Today*, 109: 24–32, 2005. With permission.)

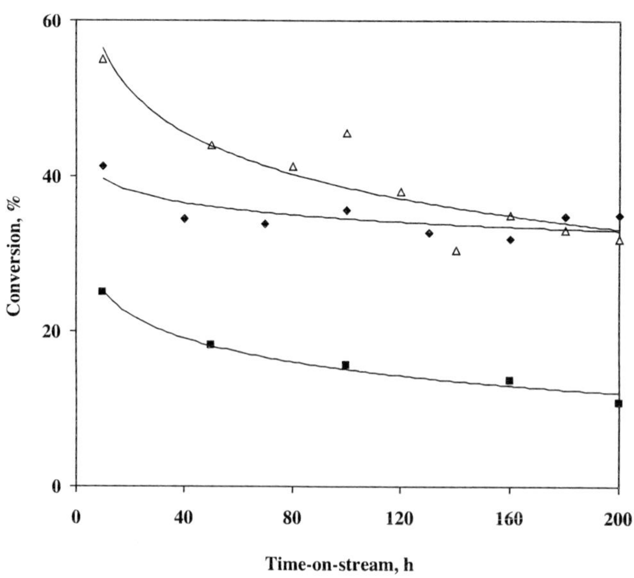

FIGURE 7.16 Catalytic activities of a $NiMo/TiO_2$-Al_2O_3 catalyst with time-on-stream during hydroprocessing of hydrotreated Maya crude with a bench-scale reactor: (◆) HDS, (■) HDM, (Δ) HDAs. (From Rana, M.S. et al., *Catal. Today*, 109: 24–32, 2005.With permission.)

shows much lower activity for the hydrogenation of asphaltenes than for HDM and HDS. However, the decay in activity with time-on-stream for the three reactions is very similar, which indicates that the catalytic sites remain constant or vary similarly with time for all catalytic activities. Surprisingly, we observed a more or less similar deactivation trend for the microflow and bench-scale reactors, while the compositions of feedstocks were different (Rana et al., 2005a). This consistency of activity with time-on-stream could be related to the porosity of the catalyst, in which hydrotreating of heavy oil is directly proportional to the stability of the catalysts (Rana et al., 2004a). On the other hand, the stability of the NiMo catalyst is not similar for all the functionalities, for example, the conversion of HDAs decreases faster than HDS and HDM. Obviously, in this case HDAs activity is higher than that of HDS and HDM, which further reveals that the NiMo catalyst enhances the hydrogenation of asphaltenes. The unusual deactivation for HDAs in the case of NiMo/TiO_2-Al_2O_3 indicates a role of catalyst acidity, which is higher in the case of the NiMo/TiO_2-Al_2O_3 catalyst (Figure 7.5). Thus, the faster deactivation can be understood by the presence of acidic sites generated due to the mixing of TiO_2 and Al_2O_3, which favors the cracking of asphaltene molecules and the subsequent deactivation by coke formation. One more plausible explanation in this respect is the stability for HDS and HDM activities with TOS, which indicates that the metallic sites (CUS) in the hydrotreating catalyst are more stable than the acidic site of the support or active phase (–SH groups). However, catalyst deactivation during heavy oil processing is not only due to the carbon deposition; it is equally possible that metallic sites (CoMoS) have been poisoned by Ni and V deposition during the HDM, although the acidic sites will be deactivated by the coke deposition. The deposition of metals or carbon is most likely at the entrance of the pores, which plugs the pore mouth or restricts diffusion of big, complex asphaltene molecules into the pores (Toulhoat et al., 1990).

7.5 DEACTIVATION AND CHARACTERIZATION OF SPENT CATALYSTS

7.5.1 Catalyst Deactivation

The major concern for heavy oil hydroprocessing catalysts is the deactivation, which decreases catalytic activity with time-on-stream, which means loss of money and time. The expected deactivation is mainly due to the metals (Ni, V, and so on) and coke deposition on the surface of the catalyst (Absi-Halabi et al., 1991; Muegge and Massoth, 1991; Furimsky and Massoth, 1993; Kim and Massoth, 1993; Chu et al., 1994; Ancheyta et al., 2005b; Rana et al., 2005d), making it necessary to raise the severity of the reactor (that is, increase the reactor temperature) to maintain the conversion, and ultimately terminate the operation and replace the catalyst. Thus, catalyst deactivation or catalytic activity loss over time is a continuing concern in heavy oil hydroprocessing. In general, the deactivation is by either the carbon coke or metals deposition over the active sites or at the pore mouth. It is likely that the deactivation by carbon deposition is due to the various types of carbon deposits on the catalyst's surface, which accumulate and adhere to catalyst surfaces during hydrocarbon processing. On the other hand, deactivation by metal sulfides is believed to proceed via blockage of surface active sites or pore plugging. Both coke and metal sulfides contribute to the loss in activity, either burying the catalytic sites or causing diffusion limitations. Therefore, in order to determine the rate and mechanisms of deactivation, it is essential to characterize the spent catalysts for coke and metal deposition. The deactivation can also occur due to the competitive adsorption of organic species on the catalytic sites. It is also well known that spent catalysts contain appreciable amounts of nitrogen compounds, which appear to be strongly adsorbed (Zeuthen et al., 1991). Thus, catalyst deactivation occurs by three simultaneous mechanisms: coking, metals deposition, and substrate interactions. The specific rate of deactivation for each mechanism and each process depends to a large extent on the properties of the catalyst, hydrogen partial pressure, characteristics of the feedstock, and process conditions.

7.5.2 Spent Catalyst Characterization

Spent catalyst characterization showed that carbon is deposited at around 7 to 16 wt% as coke on the catalyst. Metals depend on the feed composition, that is, Ni at around 0.3 wt%, while V is 1.4 wt% after 200 TOS. With variation of feed composition, the metal and carbon deposition varies, as shown in Table 7.10.

TABLE 7.10
Analysis of Spent Catalysts

	wt%						
Catalysts	Mo	Ni	V	C	S	Fe, wppm	S/Mo (mol/mol)
			HDS Feed				
CoMo/-Al_2O_3-ac	4.9	0.05	0.19	11.3	5.19		3.1
NiMo/-Al_2O_3-am	5.4	2.01	0.13	11.0	4.64		2.5
NiMo/TiO_2-Al_2O_3	4.8	3.86	0.06	7.10	4.42		2.72
			HDM Feed				
CoMo/-Al_2O_3-u	5.3	0.11	0.69	13.9	5.03		2.8
CoMo/-Al_2O_3-acs	4.7	0.11	0.65	16.1	6.23		3.9
CoMo/-Al_2O_3-am	5.4	0.08	0.43	10.5	4.45		2.4
CoMo/Al_2O_3	4.7	0.140	0.558	8.66	5.10	201.5	3.25
			Maya HDT Feed				
NiMo/TiO_2-Al_2O_3	3.8	3.8	0.171	7.53	4.30	320.1	3.34
			Maya Feed				
CoMo/Al_2O_3	4.301	0.302	1.393	13.3	3.81	804.5	2.66

Source: Adapted from Rana, M.S. et al., *Catal. Today*, 98: 151–160, 2004; Rana, M.S. et al., *Catal. Today*, 109: 24–32, 2005. (With permission.)

These results are determined over spent catalyst, which are refluxed (washed) with toluene and dried at 200°C. However, the effect of washing was very small, which slightly decreased the carbon and S/Mo ratio, possibly due to the loss of soft coke and the exchange of S atoms with oxygen (Rana et al., 2005b). In a fixed-bed reactor, the metal and carbon deposition over the catalyst is generally affected by the bed length of the reactor; broadly, the maximum deposition of Ni and V occurs at around 15 to 20% bed depth, while the carbon deposition profile is relatively uniform or slightly increases along the reactor depth (Fleisch et al., 1984). The coke and metal contaminants on the spent catalyst also depend on the catalyst, as well as on the scale and type of reactor (Thakur and Thomas, 1985).

Pore volume and pore size distribution are important properties of heavy oil hydroprocessing catalysts. As illustrated in Figure 7.13, HDM and HDAs activities depend on pore diameter, while HDS activity is distinctively different, appearing to be more likely dependent on metal dispersion. Not only catalytic activities, but also metal retention capacity increases with large pore diameter of the catalyst (Siroto et al., 1983; Kressmann et al., 1998). The deposited foulant significantly alters the textural properties of the catalyst, for example, surface area and pore volume decrease by 30 to 50% and 40 to 70%, respectively; this decrease depends on composition of feedstock and total time-on-run. The loss of surface area and pore volume is due to the decrease in micro- and lower meso- (<10 nm) pores, while pores with diameters greater than 25 nm remain unaffected after 200 h TOS, as shown in Figure 7.17. However, carbon deposition increases

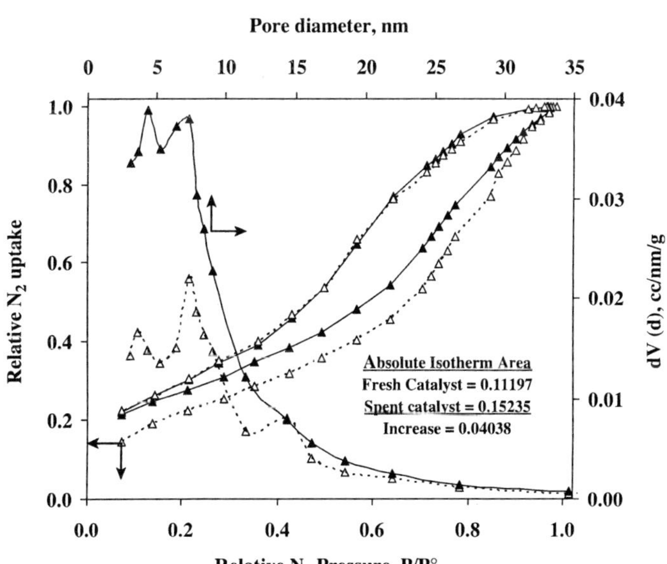

FIGURE 7.17 Comparison of pore size distributions and adsorption–desorption isotherms for $CoMo/Al_2O_3$ fresh (▲) and spent (Δ) catalysts.

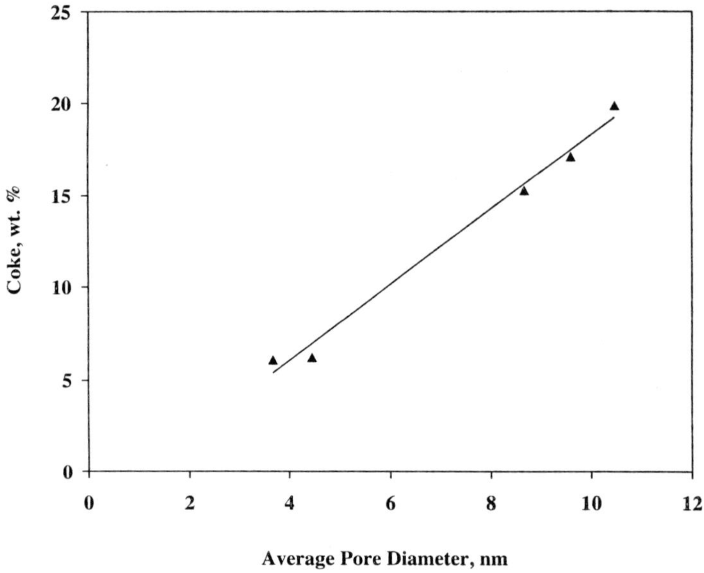

FIGURE 7.18 Relationship between average pore diameter (APD) of fresh catalysts and coke formation (60 h TOS) during Maya crude hydrotreating. (From Maity, S.K. et al., *Appl. Catal. A Gen.*, 244: 141–153, 2003. With permission.)

with the average pore diameter of the fresh catalyst, as shown in Figure 7.18 (Maity et al., 2003a). It is difficult to distinguish the deposited species by characterizing the textural properties, but it is confirmed by several researchers that initial decline in activity is attributed to coke formation (Yoshimura et al., 1991; Thakur and Thomas, 1984, 1985) (Figure 7.15).

Apart from coke deposition, metals such as V, Ni, Fe, and so on, were deposited and modified the surface of the catalyst (CoMoS). Figure 7.19 shows that the presence of deposited Ni and V metals has been verified by their Kα x-ray emission Scanning Electron Microscopy–energy dispersive x-ray microanalysis (SEM–EDAX) line at 7.47 and 4.95 keV, respectively (Rana et al., 2005a). The deposited metals exist as sulfides of V_3S_4, N_3S_2, which is confirmed by using x-ray diffraction (XRD) of spent catalyst, as shown in Figure 7.20 (Rana et al., 2004a). The XRD result indicates that the crystal size of deposited metal sulfide phases is larger than 4 nm. The deposition of the Ni sulfide phase distributed homogeneously, at the inner and outer surfaces of the catalyst. The fact that this was not visible by XRD may be due to the small crystallite size (<4 nm). In contrast, V sulfides were distributed mostly at the outer surface of the extrudate (Rana and Ancheyta, unpublished results). This may be due to the formation of large crystallites, which were detected by XRD. Similar results were observed by other researchers (Kawada et al., 1975; Takeuchi et al., 1985; Toulhoat et al., 1987, 1990; Eijsbouts, 1999). They have reported that vanadium sulfide crystallite size

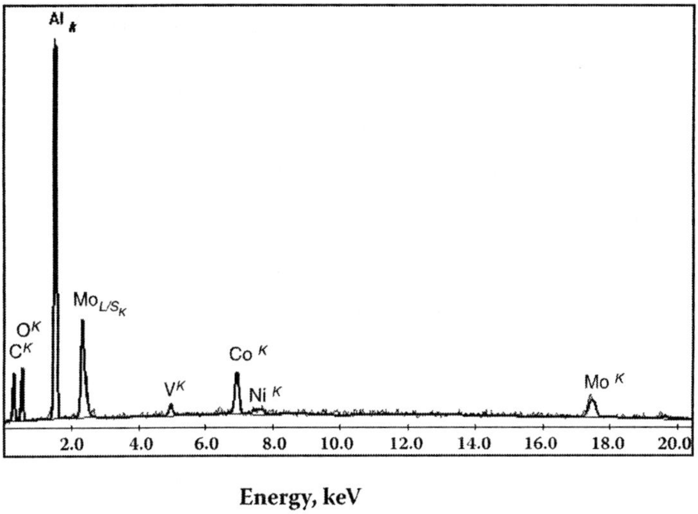

FIGURE 7.19 SEM-EDAX microanalysis of a $CoMo/Al_2O_3$ spent catalyst. (From Rana, M.S. et al., *Catal. Today*, 109: 24–32, 2005. With permission.)

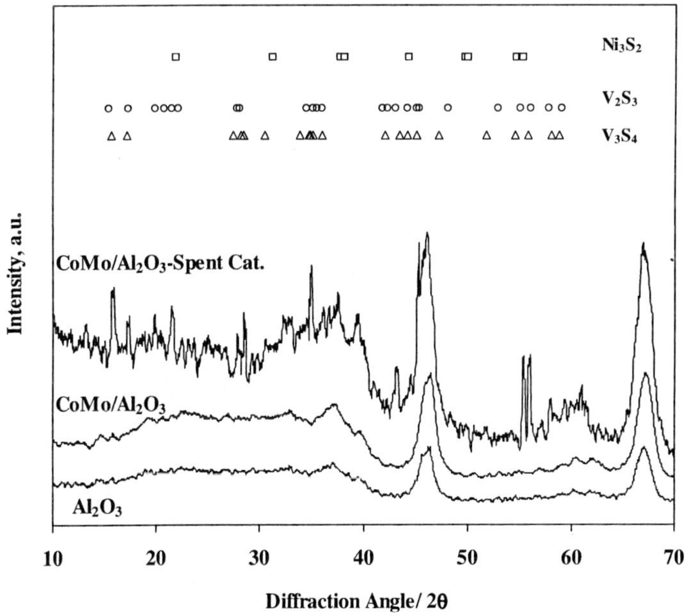

FIGURE 7.20 X-ray diffraction patterns for support, fresh, and spent $CoMo/Al_2O_3$ catalysts. Deposited Ni and V sulfides intensity is compared with JCPDS-ASTM data: (Δ) V_3S_4, (O) V_2S_3, and (□) Ni_2S_3. (From Rana, M.S., *Catal. Today*, 98: 151–160, 2004a. With permission.)

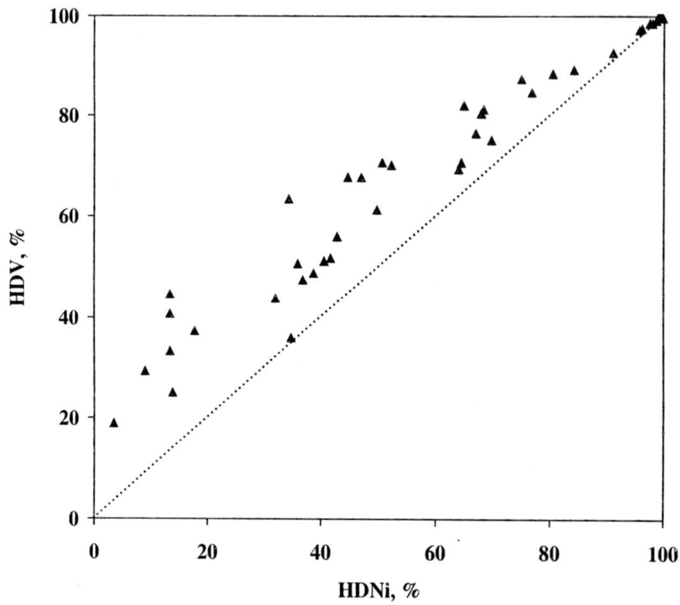

FIGURE 7.21 Selectivity for vanadium and nickel removal over $CoMo/TiO_2$-Al_2O_3 catalysts and Maya + naphtha feedstock. (From Maity, S.K. et al., *Appl. Catal. A Gen.*, 244: 141–153, 2003. With permission.)

ranged from 5 to 30 nm, and that the crystallites grew up perpendicular to the support surface. Detailed spent catalyst spectroscopic characterization studies were reported (Smith and Wei, 1991a, 1991b, 1991c; Rana et al., 2005d) using TEM, EDX, STEM, and x-ray photoelectron spectroscopy (XPS) for model molecules such as nickel etioporphyrin and vanadyl etioporphyrin, and concluded that nickel and vanadium sulfides deposited on HDM catalyst were in the form of relatively large spatially dispersed crystallites. The difference between Ni and V depositions could be due to the more refractory nature of the Ni molecule compared with the V molecule, as shown in Figure 7.21 (Maity et al., 2003a). Since most of the metalloporphyrins are associated with the asphaltene molecules, it is plausible that some of the asphaltene molecules do not diffuse into the catalyst pore. Consequently, their conversion took place at the entrance of the pore mouth due to pore diffusion limitation. Furthermore, these results are confirmed by quantitative analysis of V metal as a function of extrudate radial distribution zones, showing an enrichment of V and carbon in the outer surface of the catalyst particle, as illustrated in the Figure 7.22 (Ancheyta et al., 2002b, 2003a; Rana et al., 2007c).

7.5.3 Stability of Catalysts

Catalyst stability is an important factor that has an immense impact on the catalyst development and is defined by the variation of conversion as a function

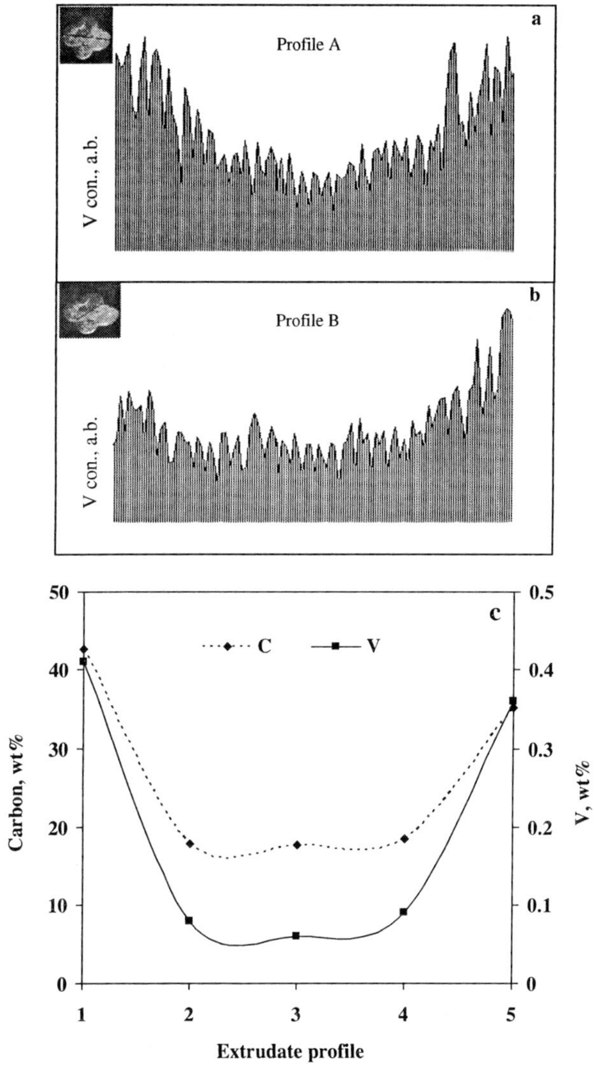

FIGURE 7.22 (a and b) SEM-EDAX analysis of V profiles in tetralobe spent commercial catalyst. (From Ancheyta, J. et al., *Energy Fuels*, 17, 462–467, 2003; Ancheyta, J. et al., *Energy Fuels*, 16, 1438–1443, 2002.) (c) Deposited V metal and carbon concentration profiles obtained in radial distribution of cylindrical extrudate. (From Rana, M.S. et al., *Petrol. Sci. Technol.* 25, 2007c. With permission.)

of time-on-stream. Heavy oil hydroprocessing catalysts finish operation either by pressure drop across the catalyst bed or when the operating temperature reaches its limiting value. The stability usually is affected by the pore diameter of the catalyst. For heavy feedstocks, it is reported that catalyst stability is directly proportional to the total pore volume and mean pore radius (Nomura et al., 1980).

The smaller the pore diameter, the shorter is the life of catalysts. Thus, the stability of the catalyst can be drawn using the equation

$$s = a_1\, V\, r + b_1 \text{ (cc/nm/g)}$$

where s is the catalyst stability, V the total pore volume, r the pore radius, and a_1 and b_1 constants. This proportionality between the stability of catalysts has been confirmed by modeling with pore volume and pore radius. The stability phenomena indicate that catalysts with large pores and pore volume provide easy access and high capacity for inorganic metal and asphaltene molecules. The life of the catalyst is closely dependent on the metal content in the feed, operating conditions, and textural properties of the catalyst, since small pores, severe operating temperatures, and low pressure provoke stronger deactivations (Howell et al., 1985; Pazos et al., 1983). The effect of feed metals level for residuum hydrotreating run length is shown in Figure 7.23 (Howell et al., 1985). Maya atmospheric residue has a sulfur content similar to that of an Arabian heavy, but the former has four times higher metal content. The greater metal contamination reduces the run length by a factor of four due to the pore plugging of the catalyst, which prevents reactive molecule diffusion to the catalytic sites. Another important

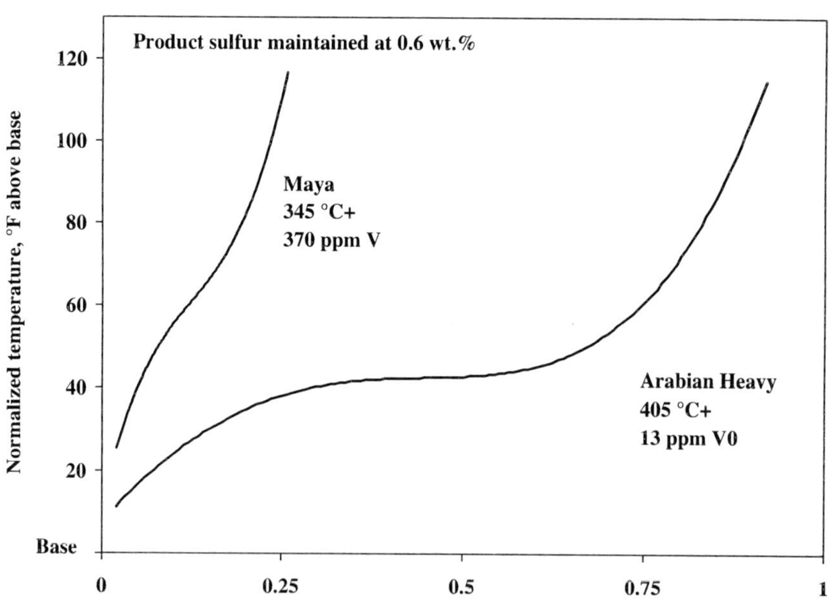

FIGURE 7.23 Effect of feed composition on catalyst run length. (From Howell, R.L. et al., *Oil Gas J.*, 29: 121–128, 1985. With permission.)

parameter for the stability is the temperature of the reactor. At low temperature the deactivation is slow; increasing temperature leads to sediment formation, which has an adverse effect on the catalyst life. Thus, for adequate conversion, textural properties and the shape and size of the catalyst appear to be more important than chemical composition. The stability of the catalyst is also sensitive to the acidic function of the catalyst, which is generally high for mixed-oxide-supported catalysts.

The activity of a catalyst having around 40% HDM conversion does not further decrease with time (Figure 7.15), which indicates that deposited metal sulfides (NixSy or VxSy) may have some role in the catalytic sites, that is, *auto catalysis*. Generally, the metal loading of a heavy oil hydroprocessing catalyst is low, and some parts of the support surface are not covered with the active metals (Mo, Co, or Ni). During hydrodemetallization, metal sulfides coming from the feedstock are deposited on a bare surface of support as part monolayer along MoS_2. It has been reported for regenerated catalysts that nickel and vanadium sulfides inhibit the HDS reaction, while they increase HDM activity (Guibard et al., 2000). In this regard, an enhancement in HDM activity with added H_2S partial pressure (Rana et al., 2007b) may also be attributed to the deposited metals sulfides; those increase the sulfhydryl groups' concentration, which further demonstrates high hydrogenation of the porphyrin ring or hydrodemetallization. Similarly, the sulfur content on spent catalysts slightly increases with average pore diameter; however, the mole ratio of S/Mo shows that the higher the average pore diameter, the greater the S/Mo ratio, which again indicates that greater stability and metal retention capacity are obtained with larger average pore diameters.

7.5.4 Pore Mouth Plugging (Semiquantitative Analysis of Pore Deactivation)

Nitrogen adsorption–desorption isotherms of fresh and spent catalysts are shown in Figure 7.17 and Figure 7.24. An increase in the hysteresis loop of spent catalyst indicates that deactivation has occurred at the pore mouth. In fact, pores of fresh catalysts are considered cylindrical in nature; however, due to the metal and carbon deposition at the pore mouth, these cylindrical pores transform into ink bottle-types of pores in the spent catalyst. The deactivation at the pore mouth is also complementary to a decrease in specific surface area and total pore volume. These deactivated catalyst isotherms display a large increase in hysteresis. The isotherm absolute areas are shown in Table 7.11 and Table 7.12, which were calculated by using the trapezoidal rule integration method. The increase in the isotherm loop and the reduction in total pore volume were considered representative of deactivation via pore mouth coking. These experimental results are also supported by a demetallation diffusion controlled reaction. The larger the pore diameter, the higher the metal retention capacity with even distribution of metal (V and Ni) on a cylindrical-shape catalyst (Tamm et al., 1981). It has also been found that catalysts with large pores have a more even distribution of contaminant metals than small-pore catalysts (Green and Broderick, 1981).

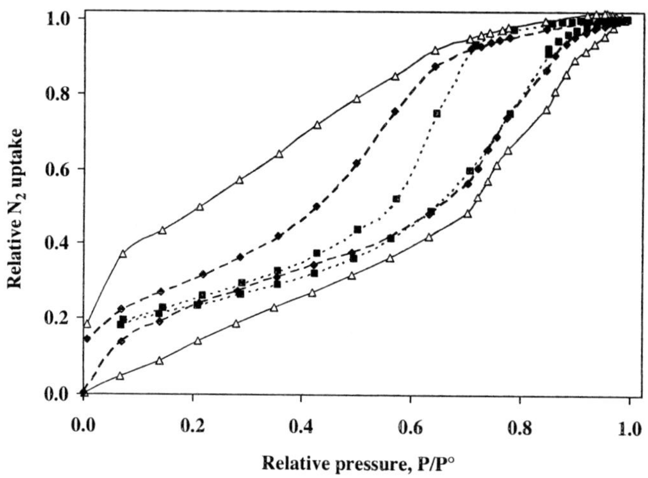

FIGURE 7.24 N_2 adsorption–desorption isotherms of NiMo/Al_2O_3-TiO_2 catalysts: (■) fresh catalyst, (◆) spent catalyst (120 h TOS) microflow reactor, (Δ) spent catalyst (200 h TOS) bench-scale reactor. (From Rana, M.S. et al., *Catal. Today*, 109: 24–32, 2005. With permission.)

TABLE 7.11
N_2 Adsorption–Desorption Measurements of Pore Mouth Plugging

Catalyst	Area Isotherm (Absolute Value)	D/A	D/BET	A/BET
CoMo/Al_2O_3-acs	0.0274	1.0	1.1	1.1
CoMo/Al_2O_3-acs-BU	—	1.3	1.2	0.9
CoMo/Al_2O_3-acs-U	0.0990	1.6	1.5	0.9
CoMo/Al_2O_3-am	0.0564	1.1	1.1	1.0
CoMo/Al_2O_3-am-U	0.4262	1.6	1.9	1.1
TA-1-WW		2.60	4.11	1.58
AT-1		2.41	2.96	1.23
AT-1-R		1.30	1.4	1.07
AT-2		1.37	1.78	1.31
AT-3		1.33	1.69	1.27
AT-4		1.35	1.70	1.20
AT-5		1.23	1.52	1.24
AT-6		1.35	1.39	1.03

Note: A = pore area calculated from BJH adsorption data; D = pore area calculated from desorption data; BET = total BET surface area; AT = Al_2O_3-TiO_2; U = spent catalyst (fixed bed, 60 h TOS); BU = spent catalyst (batch reactor, 6 h TOS); AT-1-WW = catalyst without washing; AT-1-R = catalyst regenerated (spent catalyst, 120 h TOS).

Source: Adapted from Rana, M.S. et al., *Catal. Today*, 109: 61–68, 2005; Rana, M.S. et al., *Catal. Today*, 104: 86–93, 2005. (With permission.)

TABLE 7.12
N_2 Adsorption–Desorption Hysteresis Loop Area Analysis

Catalyst	Absolute Hysteresis Loop Area		
	Fresh Catalyst	Spent Catalyst	% Increase
AT-1-WW		0.3173	19.8
AT-1	0.1191	0.2765	15.7
AT-1-R		0.1809	6.2
AT-2	0.1067	0.2227	11.6
AT-3	0.1144	0.2358	12.1
AT-4	0.1202	0.2502	12.9
AT-5	0.1248	0.2558	13.1
AT-6	0.1254	0.2509	12.5
CoMo/Al_2O_3-MFR	0.0865	0.101	
CoMo/Al_2O_3-BSR		0.139	
NiMo/Ti/Al_2O_2-MFR	0.0685	0.164	
NiMo/Ti/Al_2O_2-BSR		0.338	

Note: AT = Al_2O_3-TiO_2; MFR = microflow reactor; BSR = bench-scale reactor; AT-1-WW = catalyst without washing; AT-1-R = catalyst regenerated.

Source: Adapted from Rana, M.S. et al., *Catal. Today*, 98: 151–160, 2004; Rana, M.S. et al., *Catal. Today*, 104: 86–93, 2005. (With permission.)

We have tried to calculate the BJH adsorption and desorption surface areas to find another explanation for the formation of ink bottle-type pores due to the deposition at the pore mouth. Table 7.11 is the compilation of adsorption–desorption area of the textural properties of fresh and spent catalysts. The adsorption data measure the true pore size distribution, which is not affected by the pore mouth deactivation (Fleisch et al., 1984). Thus, adsorption data are the true values while effective pore distribution is close to desorption values. The constant adsorption/desorption pore area indicated that fresh catalyst pores are cylindrical in nature. On the other hand, the spent catalyst desorption area/BET-SSA ratio increases as pore plugging increases. We conclude that:

1. For D/A close to unity, the pores are cylindrical in nature.
2. For D/BET-SSA greater than 1, more deactivation occurs at the pore mouth.
3. The A/BET-SSA ratio should not be affected by the pore plugging and catalyst deactivation.

Therefore, the BJH desorption area is affected by the geometry of pores, while the adsorption area looks unaffected by the pore geometry and catalyst deactivation. The regenerated catalyst also followed the same principle of deactivation and indicates that around 60% of pore plugging can be recovered at the regeneration conditions used (Rana et al., 2005b).

Indeed, the ink bottle-type of pore is expected to present difficulties to desorb until the relative pressure is reasonably low. It might be possible that adsorbed nitrogen in such a type of pore exists as liquid, which is comparatively difficult to desorb and as a result, the evaporation process takes place progressively rather than abruptly. It is pertinent to mention that the same effect has been also predicted theoretically for tubular structures of low-connectivity samples (Rojas et al., 2002). Therefore, the greater desorption/adsorption (BET-D/BET-A) area can be observed in the case of spent catalysts, and this is more obvious in the case of $CoMo/Al_2O_3$-6.5, in which the APD is 6.5 nm. The difference in adsorption–desorption surface area is another reason to confirm pore mouth plugging. These results further indicate that true pore size distribution should be calculated using adsorption data; however, in the case of effective pore size distribution, desorption data are considered more reliable.

7.6 CONCLUDING REMARKS

The petroleum refining industry has entered into a significant era and important changes have occurred over the past few years. The worldwide crude oil scenario is changing every day with respect to crude oil price as well as origin; many of the world oil reserves are becoming heavy or extra-heavy petroleum. At the same time, the demand for lighter products and middle distillates has grown steadily due to the demand of transportation fuels. The main goal of Maya heavy crude oil hydroprocessing is to prepare *synthetic crude oil* with low metal content. The processes bring down the level of the impurities present in Maya crude, providing additional quantities of better-quality feedstocks for *FCC* and *hydrocracking* processes, which are the principal processes to convert the heavy fraction into the middle distillates or lighter fractions. The second target is to find a catalyst that satisfies the stringent environmental regulations and that has sufficient stability with time-on-stream.

Since catalyst function is crucial for these processes, one should optimize the catalyst properties. A promising catalyst requires porous texture (bimodal porosity) specially designed for large molecules of sulfur, metals, asphaltene, and so forth, along with an optimum number of catalytic sites. The effect of support on the different catalytic activities is an intriguing topic in heavy oil hydroprocessing. In spite of the intensive research knowledge about hydrotreating (model or middle distillate) catalysis, there are several questions that are not conceded for heavy oil processing. This is partly because of the complexity of the feedstock and the varied nature of the supports and limitations of the techniques employed to understand the role of the metals and asphaltene on various activities (hydrogenolysis, hydrogenation, and hydrocracking) that are induced by Mo-, CoMo-, and NiMo-supported systems. Asphaltene, in which major metal tends to concentrate, requires more attention with respect to catalyst pore diameter and its diffusion to active sites. The catalytic activity no doubt depends on the number of catalytic sites, but accessing these sites is a matter of diffusing complex reactant molecules and adsorbing them on the catalytic site. Therefore, an accurate description of the active site under dynamic conditions is necessary.

Knowledge about the nature of active sites in spent catalysts will not only help one to know the role of the support, but also provide clues to preparing improved catalysts. The role of Ti-containing support and the promotional effects are noticeable, but γ-Al_2O_3 remains the main support for heavy oil hydroprocessing catalysts. Understanding the deactivation behavior or catalyst stability is a subject of detailed characterization of spent catalysts.

ACKNOWLEDGMENT

We thank Prof. Jorge Ramírez, UNAM, Mexico, for his suggestions and fruitful discussion during the compilation of this work.

REFERENCES

Absi-Halabi, M., Stanislaus, A., and Al-Zaid, H. 1993. Effect of acidic and basic vapors on pore size distribution of alumina under hydrothermal conditions. *Appl. Catal. A Gen.* 101:117–128.

Absi-Halabi, M., Stanislaus, A., and Trimm, D.L. 1991. Coke formation on catalysts during the hydroprocessing of heavy oils. *Appl. Catal.* 72:193–215.

Agudo, A.L., Llambias, F.J.G., Tascon, J.H.D., and Fierro, J.L.G. 1984. Characterization of sulfided NiMo/-Al_2O_3 catalysts by O_2 and NO chemisorption: influence of the method of preparation. *Bull. Soc. Chim. Belg.* 93:719–726.

Ancheyta, J., Betancourt, G., Centeno, G., and Marroquín, G. 2003a. Catalyst deactivation during hydroprocessing of Maya heavy crude oil. II. Effect of temperature during time-on-stream. *Energy Fuels* 17:462–467.

Ancheyta, J., Betancourt, G., Centeno, G., Marroquín, G., Alonso, F., and Garciafigueroa, E. 2002b. Catalyst deactivation during hydroprocessing of Maya heavy crude oil. I. Evaluation at constant operating conditions. *Energy Fuels* 16:1438–1443.

Ancheyta, J., Betancourt, G., Marroquín, G., Centeno, G., Alonso, F., and Muñoz, J.A. 2006. Process for the Catalytic Hydrotreatment of Heavy Hydrocarbons of Petroleum. U.S. Patent, pending.

Ancheyta, J., Centeno, G., and Trejo, F. 2004. Effects of catalyst properties on asphaltenes composition during hydrotreating of heavy oils. *Petrol. Sci. Technol.* 22:219–225.

Ancheyta, J., Centeno, G., Trejo, F., and Marroquín, G. 2003b. Changes in asphaltene properties during hydrotreating of heavy crudes. *Energy Fuels* 17:1233–1238.

Ancheyta, J., Centeno, G., Trejo, F., Marroquín, G., Garcia, J.A., Tenorio, E., and Torres, A. 2002a. Extraction and characterization of asphaltenes from different crude oils and solvents. *Energy Fuels* 16:1121–1127.

Ancheyta, J., Maity, S.K., Betancourt, G., Centeno, G., Rayo, P., and Gómez-Pérez, Ma.T. 2001. Comparison of different Ni-Mo/alumina catalysts on hydrodemetallization of Maya crude oil. *Appl. Catal. A Gen.* 216:195–208.

Ancheyta, J., Rana, M.S., and Furimsky, E. 2005a. Hydroprocessing of heavy oil fraction. *Catal. Today* 109:1–2.

Ancheyta, J., Rana, M.S., and Furimsky, E. 2005b. Hydroprocessing of heavy petroleum feeds: tutorial. *Catal. Today* 109:3–15.

Ancheyta, J., Trejo, F., Centeno, G., and Speight, J.G. 2005c. Asphaltenes characterization as function of time-on-stream during hydroprocessing of Maya crude. *Catal. Today* 109:162–166.

Armor, J.N. 2005. Do you really have a better catalyst? *Appl. Catal. A Gen.* 282:1–4.

Bartholomew, C.H. 1994. Catalyst deactivation in hydrotreating of residua: a review. In *Catalytic Hydroprocessing of Petroleum and Distillates*, M.C. Oballa and S.S. Shih, Eds. Marcel Dekker, New York.

Benbow, J.J. and Bridgewater, J. 1987. The influence of formulation on extrudate structure and strength. *Chem. Eng. Sci.* 42:753–766.

Benbow, J.J., Oxley, E.W., and Bridgewater, J. 1987. The extrusion mechanics of paste: the influence of paste formulation on extrusion parameter. *Chem. Eng. Sci.* 42:2151–2162.

Bonné, R.L.C., van Steenderen, P., and Moulijin, J.A. 2001. Hydrogenation of nickel and vanadyl tetraphenylporphyrin in absence of a catalyst: a kinetic study. *Appl. Catal. A Gen.* 206:171–181.

Branco, V.A.M., Mansoori, G.A., Xavier, L.C. de A., Park, J.S., and Manafi, H. 2001. Asphaltene flocculation and collapse from petroleum fluids. *J. Petrol. Sci. Eng.* 32:217–230.

Breysse, M., Furimsky, E., Kasztelan, S., Lacroix, M., and Perot, G. 2002. Hydrogen activation by transition metal sulfides. *Catal. Rev. Sci. Eng.* 44:651–735.

Caloch, B., Rana, M.S., and Ancheyta, J. 2004. Improved hydrogenolysis (C-S, C-M) function with basic supported hydrodesulfurization catalysts. *Catal. Today* 98:91–98.

Chu, K.-S., Hanson, F.V., and Massoth, F.E. 1994. Effect of bitumen-derived coke on deactivation of an HDM catalyst. *Fuel Process. Technol.* 40:79–95.

Chuah, G.K., Jaenicke, S., and Xu, T.H. 2000. The effect of digestion on the surface area and porosity of alumina. *Microporous Mesoporous Mater.* 37:345–353.

Eijsbouts, S. 1999. Life cycle of hydroprocessing catalysts and total catalyst management. *Stud. Surf. Sci. Catal.* 127:21–36.

Eijsbouts, S., Heinemann, J.J.L., and Elzerman, H.J.W. 1993. MoS_2 structures in high activity hydrotreating catalysts. II. Evolution of the active phase during the catalyst life cycle deactivation model. *Appl. Catal. A Gen.* 105:69–82.

Falk, J.F. 1964. *Porphyrin and Metalloporphyrins*. Elsevier Publishing Company, Amsterdam, p. 18.

Farrauto, R. and Bartholomew, C. 1997. *Fundamental of Industrial Catalytic Progresses*. Blackie Academic & Professional, London, pp. 181–183.

Fleisch, T.H., Meyers, B.L., Hall, J.B., and Ott, G.L. 1984. Multitechnique analysis of a deactivated resid demetallation catalyst. *J. Catal.* 86:147–157.

Fleischer, E.B. 1963. The structure of Ni etioporphyrine: I. *J. Am. Chem. Soc.* 85:146–148.

Fukuyama, H., Terai, S., Uchida, M., Cano, J.L., and Ancheyta, J. 2004. Active carbon catalyst for heavy oil upgrading. *Catal. Today* 98:207–215.

Furimsky, E. and Massoth, F.E. 1993. Regeneration of hydroprocessing catalysts. *Catal. Today* 17:537–659.

Furimsky, E. and Massoth, F.E. 1999. Deactivation of hydroprocesssing catalysts. *Catal. Today* 52:381–495.

Gary, J.H. and Handwerk, G.E. 1994. *Petroleum Refining, Technology and Economics*, 3rd ed. Marcel Dekker, New York, 1994.

Gorbaty, M.L., George, G.N., and Kelemen, S.R. 1990. Direct determination and quantification of sulphur forms in heavy petroleum and coals. 2. The sulphur K edge x-ray absorption spectroscopy approach. *Fuel* 69:945–949.

Gosselink, J.W. 1998. Sulfide catalysts in refineries. *CatTech* 2:127–144.

Green, D.C. and Broderick, D.H. 1981. Residuum hydroprocessing in the 80s. *Chem. Eng. Prog.* 77:33–39.

Guibard, I., Kressmann, S., Morel, F., Harle, V., and Dufresne, P. 2000. Performance of fresh and regenerated catalyst for residua hydrotreatment. *Hydrocarbon Eng.* 5:54–58.

Howell, R.L., Hung, C., Gibson, K.R., and Chen, H.C. 1985. Catalyst selection important for residuum hydroprocessing. *Oil Gas J.* 121–128.

Huang, Y., White, A., Walpole, A., and Trimm, D.L. 1989. Control of porosity and surface area in alumina. I. Effect of preparation conditions. *Appl. Catal.* 56:177–186.

Inoue, S., Asaoka, S., and Nakamura, M. 1998. Recent trends of industrial catalysts for resid hydroprocessing in Japan. *Catal. Surv. Jpn.* 2:87–97.

Johnson, M.F.L. and Mooi, J. 1968. The origin and types of pores in some alumina catalysts. *J. Catal.* 10:342–354.

Kaloidas, V., Thanos, A.M., Tsamatsoulis, D.C., and Papayannakos, N.G. 2000. Preparation of Al_2O_3 carriers in an integrated mini pilot unit. *Chem. Eng. Process.* 39:407–416.

Kasztelan, S., Payen, E., Toulhoat, H., Gaimblot, J., and Bonnelle, J.P. 1986. Industrial MoO_3-promorter oxide-γ-Al_2O_3 hydrotreating catalysts: genesis and architecture description. *Polyhedron* 5:157–167.

Kawada, I., Onoda, M.N., Ishii, M., and Nakahira, M. 1975. Crystal structures of V_3S_4 and V_5S_8. *J. Solid State Chem.* 15:246–252.

Kelemen, S.R., George, G.N., and Gorbaty, M.L. 1990. Direct determination and quantification of sulphur forms in heavy petroleum and coals. 1. The x-ray photoelectron spectroscopy (XPS) approach. *Fuel* 69:939–944.

Kim, C.-S. and Massoth, F.E. 1993. Deactivation of a Ni/Mo hydrotreating catalyst by vanadium deposits. *Fuel Process. Technol.* 35:289–302.

Kressmann, S., Morel, F., Harlé, V., and Kasztelan, S. 1998. Recent developments in fixed-bed catalytic residue upgrading. *Catal. Today.* 43:203–215.

Kushiyama, S., Aizawa, R., Kobayashi, S., Koinuma, Y., and Uemasu, I. 1990. Effect of addition of sulphur and phosphorus on heavy oil hydrotreatment with dispersed molybdenum-based catalysts. *Appl. Catal.* 63:279–292.

Le Page, J.-F., Cosyns, J., Courty, P., Freund, E., Franck, J.-P., Jscquin, Y.J., Marcilly, B.C., Martino, G., Miquel, J., Montarnal, R., Sugier, A., and Landeghem van H. 1987. *Applied Heterogeneous Catalysis, Design, Manufacture, Use of Solid Catalysts.* Technip, Paris.

López-Salinas, E., Espinosa, J.G., Hernández-Cortez, J.G., Sánchez-Valente, J., and Nagira, J. 2005. Long-term evaluation of NiMo/alumina–carbon black composite catalysts in hydroconversion of Mexican 538°C+ vacuum residue. *Catal. Today* 109:69–75.

Maity, S.K., Ancheyta, J., Alonso, F., and Rana, M.S. 2004. Preparation, characterization and evaluation of Maya crude hydroprocessing catalysts. *Catal. Today* 98:193–199.

Maity, S.K., Ancheyta, J., and Rana, M.S. 2005a. Support effects on hydroprocessing of Maya heavy crude. *Energy Fuel* 19:343–347.

Maity, S.K., Ancheyta, J., Rana, M.S., and Rayo, P. 2005b. Effect of phosphorus on activity of hydrotreating catalyst of Maya heavy crude. *Catal. Today* 109:42–48.

Maity, S.K., Ancheyta, J., Rana, M.S., and Rayo, P. 2006. Alumina-titania mixed oxide used as support for hydrotreating catalysts of Maya heavy crudes: effect of support preparation methods. *Energy Fuels* 20:427–431.

Maity, S.K., Ancheyta, J., Soberanis, L., and Alonso, F. 2003b. Alumina-silica binary mixed oxide used as support of catalysts for hydrotreating of Maya heavy crude. *Appl. Catal. A Gen.* 250:231–238.

Maity, S.K., Ancheyta, J., Soberanis, L., and Alonso, F. 2003c. Catalysts for hydroprocessing of Maya heavy crude. *Appl. Catal. A Gen.* 253:125–134.

Maity, S.K., Ancheyta, J., Soberanis, L. Alonso, F., and Llanos, M.E. 2003a. Alumina-titania binary mixed oxide used as support of catalysts for hydrotreating of Maya heavy crude. *Appl. Catal. A Gen.* 244:141–153.

Mansoori, G.A. 1997. Modeling of asphaltene and other heavy organic depositions. *J. Petrol. Sci. Eng.* 17:101–111.

Marroquín, G., Ancheyta, J., and Díaz, J.A.I. 2004. On the effect of reaction conditions on liquid phase sulfiding of a NiMo HDS catalyst. *Catal. Today* 98:75–81.

McCulloch, D.C. 1983. Catalytic hydrotreating in petroleum refining. In *Applied Industrial Catalysis*, B.E. Leach, Ed. Academic Press, New York.

Mehrotra, R.P. 2006. What we need to consider for a successful development of catalysts. *Bull. Catal. Soc. India* 5:33–41.

Muegge, B.D. and Massoth, F.E. 1991. Basic studies of deactivation of hydrotreating catalysts with anthracene. *Fuel Process. Technol.* 29:19–30.

Mullins, O.L. 1995. Sulfur and nitrogen molecular structures in asphaltenes and related materials quantified by XANES spectroscopy. In *Asphaltenes: Fundamentals and Applications*, E.Y. Sheu and O.C. Mullins, Eds. Plenum Press, New York, chap. II, pp. 53–96.

Nomura, H., Sekido, Y., and Oguchi, Y. 1980. Hydrodesulfurization reactions of residual oils. Part 3. Effects of various factors on catalyst life. *Sekiyu Gakkaishi* 23:321–327.

Ono, T., Ohguchi, Y., and Togari, O. 1983. In *Preparation of Catalyst*, Vol. III, G. Poncelet, P. Grange, and P. Jacob, Eds. Elsevier, Amsterdam, p. 631.

Papayannakos, N.G., Thanos, A.M., and Kaloidas, Y.E. 1993. Effect of seeding during precursor preparation on the pore structure of alumina catalyst supports. *Microporous Mesoporous Mater.* 1:423–430.

Pazos, J.M., Gonzalez, J.C., and Salazar-Gullen, A.J. 1983. Effect of catalyst properties and operating conditions on hydroprocessing high metals feeds. *Ind. Eng. Chem. Process Des. Dev.* 22:653–659.

Pecoraro, T.A. and Chianelli, R.R. 1981. Hydrodesulfurization catalysis by transition metal sulfides. *J. Catal.* 67:430–445.

PEMEX http://www.pemex.com/index.cfm?action=content&sectionID=11&catID. (Petroleos Mexicanos). Accessed 2007.

PEMEX http://en.wikipedia.org/wiki/Pemex#Reserves. Accessed 2007.

Quann, R.J., Ware, R.A., Hung, C.-W., and Wei, J. 1988. Catalytic hydrodemetallation of petroleum. *Adv. Chem. Eng.* 14:95–259.

Ramírez, J., Rayo, P., G.-Alejandre, A., Ancheyta, J., and Rana, M.S. 2005. Analysis of the hydrotreatment of heavy Maya petroleum with NiMo catalysts supported on TiO_2-Al_2O_3 binary oxides. Effect of the incorporation method of Ti, Ni and Mo. *Catal. Today* 109:54–60.

Rana, M.S., Ancheyta, J., Maity, S.K., and Rayo, P. 2005b. Maya crude hydrodemetallization and hydrodesulfurization catalysts: an effect of TiO_2 incorporation in Al_2O_3. *Catal. Today* 109:61–68.

Rana, M.S., Ancheyta, J., Maity, S.K., and Rayo, P. 2005d. Characteristics of Maya crude hydrodemetallization and hydrodesulfurization catalysts. *Catal. Today* 104:86–93.

Rana, M.S., Ancheyta, J., Maity, S.K., and Rayo, P. 2007a. Hydrotreating of Maya crude oil. II. Generalized relationship between hydrogenolysis and hydrodeasphaltenization (HDAs). *Petrol. Sci. Technol.*, 25(1–2): 201–214.

Rana, M.S., Ancheyta, J., Maity, S.K., and Rayo, P. 2007c. Hydrotreating of Maya crude oil. I. Effect of support composition and its pore-diameter on asphaltene conversion. *Petrol. Sci. Technol.*, 25(1–2): 187–200.

Rana, M.S., Ancheyta, J., and Rayo, P. 2005a. A comparative study for heavy oil hydroprocessing catalysts at micro-flow and bench-scale reactors. *Catal. Today* 109:24–32.

Rana, M.S., Ancheyta, J., Rayo, P., and Maity, S.K. 2004a. Effect of alumina preparation on hydrodemetallization and hydrodesulfurization of Maya crude. *Catal. Today* 98:151–160.

Rana, M.S., Ancheyta, J., Rayo, P., and Maity, S.K. 2007b. Heavy oil hydroprocessing over supported NiMo sulfided catalyst: an inhibition effect by added H_2S. *Fuel*, 86(9): 1263–1269.

Rana, M.S., Huidobro, M.L., Ancheyta, J., and Gómez, M.T. 2005c. Effect of support composition on hydrogenolysis of thiophene and Maya crude. *Catal. Today* 107/108:346–354.

Rana, M.S., Maity, S.K., Ancheyta, J., Murali Dhar, G., Prasada Rao, T.S.R. 2004c. Cumene cracking functionalities on sulfided Co(Ni)Mo/TiO_2-SiO_2 catalysts. *Appl. Catal. A Gen.* 258:215–225.

Rana, M.S., Navarro, R., and Leglise, J. 2004b. Competitive effects of nitrogen and sulfur content on activity of hydrotreating CoMo/Al_2O_3 catalysts: a batch reactor study. *Catal. Today* 98:67–74.

Rana, M.S., Ramirez, J., Alejandre, G.A., Ancheyta, J., Cedeño, L., and Maity, S.K. 2007d. Support effects in CoMo hydrodesulfurization catalysts prepared with EDTA as a chelating agent. *J. Catal.* 246:100–108.

Rana, M.S., Srinivas, B.N., Maity, S.K., Murali Dhar, G., Prasada Rao, T.S.R. 2000. Origin of cracking functionality of sulfided (Ni) CoMo/SiO_2-ZrO_2 catalysts. *J. Catal.* 195:31–37.

Rankel, L.A. 1981. Reactions of metalloporphyrins and petroporphyrins with H_2S and H_2. *Am. Chem. Soc. Prepr. Div. Petrol. Chem.* 26:689–698.

Rayo, P., Ancheyta, J., Ramírez, J., and G.-Alejandre, A. 2004. Hydrotreating of diluted Maya crude with NiMo/Al_2O_3-TiO_2 catalysts: effect of diluent composition. *Catal. Today* 98:171–179.

Rayo, P., Ancheyta, J., Ramírez, J., Maity, S.K., Rana, M.S., and Martínez, F.A. 2005. Composicion catalitica mejorada para la hidrodesulfuracion de residuos y crudos pesados. Mexican Patent PA/E/2005/066614, pending.

Redey, A., Goldwasser, J., and Hall, W.K. 1988. The surface chemistry of molybdena-alumina catalysts reduced in H_2 at elevated temperatures. *J. Catal.* 113:82–95.

Riazi, M.R. 2005. *Characterization and Properties of Petroleum Fractions*, 1st ed., ASTM manual series. ASTM, p. 156.

Rojas, F., Kornhauser, I., Felipe, C., Esparza, J.M., Cordero, S., Domínguez, A., and Ricardo, J.L. 2002. Capillary condensation in heterogeneous mesoporous networks consisting of variable connectivity and pore-size correlation. *Phys. Chem. Chem. Phys.* 4:2346–2355.

Siroto, Y., Ono, T., Asaoka, S., and Nakamura, M. 1983. Catalyst for Hydrotreatment of Heavy Hydrocarbon Oils Containing Asphaltene. U.S. Patent 4,422,960.

Smith, B.J. and Wei, J. 1991a. Deactivation in catalytic hydrodemetallation. I. Model compound kinetic studies. *J. Catal.* 132:1–20.

Smith, B.J. and Wei, J. 1991b. Deactivation in catalytic hydrodemetallation. II. Catalyst characterization. *J. Catal.* 132:21–40.

Smith, B.J. and Wei, J. 1991c. Deactivation in catalytic hydrodemetallation. III. Random-spheres catalyst models. *J. Catal.* 132:41–57.

Stanislaus, A., Abasi-Halabi, M., and Al-Dolama, K. 1988. Effect of phosphorus on the acidity of γ-alumina and on the thermal stability of γ-alumina supported nickel-molybdenum hydrotreating catalysts. *Appl. Catal.* 39:239–253.

Takatsuka, T., Wada, Y., Hirohama, S., and Komatsu, S. 1988. More Insights into the *in-situ* Liquid Phase Equilibrium in the ABC Processes. Paper presented at AIChE Annual Meeting, Washington, DC, November 27–December 2, 1988.

Takeuchi, C., Asaoka, S., Nakata, S., and Shiroto, Y. 1985. Characteristics of residue hydrodemetallation catalyst. *Am. Chem. Soc. Prepr. Div. Petrol. Chem.* 30:96–107.

Tamm, P.W., Harnsberger, H.F., and Bridge, A.G. 1981. Effects of feed metals on catalyst aging in hydroprocessing residuum. *Ind. Eng. Chem. Process Des. Dev.* 20:262–273.

Tanaka, R., Hunt, J.E., Winans, R.E., Thiyagarajan, P., Sato, S., and Takanohashi, T. 2003. Aggregates structure analysis of petroleum asphaltenes with small-angle neutron scattering. *Energy Fuels*, 17:127–134.

Thakur, D.S. and Thomas, M.G. 1984. Catalyst deactivation during direct coal liquefaction: a review. *Ind. Eng. Chem. Prod. Res. Dev.* 23:349–360.

Thakur, D.S. and Thomas, M.G. 1985. Catalyst deactivation in heavy petroleum and synthetic crude processing: a review. *Appl. Catal.* 15:197–225.

Tischer, R.E. 1981. Preparation of bimodal aluminas and molybdena/alumina extrudates. *J. Catal.* 72:255–265.

Tischer, R.E., Narain, N.K., Stiegel, G.J., and Cillo, D.L. 1985. Large-pore Ni-Mo/Al_2O_3 catalysts for coal-liquids upgrading. *J. Catal.* 95:406–413.

Topsøe, H. and Clansen, B.S. 1984. Importance of CoMoS type structure in hydrodesulfurization. *Catal. Rev. Sci & Eng.* 26:395-420.

Topsøe, H., Clausen, S., and Massoth, F.E. 1996. *Hydrotreating Catalysis Science and Technology.* Springer-Verlag, New York.

Topsøe, N.-Y. and Topsøe, H. 1993. FTIR studies of Mo/Al_2O_3-based catalysts. II. Evidence for the presence of SH groups and their role in acidity and activity. *J. Catal.* 139:641–651.

Toulhoat, H., Hudebine, D., Raybaud, P., Guillaume, D., and Kressmann, S. 2005. THERMIDOR: a new model for combined simulation of operations and optimization of catalysts in residues hydroprocessing units. *Catal. Today*, 109:135–153.

Toulhoat, H., Plumail, J.C., Houpert, C., Szymanski, R., Bourseau, P., and Muratet, G. 1987. Modelling RDM catalyst deactivation by metal sulfides deposit: an original approach supported by HRTEM investigation and pilot test results. *Am. Chem. Soc. Prepr. Div. Petrol. Chem.* 32:463–464.

Toulhoat, H., Szymanski, R., and Plumail, J.C. 1990. Interrelations between initial pore structure, morphology and distribution of accumulated deposits, and lifetimes of hydrodemetallization catalysts. *Catal. Today* 7:531–568.

Trejo, F. and Ancheyta, J. 2005a. Kinetics of asphaltenes conversion during hydrotreating of Maya crude. *Catal. Today* 109:99–103.

Trejo, F., Ancheyta, J., Centeno, G., and Marroquín, G. 2005b. Effect of hydrotreating conditions on Maya asphaltenes composition and structural parameters. *Catal. Today* 109:178–184.

Trejo, F., Centeno, G., Ancheyta, J., and Marroquín, G. 2004. Precipitation, fractionation and characterization of asphaltenes from heavy and light crude oils. *Fuel* 83:2169–2175.

Trimm, D.L. 1980. *The Design of Industrial Catalysts*. Elsevier/North-Holland, Amsterdam.

Trimm, D.L. and Stanislaus, A. 1986. The control of pore size in alumina catalyst supports: a review. *Appl. Catal.* 21:215–238.

Walendziewski, J. and Trawczynski, J. 1993. Preparation of large pore alumina supports for hydrodesulfurization catalysts. *Appl. Catal. A Gen.* 96:163–174.

Walendziewski, J. and Trawczynski, J. 1994. Influence of the forming method on the pore structure of alumina supports. *Appl. Catal. A Gen.* 119:45–58.

Ware, R.A. and Wei, J. 1985a. Catalytic hydrodemetallation of nickel porphyrins. II. Effects of pyridine and of sulfiding. *J. Catal.* 93:122–134.

Ware, R.A. and Wei, J. 1985b. Catalytic hydrodemetallation of nickel porphyrins. III. Acid-base modification of selectivity. *J. Catal.* 93:135–151.

White, A., Walpole, H.Y., and Trimm, D.L. 1989. Control of porosity and surface area in alumina. II. Alcohol and glycol additives. *Appl. Catal.* 56:187–196.

Yoshimura, Y., Endo, S., Yoshitomi, S., Sato, T., Shimada, H., Matsubayashi, N., and Nishijima, A. 1991. Deactivation of hydrotreating molybdate catalysts by metal deposition. *Fuel* 70:733–739.

Zeuthen, P., Blom, P., and Massoth, F.E. 1991. Characterization of nitrogen on aged hydroprocessing catalysts by temperature-programmed oxidation. *Appl. Catal.* 78:265–276.

Zmierczak, W., Murali Dhar, G., and Massoth, F.E. 1982. Studies on molybdena catalysts. XI. Oxygen chemisorption on sulfided catalysts. *J. Catal.* 77:432–438.

8 Effect of Feedstock Composition on the Performance of Hydroconversion Catalysts

Esteban López-Salinas and Jaime S. Valente

CONTENTS

8.1 INTRODUCTION

Heavy oils are defined as residual fractions from atmospheric or vacuum distillation of petroleum, and certain ultra-heavy crude oils, produced mainly in Venezuela, Mexico, and Canada. The typical properties common to these heavy oils are high specific gravities, low H/C ratios, high carbon residues, and high concentrations of asphaltenes, S, N, V, and Ni. To illustrate these conditions, Table 8.1 shows examples of properties of vacuum residua from crude oils of indicated countries. The yields of residua from these crude oils are high and typically below 10° API. It is well known that heavy oils with asphaltenes and metals contents above 5% and 150 wt-ppm, respectively, will generate coke and metal deposits on catalyst beds and will cause serious catalyst deactivation and plugging problems.

TABLE 8.1
Properties of Various Petroleum Vacuum Residua

Feedstock	Yield on Crude (vol%)	Gravity (°API)	H/C Atomic Ratio	Asphaltenes (wt%)	CCR (wt%)	S (wt%)	N (wt%)	V/Ni (wt-ppm)
Middle East								
Kuwait	21	7.4	1.48	4.9	19.8	4.78	0.41	103/25
Gach Saran	32	8.8	1.49	6.1	17.5	3.27	0.62	227/72
Khafji	31	4.5	1.42	12.6	23.1	5.55	0.48	152/50
Basrah Heavy	37	3.1	1.38	16.1	25.6	6.12	0.40	211/71
Venezuela								
Tia Juana	33	8.7	1.48	12.5	21.3	3.00	0.60	669/66
Bachaquero	59	6.8	1.45	10.3	19.3	3.34	0.65	614/82
Orinoco	57	3.0	1.40	16.9	36.2	4.31	0.99	625/147
Canada								
Lloydminster	43	3.4	1.42	13.8	23.9	5.72	0.62	230/117
Mexico								
Maya	45	4.2	1.41	18.1	24.5	4.99	0.78	487/100
Russia								
Arlanskaya	35	5.9	1.44	7.8	20.5	4.53	0.72	335/127

Note: CCR = carbon Conradson.

Residue hydroconversion is becoming increasingly important in the world today because of several market and economic factors. The heavy oil demand continues to decline, while at the same time there is an increasing demand for motor fuels. On the other hand, the crude oil production of light crude oils is declining, and thus the relative availability of heavy crude oils is expected to increase. The residue of heavy crude oils can be upgraded through various existing processes (Gray, 1994). The most deleterious components of heavy crude oils are asphaltenes and metals (for example, V and Ni), which concentrate in oil residue.

The typical catalysts in fixed-bed or ebullated-bed processes are CoMo or NiMo sulfides (in their final condition) supported on γ-alumina, containing 11 to 14 wt% Mo and 2 to 3 wt% Ni or Co. The alumina support has a total pore volume between 0.5 and 0.9 cm^3/g, and the catalyst is formed as extrusion pellets in shapes such as cylinders (ca. 1 to 2 mm diameter), tri- or tetralobed cylinders, or rings. The catalyst pellets must have high mechanical strength (typically $\geq$ 0.7 kg/mm) and attrition resistance in order to withstand pressure from stacking in a fixed-bed reactor or turbulent and collision conditions within an ebullated-bed reactor.

This chapter focuses on the effect of the various residue components on the performance of the catalysts used in fixed-bed or ebullated-bed processes.

8.2 ASPHALTENES

Asphaltenes constitute a general class of polyaromatic- and polyheteroaromatic-type substances, defined on the basis of their solubility. Asphaltenes are soluble in toluene, but insoluble in light alkanes such as n-pentane and n-heptane. Asphaltene molecules carry a core of stacked, flat sheets of condensed (fused) aromatic rings linked at their edges by chains of alipathic or naphthenic-aromatic ring systems. The condensed sheets contain NSO atoms and probably vanadium and nickel complexes. Great effort has been invested to reveal the complex molecular structure of asphaltenes. A hypothetical representation of an asphaltene molecule is presented in Figure 8.1.

The present trend in the petroleum industry is an increasing demand for light products. In order to meet the market demand, refineries convert a portion of their residuals into light fractions. This conversion results in the production of modern heavy fuels that contain a greater concentration of sulfur, vanadium, and asphaltenes.

Asphaltenes are considered part of the "bottom of the barrel." They constitute the nonvolatile, high molecular weight fraction of petroleum. In addition, since they are nonsoluble in heptane, they remain in solid form in the crude as well. The condensed aromatic rings exist in the form of nonhomogeneous flat sheets.

In the crude, the asphaltene sheets remain dispersed. However, they have the tendency to be attracted toward each other, thus resulting in the formation of an agglomeration. The structure of the agglomeration is similar to that of a book: a compact stack of thin sheets.

Asphaltenes, which exist in crudes in a dispersed state, are held in this condition by resins. They have three characteristics that make them problematic to a refinery system:

1. They constitute the largest aromatic fraction in petroleum as well as being the highest molecular weight component.
2. They have no definite melting point and therefore remain in solid form, thus contributing to carbon residue.
3. They agglomerate to form a book-like structure.

Compared with the large molecular size of such heavy hydrocarbon fractions, ranging from 2.5 to 15 nm (Dai et al., 1990; Ruckenstein and Tsai, 1981; Pfeiffer and Saal, 1990; Yen et al., 1961), the average pore diameter of the supported hydrotreating catalyst, residual oil treatment over a supported catalyst is strongly influenced by pore diffusion limitation, resulting in poor performance.

The asphaltene fraction in vacuum residua concentrates most of the heteroatoms and metals, as shown in Table 8.2.

In hydroconversion processes, such as LC-Fining and H-Oil, where NiMo or CoMo catalysts supported on alumina are used, it is crucial that the pore network

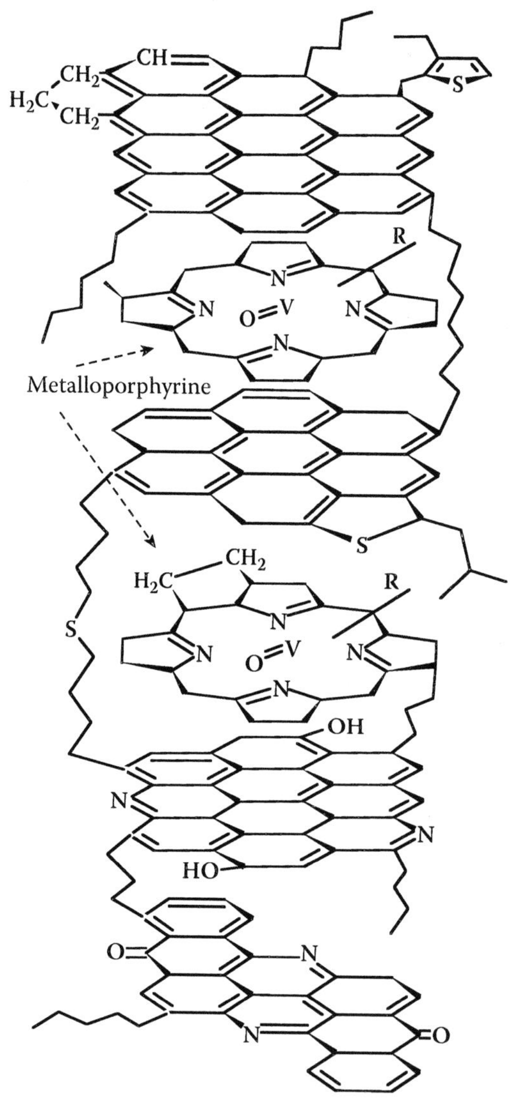

FIGURE 8.1 Hypothetical molecular structure of asphaltenes.

of the support offers the maximum accessibility to asphaltenes. To achieve high lighter hydrocarbon fractions, asphaltenes should diffuse into the pore channels and then reach the active sites. The first active sites where an asphaltene molecule touches in a pore system are in the pore mouth, where plugging by metal or coke deposition occurs. Since asphaltenes are huge molecules, their effective diffusion coefficient in the pores of a catalyst is usually small, on the order of 10^{-7} cm^2/sec.

TABLE 8.2
Properties of Asphaltenes from Various Petroleum Vacuum Residua

Feedstock	H/C Atomic Ratio	Average Molecular Weight	CCR (wt%)	S (wt%)	N (wt%)	V/Ni (wt-ppm)
Khafji	1.10	5300	58.7	7.54	0.81	608/123
Bachaquero	1.11	2700	54.2	4.42	1.72	2880/388
Lloydminster	1.17	4200	48.9	7.47	1.33	624/317
Maya	1.09	3800	53.6	6.92	1.71	1750/347
Arlanskaya	1.09	2700	54.7	6.30	1.48	1410/500

Note: CCR = carbon Conradson.

Hence, reaction rates of asphaltenes conversion and demetallation could be controlled by the pore diffusion of reacting substances (Shimura et al., 1982).

During hydroprocessing, asphaltenes undergo a multitude of reactions involving both cracking and hydrogenation, which change their initial structure (Zou and Liu, 1994; Bartholdy and Andersen, 2000; Tojima et al., 1998; Buch et al., 2000; Seki and Kumata, 2000; Callejas and Martinez, 2000; Kodera et al., 2000; Ancheyta et al. 2003; Bartholdy et al., 2001). The reaction chemistry is affected by the hydroprocessing conditions, with temperature being one of the main parameters (Zou and Liu, 1994; Bartholdy and Andersen, 2000; Tojima et al., 1998; Buch et al., 2000; Seki and Kumata, 2000; Callejas and Martinez, 2000). Accordingly, below 370°C the reaction chemistry is hydrogenation dominated, whereas above 380°C cracking reactions predominate, causing a rupture of alkyl side chains and cracking of naphthenes (Zou and Liu, 1994). The transition between these two reactions was found in the range of 370 to 390°C (Bartholdy et al., 2001). However, transition temperature may depend on the feedstock composition. For instance, Ancheyta et al. (2003) have reported that cracking of an asphaltene molecule is dominant at temperatures above 420°C (Ancheyta et al., 2003). Hydrocracking of asphaltenes in different feedstocks indicates that the number of unit sheets decreases as conversion increases, but the structure of the main unit remains unchanged (Zou and Liu, 1994). On the other hand, aromaticity of asphaltenes increases along conversion, and the average molecular weight of asphaltene falls (Zou and Liu, 1994). The increase in aromaticity and drop of molecular weight have been attributed to dealkylation rather than to any modification in the aromatic rings (Zou and Liu, 1994; Bartholdy and Andersen, 2000; Seki and Kumata, 2000; Buenrostro-Gonzalez et al., 2001).

In a recent study in the hydroconversion of a Middle East vacuum residue (asphaltene content, 13.2 wt%) using a NiMo/γ-Al_2O_3 catalyst and near H-Oil conditions, it was found that 89% conversion of asphaltenes can be achieved, indicating that unconverted asphaltenes dissociate into smaller aggregates at about

50% conversion (Merdrignac et al., 2006). The remaining asphaltenes were more aromatic than asphaltenes in the feedstock, which is due mainly to dealkylation (Merdrignac et al., 2006).

8.3 SOLUBILITY VARIATIONS

Another aspect that greatly limits asphaltene's transformation into lighter fractions is its metastable solubility. Asphaltenes in solution exhibit self-assembly and colloidal behavior, depending on the operation conditions. Several models have been proposed in the literature to characterize such complex structures (Pfeiffer and Saal, 1990; Yen et al., 1961). Molecular associations are initiated by aromatic region staking (by four or five units) to form the elementary particles, which further self-associate to form colloidal aggregates. Hydrogen and π–π bonds are the main interactions involved in such association mechanisms. The archipelago model describes asphaltenes as polyaromatic molecules linked by alkanes (Murgich et al., 1996).

Given the complex chemical structure of asphaltenes, any modification in their constitutional groups is very likely to alter their solubility, and precipitation may occur within the catalyst pore system. At higher processing temperatures, cracking becomes dominant, yielding asphaltene molecules with lower solubility. Concomitantly, the chemical structure of maltenes becomes more aliphatic, and thus they are poor solvents for asphaltenes. The chemical composition of asphaltenes affects the stability of hydroprocessed oil (Speight, 2000) Asphaltenes with low hydrogen content and high aromaticity aggregate at lower concentrations than do asphaltenes with a high hydrogen content and low aromaticity (León et al., 2000).

It is believed that self-aggregation of asphaltenes is the first step in the formation of particles precipitating from oil (Andersen and Birdi, 1991). The most reliable model for the interaction between asphaltenes and the lighter oil fractions accounts for both solubility and colloidal condition (Park et al., 1994). In this model, asphaltenes are partly dissolved and partly in a colloidal state. Resins reduce the self-aggregation tendency of asphaltenes, and thus are important for solubilization.

Variations in solubility within asphaltene fractions have been recognized. The soluble and insoluble fractions of the asphaltene show neither uniform chemical compositions nor constant molecular weight (Anderson and Speight, 1999). The insoluble portion has a strong tendency to aggregate. Fractionated asphaltenes, from hydroprocessed oil, have been separated into two different solubility classes, indicating that the concentration of the heavy and less soluble fraction increases with the processing temperature, while the stability of the product decreases (Tojima et al., 1998). Groenzin and Mullins (2003) reported that the solubility of asphaltene subfractions affects the distribution of the molecules. Asphaltenes with a broader chemical composition and molecular weight are more stable than narrowly distributed fractions (Wang and Buckley, 2003). Even though only the least soluble asphaltenes precipitate at the flocculation onset, chemical transformations occur all over asphaltene molecules (Rogel et al., 2003).

8.4 COKE FORMATION

Asphaltenes are known to be directly associated with coke formation (Lulic et al., 1990). Catalysts deactivate very fast with time in the first contact period, where up to 25 wt% of coke is laid and the active surface area reduces by about 50% (Gary and Handwerk, 1984). Deactivation and coking then slow down, but significant quantities of metals and coke are deposited on the catalyst (Gary and Handwerk, 1984). The initially high rate of coking appears to be associated with the adsorption of asphaltenes on the most acidic sites of the catalysts (Absi-Halabi et al., 1991). The enhanced adsorption arises from interaction with nitrogen groups attached to the asphaltene molecules (Furimsky, 1978).

The coke deposited on the catalysts contains several components, which have been characterized by solubility (Mochida et al., 1978). For instance, coke generated upon a catalytic hydrocracking of vacuum residue was found to be made up of 86% hexane-soluble fraction, 7% hexane-insoluble, benzene-soluble fraction, 2% benzene-insoluble, tetrahydrofuran-soluble fraction, and 5% tetrahydrofuran-insoluble fraction. The hexane-soluble fraction consisted essentially of long-chain alkanes and alkylbenzenes, whereas the hexane-insoluble fractions were aromatic and polar. These results suggest that at least part of the coke was produced through hydrocracking of resins and light asphaltenes, resulting in lower solubility capacity and sedimentation of asphaltenes. Egiebor et al. (1989) arrived at similar interpretations after characterizing organic residues on hydroprocessing spent catalysts by ^{13}C-NMR. There are also indications that condensation–hydrogenation reactions may be involved in coke formation, yielding mesophase crystals (Beuther and Berrotta, 1990). Thus, coke can derive from pyrolitic carbon, asphaltene fragments precipitated from solution or colloids, and heavier molecules formed by condensation reactions. At least a portion of the coke deposit can be withdrawn from the catalyst by solvent extraction, differing from pyrolitic carbon deposited on other catalysts.

Both thermal and catalytic coke formation occurs between 400 and 430°C (Dautzenberg and De Deken, 1985). Dealkylation of asphaltenes takes place at somewhat lower temperatures, to leave large aromatic structures that deposit as coke or coke precursors (Abdul Latif, 1990).

Minimization of coke deposits involves various factors. Acidity is important in coke formation, but it is also necessary in hydrotreating catalysts; this should be carefully balanced. Another way to ameliorate asphaltene-derived coke deposits is by solvent deasphalting or by adding suitable solvents. Asphaltenes are not the only source of coke, but they are the major contributor (Absi-Halabi et al., 1991). Mochida et al. (1989) suggested the use of 1-methyl-naphthalene as a solvent for asphaltenes.

Considering the bulkiness of asphaltene molecules or their aggregates and the gradual coke accumulation that obstructs the pore network in the catalyst, optimal pore size and pore size distribution are crucial for development of an efficient catalyst for hydroprocessing of oil residua. Moreover, and given the wide dispersion of molecular weights (5000 ~ 10,000 daltons) that characterize asphaltenes, a wide pore size distribution may be advisable. Richardson and Alley (1975) showed that asphaltenes in oil at 200 ~ 600°F are excluded by catalysts with pore diameters smaller than 7 nm. An optimal pore size of about 25.0 nm is known to minimize deactivation (Trimm, 1990). However, as the pore size increases, the surface area decreases, and so does the number of active sites. There is a compromise between the increase in activity due to an increase in surface area and the decrease due to diffusional resistances. A NiMo/γ-alumina catalyst, where γ-alumina was prepared as fibers with 25.2 nm pore size and wide pore size distribution, showed better activity for hydrodemetallization (HDM) than did a commercial one, in the hydroconversion of an atmospheric oil resid (Ying et al., 1985). Catalysts based on fibrillar alumina show both high intrinsic activity for HDM and relatively long catalyst lifetime (Ying et al., 1997).

A catalyst for hydroprocessing heavy residual oil must have a pore size distribution designed so that it is not easily plugged and the rate of reaction is not controlled by the pore diffusion of the reacting compound (for example, huge asphaltene molecules). However, when a catalyst is manufactured with larger pore size, the catalyst crushing strength declines significantly, particularly when using alumina as a support (Knudsen, 1959).

An effective commercial catalyst must have a specific pore structure for processing heavy residual oil and the mechanical strength for practical use of the same. A proprietary asphaltene conversion catalyst was proposed in 1979 containing sepiolite, characterized by a specific chemical composition, pore structure, and crushing strength that alumina catalysts do not have (Takeuchi et al., 1983). The resistance of sepiolite catalysts to coke deposition is better than that of alumina catalysts, and it has excellent capability for asphaltene cracking and metal removal (Takeuchi et al., 1983). A possible explanation may be that sepiolite is made up of Si and Mg atoms where acidity is weak, thus being unfavorable for carbonization reactions in comparison with those in alumina. Inoue et al. (1998) reported that sepiolite and pH-swing-prepared alumina combined as a catalyst support for conventional NiMo active metals are good candidates for hydroconversion of heavy residue. A different approach was taken by combining boehmite and carbon black to obtain a bimodal carbon–alumina composite support impregnated with conventional Ni and Mo compounds (López-Salinas et al., 2005). These bimodal catalysts, with ca. 20% of total pore volume in the macroporous region and containing between 8 and 18 wt% carbon black, showed good side crushing strength (0.7 to 0.9 kg/mm), in spite of the well-known fragility of carbon. The preparation of these composite supports afforded bimodal catalysts with about 78 and 22% mesopores and macropores, respectively, of the total pore volume, as shown in Figure 8.2. Accordingly, the incorporation of carbon diminishes the number or strength of acid sites in alumina, which in turn, as a NiMo catalyst, resulted in lower sediments and carbon Conradson formation than a commercial bimodal alumina catalyst.

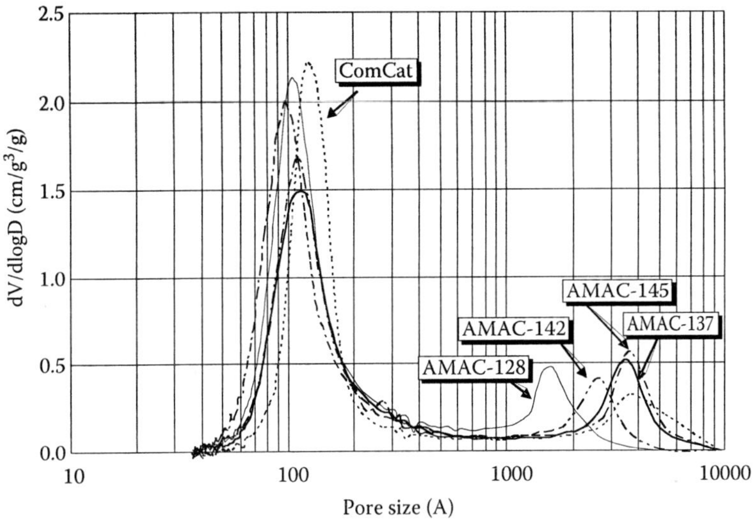

FIGURE 8.2 Pore size distribution of aluminas modified with carbon black (AMAC) and a commercial catalyst (comCat), measured by Hg porosimetry. (From López-Salinas, E. et al., *Catal. Today*, 109: 69, 2005. With permission.)

8.5 VANADIUM AND NICKEL

The metal-containing compounds in heavy feedstocks vary from a few parts per million to more than 1000 ppm. They induce many problems in petroleum refining; for instance, the catalysts employed in several processes are rapidly poisoned. Metals in fuel oils produce ash when the oil is burned. Ash deposits in engines result in abrasion of the moving parts of the engines, and the ash is injurious to the walls of the boilers and furnaces (Ali and Abbas, 2006). The common metals found are sodium, potassium, calcium, strontium, lithium, copper, silver, vanadium, manganese, tin, lead, cobalt, titanium, gold, chromium, and nickel. They are generally in combination with naphthenic acid as soaps and in the form of complex organometallic compounds such as metalloporphyrins (Vokovic, 1978; Yen, 1975).

Among these metals the most abundant and undesirable are vanadium and nickel. Depending on the origin of crude oil, the concentration of the vanadium ranges from 0.1 to 1200 ppm, while nickel commonly varies from trace to 200 ppm. Vanadium and nickel are thought to occur in petroleum in two forms: porphyrinic and nonporphyrinic. The porphyrins have been extensively studied, not only because of their detrimental effects but also for their considerable role as geochemical markers (Ali et al., 1993). Vanadium, nickel, and iron are sometimes known as heavy metals. The heavy metals in crude oil residua are agglomerated in asphaltenes in the form of phorphyrin compounds, as shown in Figure 8.3.

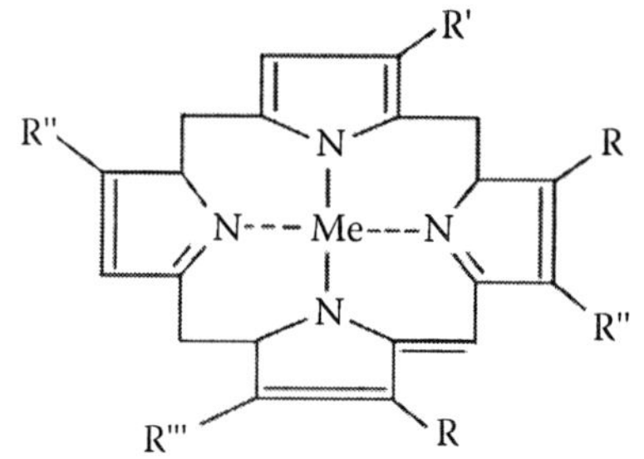

FIGURE 8.3 Phorphyrin compounds, Me: vanadium, nickel, or iron. (From Ali, M.F. and Abbas, S., *Fuel Process. Technol.*, 87: 573, 2006. With permission.)

The molecular weight of these metalloporphyrins varies between 420 and 520μ (Simanzhenkov and Indem, 2003).

In fact, the processing of heavy crude oils containing large proportions of these substances is becoming more important. Since the metals tend to accumulate in the residuum during distillation and affect its properties adversely, several methods have been developed for their removal.

Catalysts for heavy oil upgrading operate under severe process conditions that require the catalyst to be active for heteroatom removal while undergoing deposition of relatively large amounts of carbon and feed metals (Zeuthen et al., 1995). The catalyst's efficiency is estimated in terms of its activity and selectivity and its lifetime. Catalyst life may be short or long, but inevitably, the catalyst will need regeneration or replacement (Trimm, 1997). The three main reasons for catalyst deactivation are poisoning, fouling, or sintering. Loss of catalyst material by formation and volatilization of components is also occasionally observed. Deactivation caused, for example, by fouling with coke may be reversible; deactivation caused by sintering is usually irreversible (Trimm, 1997).

The deactivation's kinetics could be expressed by the following generalized power rate law (Thomas and Thomas, 1967; Bartholomew, 1984):

$$\frac{-dc}{dt} = kA_s\,[reactants,\ products]^{\propto}\eta$$

where k is the rate constant for the reaction, A_s is the surface area, is the order of reaction, and η is the effectiveness factor. Both k and η reflect the chemical interactions between the catalyst and the reactants and products. If there is a reaction between a component from the gas phase and the catalyst, then the chemical nature of the catalyst changes and both k and $\propto$ can be expected to change. Fouling results in physical blocking of the active surface of pores (Bartholomew, 1984). As a result, the main effect will be on the active surface area and the effectiveness factor. Sintering will also affect these variables, but thermal degradation as a result of, for example, solid–solid reactions can also change the nature of the catalyst and affect k and $\propto$ (Trimm, 1997). Although catalyst deactivation is an important phenomenon, it has not always received the attention that it deserves. In considering catalyst deactivation, it is useful to review the causes of deactivation, to minimize it. The safe disposal of spent catalysts should be considered, since these may contain substantial amounts of heavy metals (Trimm, 1991).

Coke formation and catalyst fouling are major problems in some liquid-phase processes, particularly those used for hydrotreating heavy oils (Absi-Halabi et al., 1991). The accumulation of coke and metals on a heavy oil hydrotreating catalyst can be very significant without necessarily having a very adverse effect on activity (Absi-Halabi et al., 1991; Rostrup-Nielsen and Trimm, 1977). However, in most cases, accumulation results in loss of activity related to surface coverage, which is a consequence of the pore blockage (Bartholomew, 1984). In order to obtain higher yields of transport fuels, it is essential to reduce the size of molecules in heavy crude oil and remove unwanted species, such as Ni and V mainly. Physical and chemical methods have been used for this end. The physical method is essentially deasphalting. In this process, the lighter oils are physically separated from heavier asphaltenes by mixing the heavy oil/residue with a very low-boiling solvent such as propane, butane, or isobutene. The chemical method includes thermal processes such as visbreaking and coking and chemical treatment. The thermal processes are based on the reshuffling of the hydrogen distribution in the residue to produce lighter products containing more hydrogen, while the asphaltenes and metals are removed in the form of coke or visbreaking residue (Ali and Abbas, 2006).

Catalytic hydroprocessing of heavy oils is a hydrogenation process used to remove compounds containing nitrogen, sulfur, oxygen, or metals from liquid petroleum fractions. A reduction in the amount of metals in the oil is accomplished by the process of hydrodemetallization (HDM), where the molecules that contain metals lose these atoms by reactions of hydrogenation.

Catalysts for treating residues are selected based on activity and cost. High-activity hydrogenation catalysts for upgrading usually consist of mixtures of cobalt and molybdenum sulfides (CoMo), nickel and molybdenum sulfides (NiMo), and nickel and tungsten sulfides (NiW), all supported on alumina or carbon (Weisser and Landa, 1973) and formed into pellets or beads. The main role of catalysts, in primary upgrading, is to enhance the uptake of hydrogen and prevent condensation and coking reactions (Miki et al., 1983).

Catalysts deactivate with time in the manner shown in Figure 8.4. During the first deactivation, up to 25% by weight of coke is deposited and the active surface area reduces by about 50% (Thakur and Thomas, 1984). Deactivation and coking then slow down, but during the slow deactivation, significant quantities of metal and coke are deposited on the catalyst (Thakur and Thomas, 1984).

The coke deposited can involve pyrolytic carbon, asphaltene fragments precipitated from the solution, and heavier molecules produced by the subsequent condensation reactions of asphaltenes. Thus, the effect of coking on catalyst

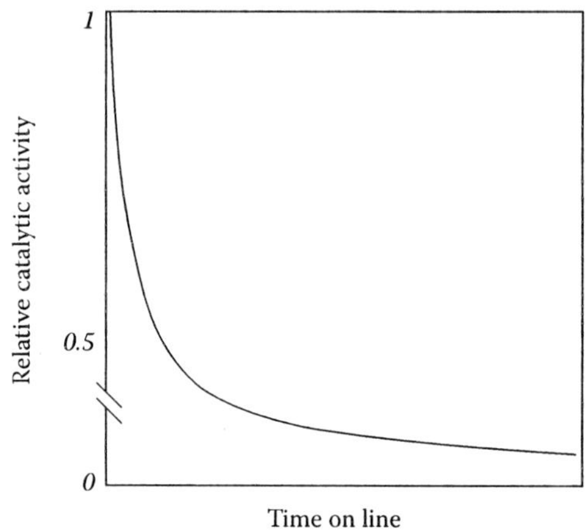

FIGURE 8.4 Catalytic activity vs. time on line.

deactivation is due to deposits coating active surfaces and blocking pores. In the case of catalytic hydroprocessing, however, the catalysts are also contaminated by metals salts originating from HDM. Catalyst deactivation appears to result more from coking than from metals, but the metals foul the surface. The evidence is that sulfur and vanadium exist in oils, resins, and asphaltenes; whereas nickel and nitrogen tend to be deposited evenly throughout the pellet, vanadium is concentrated near the pellet exterior (Tamm et al., 1981). Iron is also deposited on the pellet exterior, but tends to originate more from process equipment than from the feed. The relatively small effect on deactivation due to metal deposition can be explained by comparing the deposits with catalytic metals. One deposit is nickel, often present in the original catalyst (Weisser and Landa, 1973). Another is vanadium, known to have at least some catalytic activity in the sulfide form. The catalyst choice used for HDM is more dependent on the porosity of the support than on the activity of the catalyst itself. Since textural properties play an important role in the lifetime of the catalyst, an intended porosity allows function to continue with high loadings of vanadium.

Distribution profiles of the deposited metals (V and Ni) within spent catalyst pellets, measured by x-ray electron microprobe analysis, show that V and Ni concentrations across the pellet cross section are higher around the edges of the pellet (external surface) than in the center (Pereira et al., 1990; Absi-Halabi et al., 1996; Martínez et al., 1997; Al-Dalama and Stanislaus, 2006). A typical M-shaped profile indicates that metals tend to deposit at the exterior of the extrudates, as shown in Figure 8.5. The concentration of Ni, in comparison with that of V, tends to be more uniform across the pellet (Al-Dalama and Stanislaus, 2006). The differences in concentration profiles can be explained in terms of differences in reactivities and diffusivities of the precursor molecules bearing these metals. Accordingly, Ni-containing molecules in residual oil diffuse more rapidly within

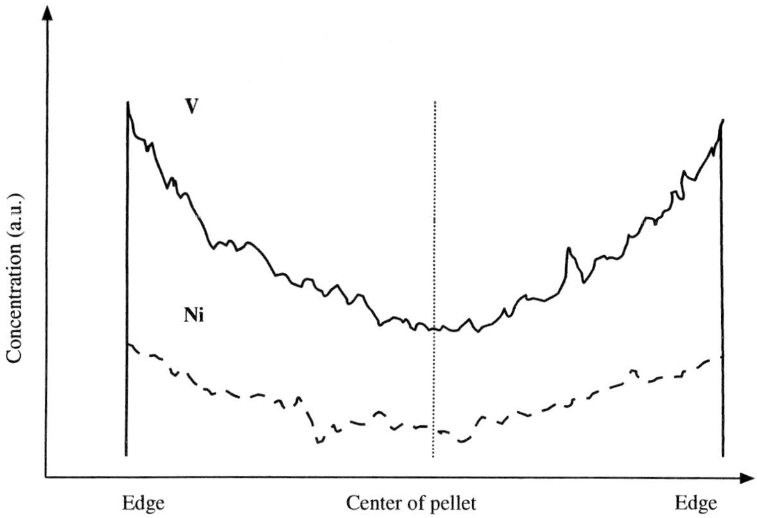

FIGURE 8.5 Typical x-ray electron microprobe analysis patterns of spent heavy oil hydroprocessing catalysts.

the catalyst pores than the V-containing molecules (Quan et al., 1988). When catalyst pellets contain macropores, V loading on the edges and in the center will be slightly different, but when the extrudate contains no macropores, the difference is considerable and the profile shows a marked M shape.

An appropriate combination of carbon black and alumina can be used to obtain bimodal NiMo catalysts with a considerable amount of macropores, without compromising mechanical strength of the final catalyst extrudate, and showing a significant resistance to metals deactivation (López-Salinas et al., 2005).

8.6 SULFUR COMPOUNDS

Total sulfur contents in oil vacuum residue feedstocks are typically from 2 ~ 7 wt% (Wenzel, 1992; Han et al., 1992). Removal of sulfur compounds in oil residue is of great importance to increase the quality of the produced lighter fractions, that is, the less sulfur they contain, (1) the less pollution they will cause when used as fuels, or (2) the less demand for hydrodesulfurization process conditions when hydrotreated downstream. The removal of sulfur from asphaltene molecules is one of the main reactions during hydroconversion processes (Zou and Liu, 1994; Takeuchi et al., 1983; Asaoka et al., 1983). Asphaltenes with high sulfur concentration are easier to transform than those with lower sulfur content. Accordingly, sulfur–carbon bonds are weaker than carbon–carbon bonds (Asaoka et al., 1983). At low reaction temperature a small quantity of sulfur can be removed from weak aliphatic fragments attached to asphaltenes, but the major part of sulfur is located in the aromatic rings (Zou and Liu, 1994; Callejas and Martinez, 2000), being more refractory. Sulfur removal from asphaltenes is negligible when the amount of sulfur removed from the feed is less than 50% of the total; asphaltene's sulfur decreases considerably when total sulfur removal is above 50% (Zou and Liu, 1994). Although sulfur concentration in asphaltenes decreases at higher reaction temperatures, the change in sulfur content is not proportional to the temperature variation (Ancheyta et al., 2003).

8.7 NITROGEN COMPOUNDS

The asphaltenes and polynuclear compounds present in residual oil feed contain heterocyclic nitrogen compounds that are more susceptible to coke formation due to strong interaction with the acid sites of the catalyst surface (Dong et al., 1997; Furimsky, 1983). Nitrogen content in vacuum residua is commonly from 0.4 ~ 1.0 wt%, and from 0.8 ~ 1.8 wt% in the asphaltene fraction. It has been reported that almost all nitrogen in asphaltenes is present in the aromatic rings located inside the asphaltene molecule (Mitra-Kirtley et al., 1993; Mullins, 1995). These studies have shown that nitrogen occurs almost entirely as pyrrolic or pyridinic structures. Nitrogen compounds are much less reactive than parent sulfur compounds, in part because only the hydrogenation pathway is observed in catalytic hydrodenitrogenation, in spite of being thermodynamically unfavorable. Hydrogenation and hydrogenolysis sites appear to be distinct on NiMo and CoMo catalysts. The hydrogenation sites are more acidic and tend to be poisoned by adsorption of basic nitrogen compounds; accordingly, simultaneous hydrodesulfurization (HDS) of dibenzothiphene and hydrodenitrogenation (HDN) of quinoline showed that the quinoline inhibited the hydrogenation of aromatic rings (Sundaram et al., 1988). The oxidation of nitrogen in spent catalysts to nitrogen oxides occurs at similar oxidation temperatures where carbon combustion takes place, which indicates that the nitrogen is associated with coke (Al-Dalama and Stanislaus, 2006). However, a temperature-programmed oxidation (TPO) study on spent catalysts revealed that nitrogen compounds preferentially adsorbed on the acid sites and beneath the coke layer require slightly higher temperatures to be oxidized (and removed from spent catalysts) than those associated with the coke deposits (Al-Dalama and Stanislaus, 2006). The oxidation of these strongly bound nitrogen compounds is likely to start after a large part of the coke is removed by combustion. Zeuthen et al. (1991) have made similar observations during TPO of nitrogen compound-fouled hydroprocessing catalysts.

8.8 REMARKS AND GUIDELINES FOR FUTURE CATALYSTS

It is clear that any upgrade on existing catalysts or development of new catalysts, for already existing processes, should take into account the cost–benefit of the final catalyst. Spent catalysts are difficult to rejuvenate or regenerate by selective leaching of fouling metals like V and Ni; thus, normally deactivated catalysts are, in the best of cases, fully oriented to metal recovery operations.

One of the key characteristics of these future supports or catalysts is the development of a high degree of macropores (20 to 30% of total pore volume) while keeping a high mechanical strength in the final pellet. Several examples in the literature cited above point out that it is possible to increase macroporosity without compromising mechanical strength by using (1) nonalumina supports, (2) additives combined with alumina, or (3) alumina with morphologies different from conventional ones. A few examples are sepiolite, fibrillar alumina, and pyrolized carbon black alumina composite as supports, among others. Recent reports in alumina preparations have disclosed new procedures to design mesostructured (Zhang and Pinnavaia, 2002), nanofibrillar (López-Salinas et al., 2004), nanotubular (Kuang et al., 2003), and nanocapsullar (Bokhimi, et al., 2005) aluminas, which are very attractive to be used as potential supports in hydroconversion catalysts. Additionally, these new-generation aluminas show greater textural properties and different surface topography than conventional ones, which may benefit CoMoS or NiMoS dispersion or hinder sintering at process conditions.

The first catalysts developed for ebullated-bed reactors were made up of unimodal mesoporous γ-alumina supports, where vanadium and carbonaceous deposits concentrated in the external surface of the extrudated catalysts, bringing about an early deactivation. But when second-generation bimodal catalysts (that is, containing ca. 80% and ca. 20% meso- and macropores, respectively, of total pore volume) were developed, a more uniform distribution of fouling metals toward the bulk of the catalyst pellet was evident, with the concomitant lifetime extension. This advancement in catalyst performance taught us the importance of facilitating the diffusion of all the components, particularly the bulkier ones, in a heavy residue through the design of an appropriate porous network. The optimum value of the mesopores' and macropores' average size will depend on the particular physicochemical properties of a given residue, specifically asphaltene content, asphaltene H/C ratio, asphaltene and average molecular weight, among other parameters.

Catalyst supports with higher macropore average size or its relative population will have a beneficial effect in the lifetime of the final catalyst, since it will resist higher loadings of deactivating metals and deposits. Besides, there are strong indications that pore mouth plugging by metal fouling, coke deposits, or sediments occlusion plays a key role in catalyst deactivation. Visualizing ways to open the mouth of outermost pores or passivate these specific pore regions by chemical methods, interconnect mesopores with macropores are still unexplored challenging fields. The problem of keeping a good catalytic activity in the presence of ongoing metals and coke deposits is a combination of chemistry and transport.

Acid sites in a given hydroprocessing catalyst play a crucial role in the scission of C–C bonds to crack bulkier molecules into valuable lighter ones, but acid sites also act as carbonization sites, where asphaltene molecules or its fragments, particularly those containing basic nitrogen, adsorb strongly and serve as carbonization seeds. On the one hand, an overall higher number or strength in acid sites in a hydroprocessing catalyst will lure us to believe that higher yields of lighter fractions will be expected as a consequence, but on the other hand, sediments will increase too, limiting the manipulation of this catalytic property. Finding ways to modulate acidity in selective locations of the catalyst pellet, that is, pore mouth or external surface, remains to be conceptualized and examined.

REFERENCES

Abdul Latif, N. 1990. *Stud. Surf. Sci. Catal.* 53, 283.

Absi-Halabi, M., Stanislaus, A., Qamra, A., and Chopra, S. 1996. Effect of presulfiding on the activity and deactivation of hydrotreating catalysts in processing Kuwait vacuum residue. *Stud. Surf. Sci. Catal.* 100, 243.

Absi-Halabi, M., Stanislaus, A., and Trimm, D.L. 1991. Coke formation on catalysts during the hydroprocessing of heavy oils. *Appl. Catal.* 72, 193.

Al-Dalama, K. and Stanislaus, A. 2006. Comparison between deactivation pattern of catalysts in fixed-bed and ebullating-bed residue hydroprocessing units. *Chem. Eng. J.* 120, 33.

Ali, M.F. and Abbas, S. 2006. A review of methods for the demetallization of residual fuel oils. *Fuel Process. Technol.* 87, 573.

Ali, M., Pernazowski, H., and Haji, A. 1993. Nickel and vanadyl porphyrins in Saudi Arabian crude oils. *Energy Fuels* 7, 179.

Ancheyta, J., Centeno, G., Trejo, F., and Marroquin, G. 2003. Changes in asphaltene properties during hydrotreating of heavy crudes. *Energy Fuels* 17, 1233.

Andersen, S.I. and Birdi, K.S. 1991. Aggregation of asphaltenes as determined by calorimetry. *J. Colloid Interface Sci.* 142, 497.

Anderson, S.I. and Speight, J.G. 1999. Thermodynamic models for asphaltene solubility and precipitation. *J. Petrol. Sci. Eng.* 22, 53.

Asaoka, S., Nakata, S., Shiroto, Y., and Takeuchi, C. 1983. Asphaltene cracking in catalytic hydrotreating of heavy oils. 2. Study of changes in asphaltene structure during catalytic hydroprocessing. *Ind. Eng. Chem. Process. Des. Dev.* 22, 236.

Bartholdy, J. and Andersen, S.I. 2000. Changes in asphaltene stability during hydrotreating. *Energy Fuels* 14, 52.

Bartholdy, J., Lauridsen, R., Mejlholm, M., and Andersen, S.I. 2001. Effect of hydrotreatment on product sludge stability. *Energy Fuels* 15, 1059.

Bartholomew, C.H. 1984. Catalyst deactivation. *Chem. Eng.* November 12, 96.

Beuther, H. and Berrotta, A.J., 1990. *Stud. Surf. Sci. Catal.* 53, 179.

Bokhimi, X., Lima, E., and Valente, J. 2005. Synthesis and characterization of nanocapsules with shells made up of Al13 tridecamers. *J. Phys. Chem. B* 109, 22222.

Buch, L., Groenzin, H., Buenrostro-Gonzalez, E., Andersen, S.I., Lira, C., and Mullins, O.C. 2000. Molecular size of asphaltene fractions obtained from residuum hydrotreatment. *Fuels* 14, 52.

Buenrostro-Gonzalez, E., Groenzin, H., and Mullins, O.C. 2001. The overriding chemical principles that define asphaltenes. *Energy Fuels* 15, 972.

Callejas, M.A. and Martinez, M.T. 2000. Hydroprocessing of a Maya residue. 1. Intrinsic kinetics of asphaltene removal reactions. *Energy Fuels* 14, 1304.

Dai, P.S.E., Sherwood, D.E., and Matrin, B.R. 1990. Effect of diffusion on resid hydrodesulfurization activity. *Chem. Eng. Sci.* 45, 2625.

Dautzenberg, F.M. and De Deken, J.C. 1985. Modes of operation in hydrometallization. Preprint. *Div. Petrol. Chem. ACS*, 30, 8.

Dong, D., Jeong, S., and Massot, F.E. 1997. Effect of nitrogen compounds on deactivation of hydrotreating catalysts by coke. *Catal. Today* 37, 267.

Egiebor, N.O., Gary, M.R., and Cyr, N. 1989. ^{13}C-NMR characterization of organic residues on spent hydroprocessing, hydrocracking and demetallization catalysts. *Appl. Catal.* 55, 81.

Furimsky, E. 1978. Chemical origin of coke deposited on catalyst surface. *Ind. Eng. Chem. Prod. Res. Dev.* 17, 329.

Furimsky, E. 1983. Thermochemical and mechanistic aspects of removal of sulphur, nitrogen and oxygen from petroleum. *Erdöl Kohle* 36, 518.

Gary, J.H. and Handwerk, G.E. 1984. *Petroleum Refining*. Marcel Dekker, New York.

Gray, M.R. 1994. *Upgrading Petroleum Residues and Heavy Oils*. Marcel Dekker, New York.

Groenzin, H. and Mullins, O.C. 2003. Molecular size of asphaltene solubility fractions. *Energy Fuels* 17, 498.

Han, C., Liao, S., and Liu, Z. 1992. Hydrocracking for high quality oil products and petrochemical feedstocks. In *Proceedings of the International Symposium on Heavy Oil and Residue Upgrading and Utilization*, C. Han and C. His (Eds.). International Academic, Beijing, p. 67.

Inoue, S.-I., Takatsuka, V., Wada, Y., Nakata, S.-I., and Ono, T. 1998. New concept for catalysts of asphaltene conversion. *Catal. Today* 43, 225.

Knudsen, F.P. 1959. Dependence of mechanical strength of brittle polycrystalline specimens on porosity and grain size. *J. Am. Ceram. Soc.* 42, 376.

Kodera, Y., Kondo, T., Saito, I., Saito, Y., and Ukegawa, K. 2000. Continuous-distribution kinetic analysis for asphaltene hydrocracking. *Energy Fuels* 14, 291.

Kuang, D., Fang, Y., Liu, H., Frommen, C., and Fenske, D. 2003. Fabrication of boehmite AlOOH and γ-alumina via a soft solution route. *J. Mater. Chem.* 13, 660.

León, O., Rogel, E., Espidel, J., and Torres, G. 2000. Asphaltenes: structural characterization, self-association, and stability behavior. *Energy Fuels* 14, 6.

López-Salinas, E., Espinosa, J.G., Hernández-Cortez, J.G., Sánchez-Valente, J., and Nagira, J. 2005. Long-term evaluation of NiMo/alumina-carbon black composite catalysts in hydroconversion of Mexican 538°C+ vacuum residue. *Catal. Today* 109, 69.

López-Salinas, E., Muñoz-López, J.A., Hernandez-Cortez, J.G., Sanchez-Valente, J., and Toledo-Antonio, A. 2004. Properties of Alumina Nanofibers Obtained by pH-Swing Method. Paper presented at Proceedings of the International Congress on Catalysis, Paris, July 11–16.

Lulic, P., Zen-Cevics, S., Melder, H., and Derdic, D. 1990. The relation between the quality of catalyst and feedstock in the hydrotreating process. *Stud. Surf. Sci. Catal.* 53, 451.

Martínez, M.T., Jímenez, J.M., Callejas, M.A., Gómez, F.J., Rial, C., and Carbó, E. 1997. Characterization of aged catalyst from hydrotreating petroleum residue. In *Hydrotreatment and Hydrocracking of Oil Fractions*, Froment, G.F., Delmon, B., Grange, P. (Eds.). Elsevier Science, Amsterdam, pp. 311–321.

Merdrignac, I., Quoineaud, A.A., and Gauthier, T. 2006. Evolution of asphaltene structure during hydroconversion conditions. *Energy Fuels*, 20, 2028.

Miki, Y., Yamadaya, S., Oba, M., and Sugimoto, Y. 1983. Role of catalyst in hydrocracking of heavy oil. *J. Catal.* 83, 371.

Mitra-Kirtley, S., Mullins, O.C., Elp, J.V., George, S.J., Chen, J., and Cramer, S.P. 1993. Determination of the nitrogen chemical structures in petroleum asphaltenes using XANES spectroscopy. *J. Am. Chem. Soc.* 115, 252.

Mochida, I., Zhao, Y.Z., and Sakanishi, K. 1978. Catalyst deactivation during the hydrotreatment of asphaltene in an Australian brown coal liquid. *Fuel* 67, 1101.

Mochida, I., Zhao, Y.Z., Sakanishi, K., Yamamoto, I., Tokashima, H.A., and Vemura, S.J. 1989. Structure and properties of sludges produced in the catalytic hydrocracking of vacuum residue. *Ind. Eng. Chem. Res.* 28, 418.

Mullins, O.C. 1995. In *Asphaltenes: Fundamentals and Applications*, Sheu, E.Y., Mullins, O.C. (Eds.). Plenum Press, New York, pp. 53–96.

Murgich, J., Rodriguez, J.M., and Aray, Y. 1996. Molecular recognition and molecular mechanics of micelles of some model asphaltenes and resins. *Energy Fuels* 10, 68.

Park, S.J., Escobedo, J., and Amansoori, G. 1994. In *Asphaltenes and Asphalts*, Vol. 1, Yen, T.F., Chilingarian, G.V. (Eds.). Elsevier Science B.V., New York, p. 179.

Pereira, C.J., Beeckman, J.W., Cheng, W.-C., and Suarez, W. 1990. Metal deposition in hydrotreating catalysts. 2. Comparison with experiment. *Ind. Eng. Chem. Res.* 29, 520.

Pfeiffer, J.P. Saal, R.N. 1990. Asphaltic bitumen as a colloid system. *J. Phys. Chem.* 44, 139.

Quan, R.J., Wane, R.A., Hung, C.W., and Wei, J. 1988. *Adv. Chem. Eng.* 14, 95.

Richardson, R.R. and Alley, S.K. 1975. Hydrocracking and hydrotreating. In *Consideration of Catalyst Pore Structure and Asphaltenic Sulfur in the Desulfurization of Resids*, ACS Symposium Series 20. ACS, Washington, DC, p. 136.

Rogel, E., León, O., Contreras, E., Carbognani, L., Torres, G., Espidel, J., and Zambrano, A. 2003. Assessment of asphaltene stability in crude oils using conventional techniques. *Energy Fuels* 17, 1583.

Rostrup-Nielsen, J.R. and Trimm, D.L. 1977. Mechanisms of carbon formation on nickel-containing catalysts. *J. Catal.* 48, 155.

Ruckenstein, E. and Tsai, M.C. 1981. Optimum pore size for the catalytic conversion of large molecules. *AICHE J.* 27, 697.

Seki, H. and Kumata, F. 2000. Structural change of petroleum asphaltenes and resins by hydrodemetallization. *Energy Fuels* 14, 980.

Shimura, M., Shiroto, Y., and Takeuchi, C. 1982. The Effect of Catalyst Pore Structure on the Hydrotreating of Heavy Oils. Paper presented at ACS Symposium on Catalyst and Related Subjects, Las Vegas, March 30.

Simanzhenkov, V. and Indem, R. 2003. *Crude Oil Chemistry.* Marcel Dekker, New York.

Speight, J.G. 2000. *The Desulphurization of Heavy Oils and Residua.* Marcel Dekker, New York, p. 159.

Sundaram, K.M., Katzer, J.R., and Bischoff, J.B. 1988. Modeling of hydroprocessing reactions. *Chem. Eng. Comm.* 71, 53.

Takeuchi, C., Fukui, Y., Nakamura, M., and Shiroto, Y. 1983. Asphaltene cracking in catalytic hydrotreating of heavy oils. 1. Processing of heavy oils by catalytic hydroprocessing and solvent deasphalting. *Ind. Eng. Chem. Proc. Des. Dev.* 22, 236.

Tamm, P.W., Harnsberfer, H.F., and Bridge, A.G. 1981. Effects of feed metals on catalyst aging in hydroprocessing residuum. *Ind. Eng. Chem. Proc. Des. Dev.*, 20, 262.

Thakur, D.S. and Thomas, M.G. 1984. Catalyst deactivation during direct coal liquefaction: a review. *Ind. Eng. Chem. Prod. Res. Dev.* 23, 349.

Thomas, J.M. and Thomas, W.J. 1967. *Introduction to the Principles of Heterogeneous Catalysis*. Academic Press, New York.

Tojima, M., Suhara, S., Imamura, M., and Furuta, A. 1998. Effect of heavy asphaltene on stability of residual oil. *Catal. Today* 43, 347.

Trimm, D.L. 1990. In *Catalysis in Petroleum Refining 1989*, Trimm, D.L., Akashah, S., Absi-Halabi, M., Bishara, A. (Eds.). Elsevier Science, Amsterdam, p. 41.

Trimm, D.L. 1997. In *Handbook of Heterogeneous Catalysis*, Vol. 3, Ertl, G., Knözinger, H., Weitkamp, J. (Eds.). Wiley-VCH, Weinheim, Germany, chap. 7.

Tynan, E.C. and Yen, T.F. 1969. Association of Vanadium chelates in petroleum asphaltenes as studied by ESR. *Fuel* 43, 191.

Vokovic, V. 1978. *Trace Elements in Petroleum*. Petroleum Publishing Company, Oklahoma City, OK.

Wang, J. and Buckley, J.S. 2003. Asphaltene stability in crude oil and aromatic solvents: the influence of oil composition. *Energy Fuels* 17, 1445.

Weisser, O. and Landa, L. 1973. *Sulphide Catalysts, Their Properties and Applications*. Pergamon, Oxford.

Wenzel, F.W. 1992. VEBA-COMBI-cracking, a commercial route for bottom of the barrel upgrading. In *Proceedings of the International Symposium on Heavy Oil and Residue Upgrading and Utilization*, Han, C., His, C. (Eds.). International Academic, Beijing, p. 185.

Yen, T.F. 1975. *The Role of Trace Metals in Petroleum*. Ann Arbor Science Publishers, Ann Arbor, MI.

Yen, T.F., Erdman, J.G., and Pollack, S.S. 1961. Investigation of the structure of petroleum asphaltenes by x-ray diffraction. *Anal. Chem.* 33, 1587.

Ying, Z.-S., Gevert, B., Otterstedt, J.-E., and Sterte, J. 1985. Large-pore catalysts for hydroprocessing of residual oils. *Ind. Eng. Chem. Res.* 34, 1566.

Ying, Z.-S., Gevert, B., Otterstedt, J.-E., and Sterte, J. 1997. Hydrodemetallisation of residual oil with catalysts using fibrillar alumina as carrier material. *Appl. Catal. A Gen.* 153, 69.

Zeuthen, P., Blom, P., and Massoth, F.E. 1991. Characterization of nitrogen on aged hydroprocessing catalysts by temperature-programmed oxidation. *Appl. Catal.* 78, 265.

Zeuthen, P., Cooper, B.H., Clark, F.T., and Arters, D. 1995. Characterization and deactivation studies of spent resid catalyst from ebullating bed service. *Ind. Eng. Chem. Res.* 34, 755.

Zhang, Z. and Pinnavaia, T. 2002. Mesostructured γ-Al_2O_3 with a lathlike framework morphology. *J. Am. Chem. Soc.* 124, 12295.

Zou, R. and Liu, L. 1994. In *Asphaltenes and Asphalts*, Vol. 1, Yen, T.F. and Chilingarian, G.V. (Eds.). Elsevier Science B.V. New York, p. 339.

9 Hydroprocesses

James G. Speight

CONTENTS

9.1 INTRODUCTION

Hydrocracking is a refining technology that, like hydrotreating, falls under the general umbrella of *hydroprocessing*. The outcome is the conversion of a variety of feedstocks to a range of products, and units to accomplish this goal can be found at various points in a refinery.

Hydrocracking allows the refiner to produce products having a lower molecular weight with higher hydrogen content and a lower yield of coke. Furthermore, hydrocracking facilities add flexibility to refinery processing and to the product slate. Hydrocracking is more severe than hydrotreating, with the intent in hydrocracking processes being to convert the feedstock to lower-boiling products rather than to treat the feedstock for heteroatom and metal removal only (Speight, 2000 and references cited therein; 2007 and references cited therein).

The mechanism of hydrocracking is similar to that of catalytic cracking, but with concurrent hydrogenation. The catalyst assists in the production of carbonium ions via olefin intermediates, and these intermediates are quickly hydrogenated under the high hydrogen partial pressures employed in hydrocracking. The rapid hydrogenation prevents adsorption of olefins on the catalyst, and hence prevents their subsequent dehydrogenation, which ultimately leads to coke formation, so that long on-stream times can be obtained without the necessity of catalyst regeneration. Overall, coke formation is relatively low in hydrocracking since the secondary reactions and formation of the precursors to coke are suppressed by the presence of hydrogen.

When applied to heavy feedstocks (heavy oil, tar sand bitumen, and residua), the problems encountered can be directly equated to the amount of complex, higher-boiling constituents that may require pretreatment (Speight and Moschopedis, 1979; Reynolds and Beret, 1989; Gray, 1994; Speight, 2000, 2007; Speight and Ozum, 2002). Furthermore, the majority of the higher molecular weight materials produce high yields (35 to 60% by weight) of coke. It is this trend of coke formation for which hydrocracking offers some relief.

Hydrocracking processes (Speight and Ozum, 2002; Speight, 2007), like any other upgrading processes, are evaluated on the basis of liquid yield (that is, naphtha, distillate, and gas oil), heteroatom removal efficiency, feedstock conversion (FC), carbon mobilization (CM), and hydrogen utilization (HU), along with other process characteristics. The definitions of FC, CM, and HU are:

$$FC = (Feedstock_{IN} - Feedstock_{OUT})/Feedstock_{IN} \times 100$$

$$CM = Carbon_{LIQUIDS}/Carbon_{FEEDSTOCK} \times 100$$

$$HU = Hydrogen_{LIQUIDS}/Hydrogen_{FEEDSTOCK} \times 100$$

High carbon mobilization (CM < 100%) and high hydrogen utilization (HU) correspond to high feedstock conversion (FC) processes involving hydrogen addition such as hydrocracking. Since hydrogen is added, hydrogen utilization can be greater than 100%. Low carbon mobilization and low hydrogen utilization correspond to low feedstock conversion such as coking (carbon rejection) processes. Maximum efficiency from an upgrading process can be obtained by maximizing the liquid yield, and its quality by minimizing the gas (C_1 to C_4) yield, simultaneously. Under these operating conditions, the hydrogen consumption would be the most efficient, that is, hydrogen is consumed to increase the liquid yield and its quality (Towler et al., 1996; Speight, 2000).

9.2 PROCESS OPTIONS

The type of process applied and the complexity of refineries in various parts of the world are determined to a greater extent by the product distribution required. A particular feature of the hydrocracking process, as compared with its alternatives, is its flexibility with respect to product production and the relatively high quality of the products. On the whole, hydrocracking can handle a wider range of feedstocks than catalytic cracking, although the latter process has seen some recent catalyst developments that narrowed the gap. There are also examples where hydrocracking is complementary rather than alternative to the other conversion process; for example, cycle oils, which cannot be recycled to extinction in the catalytic cracker, can be processed in the hydrocracker.

The simplest form of the hydrocracking process is the *single-stage* process (Figure 9.1). In this simplest of the hydrocracker configurations, the layout of the reactor section generally resembles that of the hydrotreating unit. This configuration finds application in cases where only a moderate degree of conversion (say 60% or less) is required. It may well apply to processes where the feedstock is pretreated prior to introduction to a fluid catalytic cracking (FCC) unit.

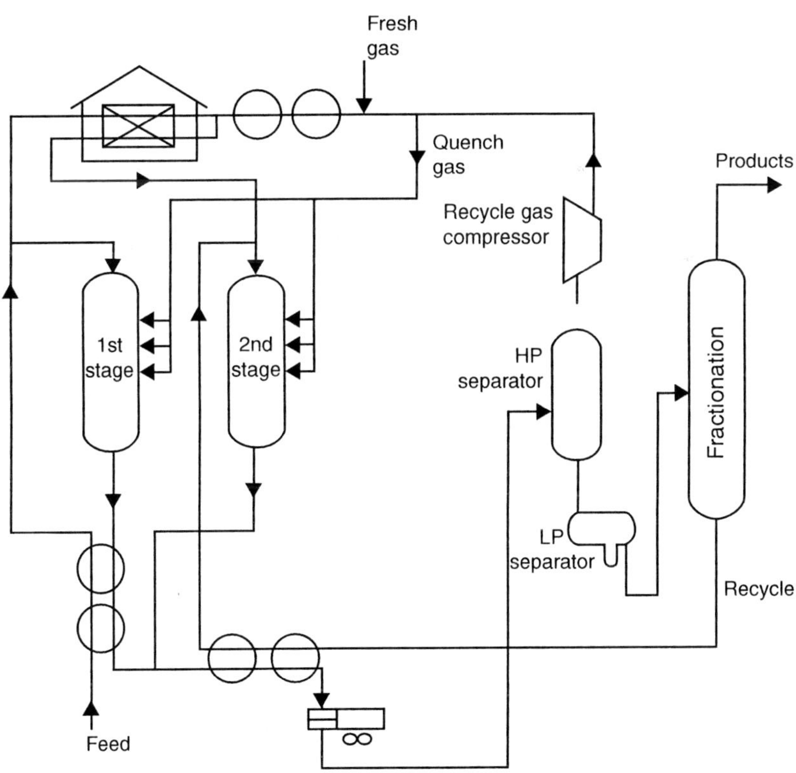

FIGURE 9.1 A single-stage or two-stage (optional) hydrocracking unit.

The catalyst used in a single-stage process comprises a hydrogenation function in combination with a strong cracking function. Sulfided metals such as cobalt, molybdenum, and nickel provide the hydrogenation function. An acidic support, usually alumina, attends to the cracking function. Nitrogen compounds and ammonia produced by hydrogenation interfere with acidic activity of the catalyst. In the cases where high/full conversion is required, the reaction temperatures and run lengths of interest in commercial operation can no longer be adhered to. Moreover, the extent of the conversion reaches asymptotes with increasing hydrogen pressure (Speight, 2007 and references cited therein), so increasing the partial pressure of the hydrogen in the reactor is a guarantee of additional feedstock conversion. In fact, it becomes necessary to switch to a different reactor bed system or to a multistage process, in which the cracking reaction mainly takes place in an added reactor.

Another form of hydrocracking process for heavier feedstocks is a *two-stage* operation. Generally, the first stage of the two-stage plant resembles a *single-stage once-through* (SSOT) unit. This flow scheme has been very popular, since it can be used to maximize the yield of transportation fuels and is an attempt to combat the adverse effect of ammonia and nitrogen compounds on catalyst activity. Similarly, the *series flow* version of the multistage hydrocracker has also been developed.

In the two-stage configuration (Figure 9.1), fresh feed is preheated by heat exchange with effluent from the first reactor. It is combined with part of a hot fresh gas/recycle gas mixture and passes through a first reactor for a denitrogenation step. These reactions, as well as those of hydrocracking, which occurs to a limited extent in the first reactor, are exothermic. The catalyst inventory is therefore divided among a number of fixed beds. Reaction temperatures are controlled by introducing part of the recycle gas as a quench medium between beds. The ensuing liquid is fractionated to remove the product made in the first reactor. Unconverted material with low nitrogen content and free of ammonia is taken as a bottom stream from the fractionation section. After heat exchange with reactor effluent and mixing with heated recycle gas, it is sent to the second reactor, where most of the hydrocracking reactions occur. A strongly acidic catalyst with a relatively low hydrogenation activity (metal sulfides on, for example, amorphous silica-alumina) is usually applied. As in the first reactor, the exothermic nature of the process is controlled by recycle gas as the quench medium of the catalyst beds. Effluent from the second reactor is cooled and joins first-stage effluent for separation from recycle gas and fractionation. The part of the second reactor feed that remains unconverted is recycled to the reactor.

In the *series flow* configuration, the principal difference is the elimination of first-stage cooling and gas/liquid separation and the ammonia removal step. The effluent from the first stage is mixed with more recycle gas and routed directly to the inlet of the second reactor. In contrast with the amorphous catalyst of the two-stage process, the second reactor in series flow generally has a zeolite catalyst, based on crystalline silica-alumina. As in the two-stage process, material not converted to the product boiling range is recycled from the fractionation section.

A *single-stage recycle* (SSREC) unit converts heavy oil completely into light products with a flow scheme resembling the second stage of the two-stage plant. Such a unit maximizes the yield of naphtha, jet fuel, or diesel depending on the recycle cut point used in the distillation section. Commercial plants have operated to produce low-pour-point diesel fuel and jet fuel.

Building on the theme of one- or two-stage hydrocracking, the *once-through partial conversion* (OTPC) concept evolved and offers the means to convert heavy feedstocks (usually vacuum gas oil) into high-quality gasoline, jet fuel, and diesel products by a partial conversion operation. Because total conversion of the higher molecular weight compounds in the feedstock is not required, once-through hydrocracking can be carried out at lower temperatures and, in most cases, at lower hydrogen partial pressures than in recycle hydrocracking, where total conversion of the feedstock is normally an objective.

Recycle hydrocracking units are designed to operate at hydrogen partial pressures from about 1200 to 2300 psi (8274 to 15,858 kPa), depending on the type of feed processed. Hydrogen partial pressure is set in the design in part depending on required catalyst cycle length, but also to enable the catalyst to convert high molecular weight polynuclear aromatic and naphthene compounds that must be hydrogenated before they can be cracked. Hydrogen partial pressure also affects properties of the hydrocracked products that depend on hydrogen uptake, such as jet fuel aromatics content and smoke point and diesel cetane number. In general, the higher the feed endpoint, the higher the required hydrogen partial pressure necessary to achieve satisfactory performance of the plant.

One disadvantage of once-through hydrocracking compared to a recycle operation is a somewhat reduced flexibility for varying the ratio of gasoline to middle distillate that is produced. A greater quantity of naphtha can be produced by increasing conversion and production of jet fuel, plus diesel can also be increased. But selectivity for higher-boiling products is also a function of conversion. Selectivity decreases as once-through conversion increases. If conversion is increased too much, the yield of desired product will decrease, accompanied by an increase in light ends and gas production. Higher yields of gasoline or jet fuel plus diesel are possible from a recycle than from a once-through operation.

The goals of *heavy oil and residuum hydroconversion* processes are to (1) desulfurize feedstocks to supply low-sulfur fuel oils and (2) pretreat feedstocks for residuum fluid catalytic cracking processes. Some of the processes available for hydroprocessing heavy feedstocks are presented below (listed in alphabetical order with no other preference in mind).

9.2.1 Asphaltenic Bottom Cracking (ABC) Process

The ABC process can be used for distillate production, hydrodemetallization, asphaltene cracking, and moderate hydrodesulfurization as well as sufficient resistance to coke fouling and metal deposition using such feedstocks as vacuum residua, thermally cracked residua, solvent-deasphalted bottoms, and bitumen with fixed catalyst beds (Takeuchi et al., 1982). The process can be combined with solvent deasphalting for complete or partial conversion of the residuum, or hydrodesulfurization to promote the conversion or hydrovisbreaking.

In the process, the feedstock is pumped up to the reaction pressure and mixed with hydrogen. The mixture is heated to the reaction temperature in the charge heater after a heat exchange and fed to the reactor. In the reactor, hydrodemetallization and subsequent asphaltene cracking with moderate hydrodesulfurization take place simultaneously under conditions similar to residuum hydrodesulfurization. The reactor effluent gas is cooled, cleaned up, and recycled to the reactor section, while the separated liquid is distilled into distillate fractions and vacuum residue, which is further separated by deasphalting into deasphalted oil and asphalt using butane or pentane.

In the vis-ABC process, a soaking drum is provided after the heater, when necessary. Hydrovisbroken oil is first stabilized by the ABC catalyst through hydrogenation of coke precursors, and then desulfurized by the hydrodesulfurization catalyst.

9.2.2 Aquaconversion

The aquaconversion process is a hydrovisbreaking technology that uses catalyst-activated transfer of hydrogen from water added to the feedstock. Reactions that lead to coke formation are suppressed, and there is no separation of asphaltene-type material (Marzin et al., 1998; Pereira et al., 1998, 2001).

The old visbreaking technology is limited in conversion level because of the stability of the resulting product. The aquaconversion process extends the maximum conversion level within the stability specification by adding a homogeneous catalyst in the presence of steam. This allows hydrogen from the water to be transferred to the resid when contacted in a coil-soaker system normally used for the visbreaking process. The hydrogen incorporation is much lower than that obtained when using a deep hydroconversion process under high hydrogen partial pressure. Nevertheless, it is high enough to saturate the free radicals, formed within the thermal process, which would normally lead to coke formation. With hydrogen incorporation, a higher conversion level can be reached, and thus higher API and viscosity improvements while maintaining syncrude stability.

The important aspect of the aquaconversion technology is that it does not produce coke, and it does not require any hydrogen source or high-pressure equipment. In addition, the aquaconversion process can be implanted in the production area, and thus the need for external diluent and its transport over large distances is eliminated. Light distillates from the raw crude can be used as diluent for both the production and desalting processes.

9.2.3 CANMET Hydrocracking Process

The CANMET hydrocracking process is for heavy oils, atmospheric residua, and vacuum residua (Pruden, 1978; Waugh, 1983; Ng and Rahimi, 1991; Pruden et al., 1993). Initially developed to upgrade heavy oil and tar sand bitumen as well as residua, the process does not use a catalyst but employs an additive to inhibit coke formation and allow high conversion to lower-boiling products using a single reactor.

In the process, the feedstock and recycle hydrogen gas are heated to reactor temperature in separate heaters. A small portion of the recycle gas stream and the required amount of additive are routed through the oil heater to prevent coking in the heater tubes. The outlet streams from both heaters are fed to the bottom of the reactor.

The vertical reactor vessel is free of internal equipment and operates in a three-phase mode. The solid additive particles are suspended in the primary liquid hydrocarbon phase through which the hydrogen and product gases flow rapidly in bubble form. The reactor exit stream is quenched with cold recycle hydrogen prior to the high-pressure separator. The heavy liquids are further reduced in pressure to a hot, medium-pressure separator and from there to fractionation. The spent additive leaves with the heavy fraction and remains in the unconverted vacuum residue.

The vapor stream from the hot, high-pressure separator is cooled stepwise to produce middle distillate and naphtha that are sent to fractionation. High-pressure purge of low-boiling hydrocarbon gases is minimized by a sponge oil circulation system. Product naphtha will be hydrotreated and reformed, light gas oil will be hydrotreated and sent to the distillate pool, the heavy gas oil will be processed in the FCC, and the pitch will be sold.

The additive, prepared from iron sulfate [$Fe_2(SO_4)_3$], is used to promote hydrogenation and effectively eliminate coke formation. The effectiveness of the dual-role additive permits the use of operating temperatures that give high conversion in a single-stage reactor. The process also offers the attractive option of reducing the coke yield by slurrying the feedstock with less than 10 ppm catalyst (molybdenum naphthenate) and sending the slurry to a hydroconversion zone to produce low-boiling products (Kriz and Ternan, 1994).

9.2.4 Chevron RDS Isomax and VRDS Process

The residuum desulfurizer (RDS)/vacuum residuum desulfurizer (VRDS) process is (like the Residfining process, q.v.) designed to hydrotreat vacuum gas oil, atmospheric residuum, or vacuum residuum to remove sulfur metallic constituents while part of the feedstock is converted to lower-boiling products. In the case of residua, the asphaltene content is reduced.

The process consists of a once-through operation and is ideally suited to produce feedstocks for residuum fluid catalytic crackers or delayed coking units to achieve minimal production of residual products in a refinery.

The basic elements of each process are similar and consist of a once-through operation of the feedstock coming into contact with hydrogen and the catalyst in a downflow reactor that is designed to maintain activity and selectivity in the presence of deposited metals. Moderate temperatures and pressures are employed to reduce the incidence of hydrocracking, and hence minimize production of low-boiling distillates. The combination of a desulfurization step and a VRDS is often seen as an attractive alternate to the atmospheric RDS. In addition, either the RDS or VRDS option can be coupled with other processes (such as delayed coking, fluid catalytic cracking, and solvent deasphalting) to achieve the most optimum refining performance.

9.2.5 Chevron Deasphalted Oil Hydrotreating Process

The Chevron deasphalted oil hydrotreating process is designed to desulfurize heavy feedstocks that have had the asphaltene fraction removed by prior application of a deasphalting process. The principal product is a low-sulfur fuel oil that can be used as a blending stock or as a feedstock for a fluid catalytic unit.

The process employs a downflow, fixed-bed reactor containing a highly selective catalyst that provides extensive desulfurization at low pressures with minimal cracking, and therefore low consumption of hydrogen.

9.2.6 Gulf Resid Hydrodesulfurization Process

This is a regenerative fixed-bed process to upgrade petroleum residues by catalytic hydrogenation to refined heavy fuel oils or to high-quality catalytic charge stocks.

The catalyst is a metallic compound supported on pelletized alumina and may be regenerated *in situ* with air and steam or flue gas through a temperature cycle of 400 to 650°C (750 to 1200°F). On-stream cycles of 4 to 5 months can be obtained at desulfurization levels of 65 to 75%, and catalyst life may be as long as 2 years.

9.2.7 H-Oil Process

The H-Oil process (Figure 9.2) (*Hydrocarbon Processing*, 1998, p. 86) is a catalytic process that uses a single-stage, two-stage, or three-stage ebullated-bed reactor in which, during the reaction, considerable hydrocracking takes place. The process is designed for hydrogenation of residua and other high feedstocks in an ebullated-bed reactor to produce upgraded petroleum products (*Hydrocarbon Processing*, 1996). The process is able to convert all types of feedstocks to distillate products as well as to desulfurize and demetallize residues for feed to coking units or residue fluid catalytic cracking units, for production of low-sulfur fuel oil, or for production to asphalt blending. A modification of the H-Oil process (Hy-C Cracking) converts high-boiling distillates to middle distillates and kerosene.

A wide variety of process options can be used with the H-Oil process depending on the specific operation. In all cases, a catalytic ebullated-bed reactor system is used to provide an efficient hydroconversion. The system ensures uniform distribution of liquid, hydrogen-rich gas, and catalyst across the reactor. The ebullated-bed system operates under essentially isothermal conditions, exhibiting little temperature gradient across the bed (Kressmann et al., 2000). The heat of reaction is used to bring the feed oil and hydrogen up to reactor temperature.

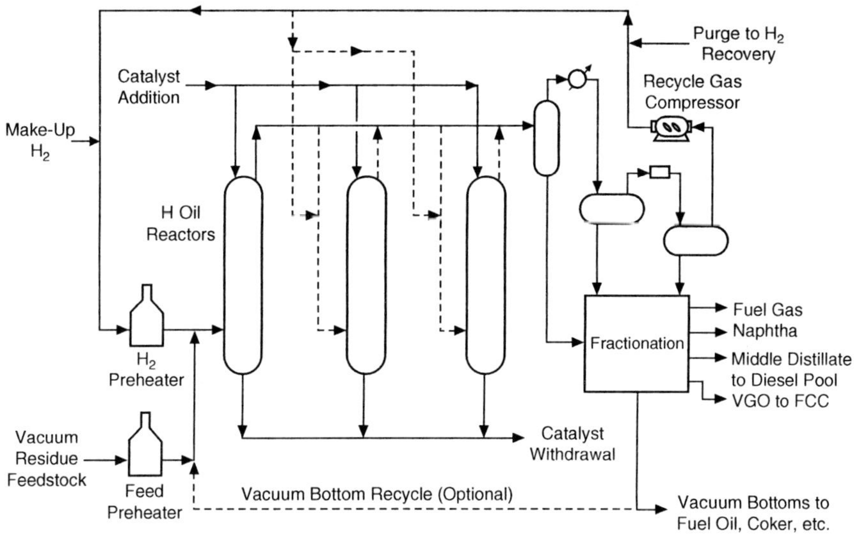

FIGURE 9.2 H-Oil process.

In the process, the feedstock (a vacuum residuum) is mixed with recycle vacuum residue from downstream fractionation, hydrogen-rich recycle gas, and fresh hydrogen. This combined stream is fed into the bottom of the reactor, whereby the upward flow expands the catalyst bed. The mixed-vapor liquid effluent from the reactor goes to either the flash drum for phase separation or the next reactor. A portion of the hydrogen-rich gas is recycled to the reactor. The product oil is cooled and stabilized and the vacuum residue portion is recycled to increase conversion.

A catalyst of small particle size can be used, giving efficient contact among gas, liquid, and solid with good mass and heat transfer. Part of the reactor effluent is recycled back through the reactor for temperature control and to maintain the requisite liquid velocity. The entire bed is held within a narrow temperature range, which provides essentially an isothermal operation with an exothermic process. Because of the movement of catalyst particles in the liquid–gas medium, deposition of tar and coke is minimized and fine solids entrained in the feed do not lead to reactor plugging. The catalyst can also be added and withdrawn from the reactor without destroying the continuity of the process. The reactor effluent is cooled by exchange and separates into vapor and liquid. After scrubbing in a lean oil absorber, hydrogen is recycled and the liquid product is either stored directly or fractionated before storage and blending.

9.2.8 Hydrovisbreaking (HYCAR) Process

Briefly, *hydrovisbreaking*, a noncatalytic process, is conducted under conditions similar to those of visbreaking and involves treatment with hydrogen under mild conditions.

The presence of hydrogen leads to more stable products (lower *flocculation threshold*) than can be obtained with straight visbreaking, which means that higher conversions can be achieved, producing a lower-viscosity product.

The HYCAR process is composed fundamentally of three parts: (1) visbreaking, (2) hydrodemetallization, and (3) hydrocracking. In the visbreaking section, the heavy feedstock (for example, vacuum residuum or bitumen) is subjected to moderate thermal cracking, while no coke formation is induced. The visbreaker oil is fed to the demetallization reactor in the presence of catalysts, which provides sufficient pore for diffusion and adsorption of high molecular weight constituents. The product from this second stage proceeds to the hydrocracking reactor, where desulfurization and denitrogenation take place along with hydrocracking.

9.2.9 Hyvahl F Process

This process is used to hydrotreat atmospheric and vacuum residua to convert the feedstock to naphtha and middle distillates (Peries et al., 1988; Billon et al., 1994; *Hydrocarbon Processing*, 1996).

The main features of this process are its dual-catalyst system and its fixed-bed swing-reactor concept. The first catalyst has a high capacity for metals (to 100% by weight of new catalyst) and is used for both hydrodemetallization (HDM) and most of the conversion. This catalyst is resistant to fouling, coking, and plugging by asphaltene constituents (as well as by reacted asphaltene constituents) and shields the second catalyst from the same. Protected from metal poisons and deposition of coke-like products, the highly active second catalyst can carry out its deep hydrodesulfurization (HDS) and refining functions. Both catalyst systems use fixed beds that are more efficient than moving beds and are not subject to attrition problems.

In the process, the preheated feedstock enters one of the two guard reactors where a large proportion of the nickel and vanadium are adsorbed and hydroconversion of the high molecular weight constituents commences. Meanwhile, the second guard reactor catalyst undergoes a reconditioning process and then is put on standby. From the guard reactors, the feedstock flows through a series of hydrodemetallization reactors that continue the metals removal and conversion of heavy ends.

The next processing stage, hydrodesulfurization, is where most of the sulfur, some of the nitrogen, and the residual metals are removed. A limited amount of conversion also takes place. From the final reactor, the gas phase is separated, hydrogen is recirculated to the reaction section, and the liquid products are sent to a conventional fractionation section for separation into naphtha, middle distillates, and heavier streams.

9.2.10 IFP Hydrocracking Process

This process features a dual-catalyst system: the first is a promoted nickel–molybdenum amorphous catalyst. It acts to remove sulfur and nitrogen and hydrogenate aromatic rings. The second catalyst is a zeolite that finishes the hydrogenation and promotes the hydrocracking reaction.

In the single-stage process, the first reactor effluent is sent directly to the second reactor, followed by the separation and fractionation steps. The fractionator bottoms are recycled to the second reactor or sold.

In the two-stage process, feedstock and hydrogen are heated and sent to the first reaction stage where conversion to products occurs (RAROP, 1991, p. 85). The reactor effluent phases are cooled and separated, and the hydrogen-rich gas is compressed and recycled. The liquid leaving the separator is fractionated, the middle distillates and lower-boiling streams are sent to storage, and the high-boiling stream is transferred to the second reactor section and then recycled back to the separator section.

9.2.11 Isocracking Process

The process has been applied commercially in the full range of process flow schemes: single-stage, once-through liquid; single-stage, partial recycle of heavy oil; single-stage extinction recycle of oil (100% conversion); and two-stage extinction recycle of oil (Bridge, 1997; *Hydrocarbon Processing*, 1998, p. 84). The preferred flow scheme will depend on the feedstock properties (Bridge, 1997).

The process uses multibed reactors, and in most applications, a number of catalysts are used in a reactor. The catalysts are dual function, being a mixture of hydrous oxides (for cracking) and heavy metal sulfides (for hydrogenation) (Bridge, 1997). The catalysts are used in a layered system to optimize the processing of the feedstock that undergoes changes in its properties along the reaction pathway. In most commercial units, the entire fractionator bottom's fraction is recycled, or all of it is drawn as heavy product, depending on whether the low-boiling or high-boiling products are of greater value. If the low-boiling distillate products (naphtha or naphtha/kerosene) are the most valuable, the higher-boiling-point distillates (like diesel) can be recycled to the reactor for conversion rather than drawn as a product (RAROP, 1991, p. 83).

9.2.12 LC-Fining Process

LC-Fining (Figure 9.3) is a hydrocracking process capable of desulfurizing, demetallizing, and upgrading a wide spectrum of heavy feedstocks by means of an expanded-bed reactor (van Driesen et al., 1979; Fornoff, 1982; Bishop, 1990; RAROP, 1991, p. 61; Reich et al., 1993; *Hydrocarbon Processing*, 1998, p. 82). Operating with the expanded bed allows the processing of heavy feedstocks, such as atmospheric residua, vacuum residua, and oil sand bitumen. The catalyst in the reactor behaves like fluid, enabling the catalyst to be added to and withdrawn from the reactor during operation. The reactor conditions are near isothermal because the heat of reaction is absorbed by the cold fresh feed immediately owing to thorough mixing of reactors.

In the process, the feedstock and hydrogen are heated separately and then pass upwards in the hydrocracking reactor through an expanded bed of catalyst. Reactor products flow to the high-pressure, high-temperature separator.

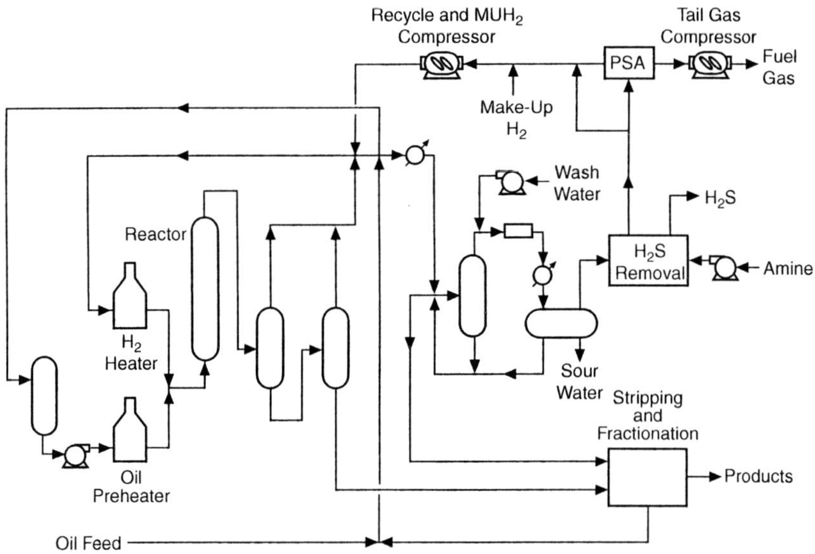

FIGURE 9.3 LC-Fining process.

Vapor effluent from the separator is let down in pressure, and then goes to the heat exchange and thence to a section for the removal of condensable products and purification.

The residence time in the reactor is adjusted to provide the desired conversion levels. Catalyst particles are continuously withdrawn from the reactor, regenerated, and recycled back into the reactor, which provides the flexibility to process a wide range of heavy feedstocks, such as atmospheric and vacuum tower bottoms, coal-derived liquids, and bitumen. An internal liquid recycle is provided with a pump to expand the catalyst bed continuously. As a result of the expanded-bed operating mode, small pressure drops and isothermal operating conditions are accomplished. Extruded catalyst particles as small as 0.8 mm (1/32 in.) in diameter can be used in this reactor.

9.2.13 Microcat-RC Process

Microcat-RC (also referred to as the M-Coke process) is a catalytic ebullated-bed hydroconversion process that is similar to Residfining (q.v.) and that operates at relatively moderate pressures and temperatures (Bearden and Aldridge, 1981; Bauman et al., 1993). The catalyst particles, containing a metal sulfide in a carbonaceous matrix formed within the process, are uniformly dispersed throughout the feed. Because of their ultrasmall size (10^{-4} in. diameter), there are typically several orders of magnitude more of these microcatalyst particles per cubic centimeter of oil than is possible in other types of hydroconversion reactors using conventional catalyst particles. This results in a smaller distance between particles and less time for a reactant molecule or intermediate to find an active catalyst site. Because of their physical structure, microcatalysts are not prone to the pore-plugging problems that tend to plague conventional catalysts.

In the process, fresh vacuum residuum, microcatalyst, and hydrogen are fed to the hydroconversion reactor. Effluent is sent to a flash separation zone to recover hydrogen, gases, and liquid products, including naphtha, distillate, and gas oil. The residuum from the flash step is then fed to a vacuum distillation tower to obtain a 565°C$^-$ (1050°F$^-$) product oil and a 565°C$^+$ (1050°F$^+$) bottom fraction that contains unconverted feed, microcatalyst, and essentially all of the feed metals.

9.2.14 Mild Hydrocracking Process

The *mild hydrocracking process* uses operating conditions similar to those of a vacuum gas oil (VGO) desulfurizer to convert a vacuum gas oil to significant yields of lighter products. Consequently, the flow scheme for a mild hydrocracking unit is virtually identical to that of a vacuum gas oil desulfurizer.

For example, in a simplified process for vacuum gas oil desulfurization, the vacuum gas oil feedstock is mixed with hydrogen makeup gas and preheated against reactor effluent. Further preheating to reaction temperature is accomplished in a fired heater. The hot feedstock is mixed with recycle gas before entering the reactor. The temperature rises across the reactor due to the exothermic heat of reaction. Catalyst bed temperatures are usually controlled by using multiple catalyst beds and introducing recycle gas as an interbed quench medium. Reactor effluent is cooled against incoming feedstock and air or water before entering the high-pressure separator. Vapors from this separator are scrubbed to remove hydrogen sulfide (H_2S) before compression back to the reactor as recycle and quench. A small portion of these gases is purged to fuel gas to prevent buildup of light ends. Liquid from the high-pressure separator is flashed into the low-pressure separator. Sour flash vapors are purged from the unit. Liquid is preheated against stripper bottoms and in a feed heater before steam stripping in a stabilizer tower. Water wash facilities are provided upstream of the last reactor effluent cooler to remove ammonium salts produced by denitrogenation of the vacuum gas oil feedstock.

The conditions for mild hydrocracking are typical of many low-pressure desulfurization units, and the process is a simple form of hydrocracking.

9.2.15 MRH Process

The Mild Resid Hydrocracking (MRH) process is designed to upgrade heavy feedstocks containing large amounts of metals and asphaltene, such as vacuum residua and bitumen, and to produce mainly middle distillates (Sue, 1989; RAROP, 1991, p. 65). The reactor is designed to maintain a mixed three-phase slurry of feedstock, fine powder catalyst, and hydrogen, and to promote effective contact.

In the process, a slurry consisting of heavy feedstock and fine powder catalyst is preheated in a furnace and fed into the reactor vessel. Hydrogen is introduced from the bottom of the reactor and flows upward through the reaction mixture, maintaining the catalyst suspension in the reaction mixture. In the upper section of the reactor, vapor is disengaged from the slurry, and hydrogen and other gases are removed in a high-pressure separator. The liquid condensed from the overhead vapor is distilled and then flows out to the secondary treatment facilities.

From the lower section of the reactor, bottom slurry oil (SLO) that contains catalyst, uncracked residuum, and a small amount of vacuum gas oil fraction is withdrawn. Vacuum gas oil is recovered in the slurry separation section, and the remaining catalyst and coke are fed to the regenerator.

9.2.16 RCD Unibon (BOC) Process

The RCD Unibon (BOC) process upgrades vacuum residua (RAROP, 1991, p. 67; Thompson, 1997; *Hydrocarbon Processing*, 1998). There are several possible flow scheme variations involving the process. It can operate as an independent unit or be used in conjunction with a thermal conversion unit. In this configuration, hydrogen and a vacuum residuum are introduced separately to the heater and mixed at the entrance to the reactor. The effluent from the reactor is directed to the hot separator. The overhead vapor phase is cooled and condensed, and the separated hydrogen is recycled to the reactor. The bottom liquid stream then goes to the vacuum column, where the gas oils are recovered for further processing, and the residuals are blended into the heavy fuel oil pool.

9.2.17 Residfining Process

Residfining is a catalytic fixed-bed process for the desulfurization and demetallization of atmospheric and vacuum residua (RAROP, 1991, p. 69; *Hydrocarbon Processing*, 1996, 1998).

In the process, liquid feedstock to the unit is filtered, pumped to pressure, preheated, and combined with hydrogen-containing gas prior to entering the reactors. A small guard reactor can be employed to prevent plugging and fouling of the main reactors. Provisions are employed to periodically remove the guard while keeping the main reactors on-line. The temperature rise associated with the exothermic reactions is controlled utilizing either a gas quench or liquid quench. The liquid product is sent to a fractionator where the product is fractionated.

Residfining is an option that can be used to reduce the sulfur, to reduce metals and coke-forming precursors, or to accomplish some conversion to lower-boiling products as a feed pretreat step ahead of a fluid catalytic cracking unit. There is also a hydrocracking option where substantial conversion of the resid occurs.

9.2.18 Residue Hydroconversion (RHC) Process

Residue hydroconversion is a high-pressure fixed-bed trickle-flow hydrocatalytic process (RAROP, 1991, p. 71). The feedstock can be atmospheric or vacuum residue.

The reactors are of multibed design with interbed cooling, and the multicatalyst system can be tailored according to the nature of the feedstock and the target conversion. For residua with a high metal content, a hydrodemetallization catalyst is used in the front-end reactors, which excels in its high metal uptake capacity and good activities for metal removal, asphaltene conversion, and residue cracking. Downstream of the demetallization stage, one or more hydroconversion stages, with optimized combination of catalysts' hydrogenation function and texture, are used to achieve desired catalyst stability and activities for denitrogenation, desulfurization, and heavy hydrocarbon cracking. A guard reactor may be employed to remove contaminants that promote plugging or fouling of the main reactors.

9.2.19 Shell Residual Oil Hydrodesulfurization

The Shell residual oil hydrodesulfurization process improves the quality of residual oils by removing sulfur, metals, and asphaltene constituents. The process is suitable for a wide range of the heavier feedstocks, irrespective of the composition and origin, and even includes those feedstocks that are particularly high in metals and asphaltene constituents.

The process centers on a fixed-bed downflow reactor that allows catalyst replacement without causing any interruption in the operation of the unit. Feedstock is introduced to the process via a filter (backwash, automatic), after which hydrogen and recycle gas are added to the feedstock stream, which is then heated to reactor temperature by means of feed-effluent heat exchangers whereupon the feed stream passes down through the reactor in trickle flow. Sulfur removal is excellent and substantial reductions in the vanadium content and asphaltene constituents also occur. In addition, a marked increase occurs in the API gravity, and the viscosity is reduced considerably.

A bunker reactor provides extra process flexibility if it is used upstream from the desulfurization reactor, especially with reference to the processing of feedstocks with a high metal content. A catalyst with a capacity for metals is employed in the bunker reactor to protect the desulfurization catalyst from poisoning by the metals. In the bunker reactor, inverted cone segments support the catalyst and are designed to allow catalyst removal.

9.2.20 Unicracking/HDS Process

Unicracking is a fixed-bed catalytic process that employs a high-activity catalyst with a high tolerance for sulfur and nitrogen compounds and can be regenerated (Reno, 1997). The design is based upon a single-stage or two-stage system with provisions to recycle to extinction (RAROP, 1991, p. 79).

In the process, a two-stage reactor system receives untreated feed, makeup hydrogen, and a recycle gas at the first stage, in which gasoline conversion may be as high as 60% by volume. The reactor effluent is separated to recycle gas, liquid product, and unconverted oil. The second-stage oil may be either once-through or recycle cracking; feed to the second stage is a mixture of unconverted first-stage oil and second-stage recycle. The process operates satisfactorily for a variety of feedstocks, and the rate of desulfurization is dependent on the sulfur content of the feedstock, as are catalyst life, product sulfur, and hydrogen consumption (Speight, 2000 and references cited therein).

In the process, the feedstock and hydrogen-rich recycle gas are preheated, mixed, and introduced into a guard reactor that contains a relatively small quantity of the catalyst. The guard chamber removes particulate matter and residual salt from the feed. The effluent from the guard chamber flows down through the main reactor, where it contacts one or more catalysts designed for removal of metals and sulfur. The catalysts, which induce desulfurization, denitrogenation, and hydrocracking, are based upon both amorphous and molecular-sieve-containing supports. The product from the reactor is cooled, separated from hydrogen-rich recycle gas, and either stripped to meet fuel oil flash point specifications or fractionated to produce distillate fuels, upgraded vacuum gas oil, and upgraded vacuum residuum. Recycle gas, after hydrogen sulfide removal, is combined with makeup gas and returned to the guard chamber and main reactors.

The process uses base-metal or noble-metal hydrogenation activity promoters impregnated on combinations of zeolites and amorphous aluminosilicates for cracking activity (Reno, 1997). The specific metals chosen and the proportions of the metals, zeolite, and nonzeolite aluminosilicates are optimized for the feedstock and desired product balance. This is effective in the production of clean fuels, especially for cases where a partial conversion unicracking unit and a fluid catalytic cracking unit are integrated.

The Advanced Partial Conversion Unicracking (APCU) process is a recent advancement in the area of ultra-low-sulfur diesel (ULSD) production and feedstock pretreatment for catalytic cracking units.

In the process, high-sulfur feeds such as vacuum gas oil and heavy cycle gas oil are mixed with a heated hydrogen-rich recycle gas stream and passed over beds of high-activity catalyst. The hydrocracked products and desulfurized feedstock for a fluid catalytic cracking unit are separated at reactor pressure in an enhanced hot separator. The overhead products for the separator are immediately hydrogenated in the integrated finishing reactor. As pretreatment severity is increased, conversion is increased in the fluid catalytic cracker.

Another development in the unicracking family is HyCycle Unicracking technology, designed to maximize diesel production for full-conversion applications.

9.2.21 Veba Combi Cracking Process

Veba Combi Cracking (VCC) is a hydrocracking/hydrogenation process for converting residua and other heavy feedstocks (Niemann et al., 1988; RAROP, 1991, p. 81; Wenzel and Kretsmar, 1993; *Hydrocarbon Processing*, 1998, p. 88).

In the process, the heavy feedstock, slurried with a small amount of fine-powdered additive and mixed with hydrogen and recycle gas, is hydrogenated (hydrocracked) using a commercial catalyst and liquid-phase hydrogenation reactor operating at 440 to 485°C (825 to 905°F) and 2175 to 4350 psi (14,996 to 29,993 kPa) pressure. The product obtained from the reactor is fed into the hot separator operating at temperatures slightly below the reactor temperature. The liquid and solid materials are fed into a vacuum distillation column, and the gaseous products are fed into a gas-phase hydrogenation reactor operating at an identical pressure (Graeser and Niemann, 1982, 1983).

However, the system operates in a trickle-flow mode, which may not be efficient for some heavy feedstocks. The separation of the liquid product from associated gases is performed in a cold separator system. The liquid product may be sent to a stabilization and fractionation unit as required, while the gases are sent to a lean oil scrubbing system for contaminant removal and are recycled.

9.3 CATALYSTS

Proper selection of the types of catalysts employed can even permit partial conversion of heavy gas oil feeds to diesel and lighter products at the low-hydrogen partial pressures for which gas oil hydrotreaters are normally designed. This so-called mild hydrocracking has been attracting a great deal of interest from refiners who have existing hydrotreaters and wish to increase their refinery's conversion of fuel oil into lower-boiling, higher-value products (Speight, 2000 and references cited therein; Speight and Ozum, 2002 and references cited therein).

Hydrocracking reactions require a dual-function catalyst with high cracking and hydrogenation activities (Katzer and Sivasubramanian, 1979). The catalyst base, such as acid-treated clay, usually supplies the cracking function of alumina or silica-alumina that is used to support the hydrogenation function supplied by metals, such as nickel, tungsten, platinum, and palladium. These highly acid catalysts are very sensitive to nitrogen compounds in the feed, which break down the conditions of reaction to give ammonia and neutralize the acid sites. Because many heavy gas oils contain substantial amounts of nitrogen (up to approximately 2500 ppm), a purification stage is frequently required. Denitrogenation and desulfurization can be carried out using cobalt-molybdenum or nickel-cobalt-molybdenum on alumina or silica-alumina.

Hydrocracking catalysts typically contain separate hydrogenation and cracking functions. Palladium sulfide and promoted group VI sulfides (nickel-molybdenum or nickel-tungsten) provide the hydrogenation function. These active compositions saturate aromatics in the feed, saturate olefins formed in the cracking, and protect the catalysts from poisoning by coke. Zeolites or amorphous silica-alumina provide the cracking functions. The zeolites are usually type Y (faujasite), ion exchanged to replace sodium with hydrogen, and make up 25 to 50% of the catalysts. Pentasils (silicalite or ZSM-5) may be included in dewaxing catalysts.

Hydrocracking catalysts, such as nickel (5% by weight) on silica-alumina, work best on feedstocks that have been hydrofined to low nitrogen and sulfur levels. The nickel catalyst then operates well at 350 to 370°C (660 to 700°F) and a pressure of about 1500 psi to give good conversion of feed to lower-boiling liquid fractions with minimum saturation of single-ring aromatics and a high ratio of *iso*-paraffin to *n*-paraffin in the lower molecular weight paraffins.

The poisoning effect of nitrogen can be offset to a certain degree by operation at a higher temperature. However, the higher temperature tends to increase the production of material in the methane (CH_4)-to-butane (C_4H_{10}) range and decrease the operating stability of the catalyst so that it requires more frequent regeneration. Catalysts containing platinum or palladium (approximately 0.5%) on a zeolite base appear to be somewhat less sensitive to nitrogen than nickel catalysts, and successful operation has been achieved with feedstocks containing 40 ppm nitrogen.

In addition to the chemical nature of the catalyst, the physical structure of the catalyst is also important in determining the hydrogenation and cracking capabilities, particularly for heavy feedstocks (Kobayashi et al., 1987; Fischer and Angevine, 1986; Kang et al., 1988; van Zijll Langhout et al., 1980). When gas oils and residua are used, the feedstock is present as liquid under the conditions of the reaction. Additional feedstock and the hydrogen must diffuse through this liquid before reaction can take place at the interior surfaces of the catalyst particle.

Catalyst operating temperature can influence reaction selectivity since the activation energy for hydrotreating reactions is much lower than for hydrocracking reactions. Therefore, raising the temperature in a residuum hydrotreater increases the extent of hydrocracking relative to hydrotreating, which also increases the hydrogen consumption (Bridge et al., 1975, 1981).

Clays have been used as cracking catalysts, particularly for heavy feedstocks, and have also been explored in the demetallization and upgrading of heavy crude oil (Rosa-Brussin, 1995). The results indicated that the catalyst prepared was mainly active toward demetallization and conversion of the heaviest fractions of crude oils.

Zeolite catalysts have also found use in the refining industry during the last two decades (Occelli and Robson, 1989; Sherman, 1998). Like silica-alumina catalysts, zeolites also consist of a framework of tetrahedrons, usually with a silicon atom or an aluminum atom at the center. The geometric characteristics of the zeolites are responsible for their special properties, which are particularly attractive to the refining industry (DeCroocq, 1984). Specific zeolite catalysts have shown up to 10,000 times more activity than the so-called conventional catalysts in specific cracking tests.

Zeolites provide the cracking function in many hydrocracking catalysts, as they do in fluid catalytic cracking catalysts. The zeolites are crystalline aluminosilicates, and in almost all commercial catalysts today, the zeolite used is faujasite. Pentasil zeolites, including silicalite and ZSM-5, are also used in some catalysts for their ability to crack long-chain paraffins selectively. Typical levels are 25 to 50 wt% zeolite in the catalysts, with the remainder being the hydrogenation component and a silica (SiO_2) or alumina (Al_2O_3) binder.

While zeolites provided a breakthrough that allowed catalytic hydrocracking to become commercially important, continued advances in the manufacture of amorphous silica-alumina made these materials competitive in certain kinds of applications. Typical catalysts of this type contain 60 to 80 wt% silica-alumina, with the remainder being the hydrogenation component. The compositions of these catalysts are closely held secrets. Over the years, broad ranges of silica/alumina molar ratios have been used in various cracking applications, but silica is almost always in excess for high acidity and stability.

In a well-designed hydrocracking catalyst system, the hydrogenation function adds hydrogen to the tarry deposits. This reduces the concentration of coke precursors on the surface. However, there is a slow accumulation of coke that reduces activity over a 1- to 2-year period. Refiners respond to this slow reduction in activity by raising the average temperature of the catalyst bed to maintain conversions. Eventually, an upper limit to the allowable temperature is reached and the catalyst must be removed and regenerated.

Catalysts carrying coke deposits can be regenerated by burning off the accumulated coke. This is done by service in rotary or similar kilns rather than leaving catalysts in the hydrocracking reactor, where the reactions could damage the metals in the walls. Removing the catalysts also allows inspection and repair of the complex and expensive reactor internals, discussed below. Regeneration of a large catalyst charge can take weeks or months, so refiners may own two catalyst loads, one in the reactor and one regenerated and ready for reload.

Catalysts used in residuum upgrading processes typically use an association of several kinds of catalysts, each of them playing a specific and complementary role (Kressman et al., 1998). The first major function to be performed is hydrodemetallization (HDM). Therefore, the hydrodemetallization catalyst must desegregate asphaltene constituents and remove as much metal (nickel and vanadium) as possible. One catalyst in particular has been developed by optimizing the support pore structure and acidity (Toulhoat et al., 1990). This catalyst allows a uniform distribution of metals deposited, and therefore a high metal retention capacity is reached. A specific hydrodesulfurization catalyst can be placed downstream of the hydrodemetallization catalyst; the main function of such positioning is to desulfurize the already deeply demetallized feedstock as well as to reduce coke precursors. Thus, the main function of the hydrodesulfurization catalyst is not the same as that of the hydrodemetallization catalyst. In addition, for fixed-bed processes, swing-guard reactors may be used to improve the protection of downstream catalysts and increase the unit cycle length. For example, the Hyvahl process (q.v.) includes two swing-guard reactors followed by conventional hydrodemetallization and hydrodesulfurization reactors (DeCroocq, 1997). The hydrodemetallization catalyst in the guard reactors may be replaced during unit operation, and the total catalyst amount is replaced at the end of a cycle.

REFERENCES

Bauman, R.F., Aldridge, C.L., Bearden, R., Jr., Mayer, F.X., Stuntz, G.F., Dowdle, L.D., and Fiffron, E. 1993. *Oil Sands: Our Petroleum Future*, Preprint. Alberta Research Council, Edmonton, Alberta, Canada, p. 269.

Bearden, R. and Aldridge, C.L. 1981. *Energy Progr.* 1: 44.

Billon, A., Morel, F., Morrison, M.E., and Peries, J.P. 1994. Converting residues with IPP's Hyvahl and Solvahl processes. *Rev. Inst. Franç. Pétrole* 49: 495.

Bishop, W. 1990. *Symposium on Heavy Oil: Upgrading to Refining*, Proceedings. Canadian Society for Chemical Engineers, p. 14.

Bridge, A.G. 1997. In *Handbook of Petroleum Refining Processes*, 2nd ed., R.A. Meyers (Ed.). McGraw-Hill, New York, chap. 7.2.

Bridge, A.G., Gould, G.D., and Berkman, J.F. 1981. *Oil Gas J.* 79: 85.

Bridge, A.G., Reed, E.M., and Scott, J.W. 1975. Paper presented at the API Midyear Meeting, May.

DeCroocq, D. 1984. *Catalytic Cracking of Heavy Petroleum Hydrocarbons*. Editions Technip, Paris.

DeCroocq, D. 1997. Major scientific and technical challenges about development of new processes in refining and petrochemistry. *Rev. Inst. Franç. Pétrole* 52: 469.

Fischer, R.H. and Angevine, P.V. 1986. *Appl. Catal.* 27: 275.

Fornoff, L.L. 1982. *Second International Conference on the Future of Heavy Crude and Tar Sands*, Proceedings. Caracas, Venezuela.

Graeser, U. and Niemann, K. 1982. *Oil Gas J.* 80: 121.

Graeser, U. and Niemann, K. 1983. Preprints. *Am. Chem. Soc. Div. Petrol. Chem.* 28: 675.

Gray, M.R. 1994. *Upgrading Petroleum Residues and Heavy Oils*. Marcel Dekker, New York.

Hydrocarbon Processing. 1996. 75: 89.

Hydrocarbon Processing. 1998. 77: 53.

Kang, B.C., Wu, S.T., Tsai, H.H., and Wu, J.C. 1988. *Appl. Catal.* 45: 221.

Katzer, J.R. and Sivasubramanian, R. 1979. *Catal. Rev. Sci. Eng.* 20: 155.

Kobayashi, S., Kushiyama, S., Aizawa, R., Koinuma, Y., Inoue, K., Shmizu, Y., and Egi, K. 1987. *Ind. Eng. Chem. Res.* 26: 2241, 2245.

Kressmann, S., Boyer, C., Colyar, J.J., Schweitzer, J.M., and Viguié, J.C. 2000. *Rev. Inst. Franç. Pétrole* 55: 397.

Kriz, J.F. and Ternan, M. 1994. U.S. Patent 5,296,130. March 22.

Marzin, R., Pereira, P., McGrath, M.J., Feintuch, H.M., and Thompson, G. 1998. *Oil Gas J.* 97: 79.

Ng, S.H. and Rahimi, P.M. 1991. *Energy Fuels* 5: 595.

Niemann, K., Kretschmar, K., Rupp, M., and Merz, L. 1988. *4th UNITAR/UNDP International Conference on Heavy Crude and Tar Sand*, Proceedings. Edmonton, Alberta, Canada, 5, p. 225.

Occelli, M.L. and Robson, H.E. 1989. *Zeolite Synthesis*, Symposium Series 398. American Chemical Society, Washington, DC.

Pereira, P., Flores, C., Zbinden, H., Guitian, J., Solari, R.B., Feintuch, H., and Gillis, D. 2001. *Oil Gas J.*

Pereira, P., Marzin, R., McGrath, M., and Thompson, G.J. 1998. *17th World Energy Congress*, Proceedings. Houston, TX.

Peries, J.P., Quignard, A., Farjon, C., and Laborde, M. 1988. Thermal and catalytic ASVAHL processes under hydrogen pressure for converting heavy crudes and conventional residues. *Rev. Inst. Franç. Pétrole* 43: 847.

Pruden, B.B. 1978. *Can. J. Chem. Eng.* 56: 277.

Pruden, B.B., Muir, G., and Skripek, M. 1993. *Oil Sands: Our Petroleum Future*, Preprints. Alberta Research Council, Edmonton, Alberta, Canada, p. 277.

RAROP. 1991. *Heavy Oil Processing Handbook*, Y. Kamiya (Ed.). Research Association for Residual Oil Processing, Agency of Natural Resources and Energy, Ministry of International Trade and Industry, Tokyo.

Reich, A., Bishop, W., and Veljkovic, M. 1993. *Oil Sands: Our Petroleum Future*, Preprints. Alberta Research Council, Edmonton, Alberta, Canada, p. 216.

Reno, M. 1997. In *Handbook of Petroleum Refining Processes*, 2nd ed., R.A. Meyers (Ed.). McGraw-Hill, New York, chap. 7.3.

Reynolds, J.G. and Beret, S. 1989. *Fuel Sci. Technol. Int.* 7: 165.

Rosa-Brussin, M.F. 1995. *Catal. Rev. Sci. Eng.* 37: 1.

Sherman, J.D. 1998. Synthetic zeolites and other microporous oxide molecular sieves. In *Colloquium on Geology, Mineralogy, and Human Welfare*, Proceedings. National Academy of Sciences, Irvine, CA.

Speight, J.G. 2000. *The Desulfurization of Heavy Oils and Residua*, 2nd ed. Marcel Dekker, New York.

Speight, J.G. 2007. *The Chemistry and Technology of Petroleum*, 4th ed. CRC Press, Taylor & Francis, Boca Raton, FL.

Speight, J.G. and Moschopedis, S.E. 1979. *Fuel Process. Technol.* 2: 295.

Speight, J.G. and Ozum, B. 2002. *Petroleum Refining Processes*. Marcel Dekker, New York.

Sue, H. 1989. *4th UNITAR/UNDP International Conference on Heavy Crude and Tar Sand*, Proceedings. Edmonton, Alberta, Canada, 5, p. 117.

Takeuchi, C., Fukui, Y., Nakamura, M., and Shiroto, Y. 1982. *Ind. Eng. Chem. Process Des. Dev.* 22: 236.

Thompson, G.J. 1997. In *Handbook of Petroleum Refining Processes*, R.A. Meyers (Ed.). McGraw-Hill, New York, chap. 8.4.

Toulhoat, H., Szymanski, R., and Plumail, J.C. 1990. *Catal. Today* 7: 531.

Towler, G.P., Mann, R., Serriere, A.J.L., and Gabaude, C.M.D. 1996. *Ind. Eng. Chem. Res.* 35: 278.

van Driesen, R.P., Caspers, J., Campbell, A.R., and Lunin, G. 1979. *Hydrocarbon Process.* 58: 107.

van Zijll Langhout, W.C., Ouwerkerk, C., and Pronk, K.M.A. 1980. *Oil Gas J.* 78: 120.

Waugh, R.J. 1983. Annual Meeting. National Petroleum Refiners Association, San Francisco.

Wenzel, F. and Kretsmar, K. 1993. *Oil Sands: Our Petroleum Future*, Preprint. Alberta Research Council, Edmonton, Alberta, Canada, p. 248.

10 Commercial Hydrotreating and Hydrocracking

Paul R. Robinson and Geoffrey E. Dolbear

CONTENTS

10.1 INTRODUCTION

Hydrotreaters are the most common process units in modern petroleum refineries. As shown in Table 10.1, the world's hydrotreating capacity is nearly half as large as the world's crude distillation capacity (Stell, 2003). In more than 700 refineries around the globe, there are more than 1300 hydrotreating units. A typical Western petroleum refinery (Figure 10.1) uses at least three hydrotreaters — one for naphtha, one or two for light gas oil, and one or two for heavy gas oil or vacuum gas oil.

Hydrocracking is far less common than hydrotreating, but the number of partial conversion mild hydrocrackers is increasing as refiners build new units to meet clean fuel regulations.

10.2 HYDROPROCESSING UNITS: SIMILARITIES AND DIFFERENCES

Process flow schemes for hydrotreating and hydrocracking are similar. Both use high-pressure hydrogen to catalytically remove contaminants from petroleum fractions. Both achieve at least some conversion, and they use the same kinds of hardware. Therefore, to avoid redundancy, we will discuss them together. As is common in the refining industry, we use the term *hydroprocessing* when a statement applies to both hydrotreating and hydrocracking.

TABLE 10.1
Worldwide Refining Process Units (as of January 1, 2004)

	Crude Distillation	Coking + Visbreaking	FCC	Catalytic Reforming	Hydrotreating	Hydrocracking
Number of units	>710	>330	360	550	1316	168
Total world capacity[a]	82.0	8.0	14.3	11.3	40.3	4.6
Average capacity[b]	114,000	45,700	39,700	20,500	30,600	27,400

[a] Million barrels per calendar day.
[b] Barrels per calendar day.

Source: Stell, J. *Oil & Gas J.* 101(49), December 22, 2003. (With permission.)

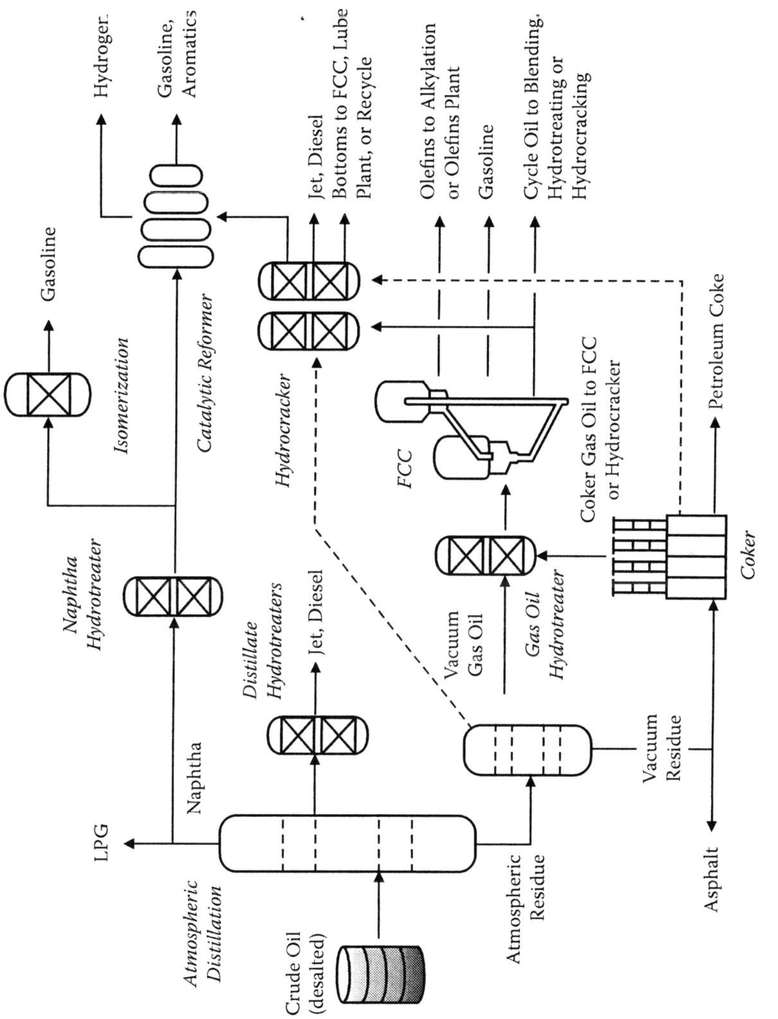

FIGURE 10.1 Layout of a typical high-conversion oil refinery.

TABLE 10.2
Hydrotreating and Hydrocracking: Ranges of H_2 Partial Pressure and Conversion

Process, Feedstock Types	H_2 Partial Pressure psig	H_2 Partial Pressure kPa	Conversion wt%
Hydrotreating			
Naphtha	250–450	1825–3204	0.5–5
LGO (kerosene)	250–600	1825–4238	0.5–5
HGO (diesel), LCO	600–800	4238–5617	5–15
VGO, VBGO, DAO, CGO, HCO	800–2000	5617–13,891	5–15
Residual oil	2000–3000	13,891–20,786	5–15
Mild Hydrocracking			
VGO, VBGO, DAO, CGO, LCO, HCO	800–1200	5617–8375	20–40
Once-Through Hydrocracking			
VGO, VBGO, DAO, CGO, LCO, HCO	1500–2000	10,443–13,891	60–90
Residual oil	2000–3000	13,891–20,786	15–25
Recycle Hydrocracking			
VGO, VBGO, DAO, CGO, LCO, HCO	1500–2000	10,443–13,891	80–99
Ebullated-Bed Hydrocracking			
VGO, VBGO, DAO, HCO	2000	13,891	80–99
Residual oil	2000–3000	13,891–20,786	>50

Note: LGO = light gas oil; HGO = heavy gas oil; LCO = FCC light-cycle oil; HCO = FCC heavy-cycle oil; VGO = vacuum gas oil; VBGO = visbreaker gas oil; DAO = deasphalted oil; CGO = coker gas oil.

As shown in Table 10.2, the extent of conversion is the most significant difference between hydrotreating and hydrocracking. In this context, the term *conversion* is defined as the amount of unconverted oil in the product divided by the amount of unconverted oil in the feed. Unconverted oil is defined as material that boils above a specified temperature. For vacuum gas oil (VGO), a typical specified temperature is 650°F (343°C). Conversion in hydrotreaters is less than 15 wt%, while conversion in hydrocrackers and mild hydrocrackers exceeds 20 wt%.

In hydrotreating units, reactions that convert organic sulfur and nitrogen into H_2S and NH_3 also produce light hydrocarbons. The removal of sulfur from dibenzothiophene (boiling point = 630°F, 332°C) generates biphenyl (492.6°F, 255.9°C). This reaction does not break any carbon-to-carbon bonds, but it does convert a molecule that boils above 600°F (315.5°C) into one that boils below 600°F (315.5°C).

Hydrotreating and hydrocracking differ in other ways. For a given amount of feed, hydrocrackers use more catalyst and operate at higher pressures. They also use different catalysts. Because they make large amounts of light products, hydrocracker fractionation sections must be more complex. In some hydrocrackers, unconverted oil from the fractionation section is recycled, either back to the front of the unit or to a separate cracking reactor.

Many mild hydrocrackers contain at least one bed of cracking catalyst, which allows them to achieve higher conversion — between 20 and 40 wt%. The unconverted bottoms can go to a fluid catalytic cracking (FCC) unit, a lube plant, or fuel oil blender. Due to its high value in other applications, the bottoms are blended into fuel oil only when there is no other feasible option.

In hydrocrackers that process vacuum gas oils or other feeds with similar boiling ranges, the typical once-through conversion exceeds 60 wt%. If the unconverted oil is recycled, the overall conversion can exceed 95 wt%. As with mild hydrocracking, the unconverted bottoms are high-value oils, which usually are sent to FCC units, lube plants, or olefin plants. For heavier feeds — atmospheric and vacuum residues — conversions are much lower, especially in fixed-bed units. In ebullated-bed units, the conversion of 1050°F+ (566°C+) residue can exceed 60 wt%.

Catalytic isomerization and dewaxing is a special kind of hydrocracking used to make high-quality lube base stocks.

10.3 PROCESS OBJECTIVES

Table 10.3 presents a list of feeds and product objectives for different kinds of hydrotreaters and hydrocrackers. In the 1950s, the first hydrotreaters were used to remove sulfur from feeds to catalytic reformers. In the 1960s, the first hydrocrackers were built to convert gas oil into naphtha.

Today, in addition to naphtha, hydrotreaters process kerosene, gas oil, vacuum gas oil, and residue. Hydrocrackers process vacuum gas oil, coker gas oil, visbreaker gas oil, FCC heavy cycle oil, and other feeds that boil between 650 and 1050°F (343 and 566°C). Most residue hydrocrackers use fluidized-bed or ebullated-bed technology.

For hydroprocessing units, product specifications are set to meet plant-wide objectives. For example, the naphtha that goes to catalytic reforming and isomerization units must be (essentially) sulfur-free. Before it can be sold as jet fuel, the aromatics content of kerosene must be low enough to meet smoke-point specifications (American Society for Testing and Materials [ASTM] D1655). Heavier distillates cannot be sold as diesel fuel unless they meet stringent sulfur specifications.

10.3.1 Clean Fuels

As mentioned in Chapter 1, on-road diesel in the United States must contain <15 wppm sulfur by 2006. The sulfur limit for nonroad diesel will be 500 wppm in 2007.

TABLE 10.3
Feeds and Products for Hydroprocessing Units

Feeds	Products from Hydrotreating	Products from Hydrocracking
Naphtha	Catalytic reformer feed	LPG
Straight-run light gas oil	Kerosene, jet fuel	Naphtha
Straight-run heavy gas oil	Diesel fuel	Naphtha
Atmospheric residue	Lube base stock, low-sulfur fuel oil, RFCC[a] feed	Naphtha, middle distillates, FCC feed
Vacuum gas oil	FCC feed, lube base stock	Naphtha, middle distillates, FCC feed, lube base stock, olefin plant feed
Vacuum residue	RFCC feed	Naphtha, middle distillates, RFCC feed
FCC light-cycle oil	Blend stocks for diesel, fuel oil	Naphtha
FCC heavy-cycle oil	Blend stock for fuel oil	Naphtha, middle distillates
Visbreaker gas oil	Blend stocks for diesel, fuel oil	Naphtha, middle distillates
Coker gas oil	FCC feed	Naphtha, middle distillates, FCC feed, lube base stock, olefin plant feed
Deasphalted oil	Lube base stock, FCC feed	Naphtha, middle distillates, FCC feed, lube base stock

[a] RFCC = residue FCC unit or reduced crude FCC unit, both of which are specially designed to process feeds that contain high concentrations of carbon-forming compounds.

The present U.S. specification for gasoline is <30 wppm sulfur. In the European Union, the sulfur content of both gasoline and diesel must be <50 wppm by 2005 and <10 wppm by 2008.

To meet clean fuel specifications, refiners in North America and Europe are increasing their hydroprocessing capabilities and adjusting operations. Two real-world examples are described below.

Example 10.1

A U.S. refinery is planning to produce diesel fuel that contains <15 wppm sulfur by June 2006. At present, the hydrocracker makes 39,000 barrels/day of middle distillate that is nearly sulfur-free. The existing 60,000 barrels/day distillate hydrotreater (DHT) gives a product with 600 to 700 wppm sulfur. Mixing the two streams yields a blend containing 425 to 485 wppm sulfur, which meets existing specifications for low-sulfur diesel fuel (per ASTM D975). To make ultra-low-sulfur diesel (ULSD), the refiner is adding a reactor and a high-pressure amine absorber to the existing DHT, enabling the unit to make a stream with 12 to 18 wppm sulfur. Blending this with distillate from the hydrocracker will give a final product containing 7 to 12 wppm sulfur.

Example 10.2

A European refiner now runs a mild hydrocracker (MHC) to maximize conversion of VGO and to pretreat the feed to its FCC unit. The plant cannot posttreat its FCC gasoline, so the sulfur content of the MHC bottoms must be less than 500 wppm to guarantee that the sulfur content of the FCC gasoline is less than 150 wppm. Other low-sulfur streams (reformate, alkylate, and hydrotreated gas oil) go into the final gasoline blend, so sulfur in the FCC gasoline can exceed the final product limit of 50 wppm.

10.3.2 The Process in Between

As shown in Figure 10.1, hydrocracking often is an in-between process. The required hydrogen comes from catalytic reformers, steam/methane reformers, or both. Liquid feeds can come from atmospheric or vacuum distillation units, delayed cokers, fluid cokers, visbreakers, or FCC units. Middle distillates from a hydrocracker usually meet or exceed finished product specifications, but the heavy naphtha from a hydrocracker usually is sent to a catalytic reformer for octane improvement. The fractionator bottoms can be recycled or sent to an FCC unit, an olefins plant, or a lube plant.

10.4 PROCESS MODELING

During the past 20 years, academic and industrial researchers have developed composition-based kinetic models with hundreds or even thousands of lumps and pure compounds. The quantitative structure-reactivity correlation (QSRC) and linear free energy relationship (LFER) lumping techniques are discussed by Klein and Hou (2006). The structure-oriented lumping (SOL) approach of Quann and Jaffe (1996) yields models rigorous enough for use in closed-loop real-time optimizers (CLRTOs), which automatically adjust setpoints for commercial process units several times each day.

In the composition-based model developed by Lapinas et al. (1991) and applied to a commercial hydrocracker by Pedersen et al. (1995) rate equations are based on the Langmuir–Hinshelwood–Hougen–Watson (LHHW) mechanism for heterogeneous reactions. In brief, the LHHW mechanism describes (1) the adsorption of reactants to acid and metal sites on a catalyst surface; (2) reactions between the reactants, including saturation, cracking, ring opening, dealkylation, hydrodesulfurization (HDS); hydrodenitrogenation (HDN); and so forth, and (3) desorption of products. Inhibition effects are modeled, too. These include the adsorption of organic nitrogen to acid sites and the inhibition of HDS reactions by H_2S.

Rigorous, flow-sheet-based models for hydrocrackers include submodels for furnaces, pumps, compressors, reactors, quench zones, flash drums, recycle gas scrubbers, fractionation towers, and — importantly — economic data. As discussed by Mudt et al. (2006) such models can comprise hundreds of reactions and hundreds of thousands of equations. The model grows when inequalities are

TABLE 10.4
Supports Used in Hydroprocessing Catalysts

Support	Major Use	Acidity
γ-Alumina	Hydrotreating catalysts	Low
Amorphous aluminosilicates	Distillate-selective hydrocracking catalysts	High
Zeolites (X, Y, or mordenite)	High-stability hydrocracking catalysts	Very high

included to ensure a feasible solution that honors process constraints. To solve such models in real time (that is, in less than 1 h), open-equation mathematics and high-powered solvers are used.

10.5 HYDROPROCESSING CATALYSTS

Recent books by Magee and Dolbear (1998) and Scherzer and Gruia (1996) are superb sources of technical information on hydroprocessing catalysts. The hydroprocessing catalyst business is big, with annual sales approaching U.S.$800 million per year. The materials most commonly used to make these catalysts are shown in Table 10.4 and Table 10.5.

In fixed-bed hydroprocessing units, the catalysts must be able to drive the desired reactions, but they also must possess a high surface area and great physical strength, enough to resist crushing under the forces imposed by rapidly flowing high-pressure fluids and the weight of the catalyst itself. A single bed can contain several hundred tons of catalyst.

Chemical reactions take place inside small pores, which account for most of the catalyst surface area. The diameters of these pores range from 75 to 85 Å for catalysts that process light and heavy gas oils. For catalysts that process residue, the average pore size ranges from 150 to 250 Å.

TABLE 10.5
Active Metals Used in Hydroprocessing Catalysts

Metals	Major Use	Activation Method	Hydrogenation Activity
CoMo	HDS	Sulfiding	Moderate
NiMo	HDN, hydrocracking	Sulfiding	High
NiW	HDN, hydrocracking	Sulfiding	Very high
Pd, Pt[a]	Hydrocracking	Reduction by H_2	Highest

[a] Pd and Pt are poisoned by sulfur and can only be used in low-H_2S environments.

10.5.1 Catalyst Preparation

The following steps may be used to prepare the supported metal catalysts used in hydrotreaters and hydrocrackers (Magee and Dolbear, 1998; Scherzer and Gruia, 1996):

- Precipitation
- Filtration (or centrifugation), washing, and drying
- Forming
- Calcining
- Impregnation
- Activation

Other steps, such as kneading, mulling, grinding, and sieving, may also be used. For some catalysts, some of the above-listed steps are eliminated or additional steps are added. For example, if mulling is used to mix active metals with a support, precipitation and impregnation may not be needed.

10.5.1.1 Precipitation

In the catalyst world, precipitation involves combining two solutions to form a desired solid. For example, mixing an aqueous solution of aluminum nitrate [$Al(NO_3)_3$] with sodium aluminate [$Na_2Al_2O_4$] yields aluminum hydroxide [$Al(OH)_3$], which forms a gelatinous solid. As the gel ages, tiny crystals grow larger and a pore structure starts to develop.

The zeolites used in hydrocracking catalysts are also prepared by precipitation. Zeolites occur naturally, but the ones used for catalysis are synthetic. Figure 10.2 outlines a common procedure for synthesizing Na-Y and H-Y zeolites.

These remarkable aluminosilicates can be used as drying agents, ion exchangers, and molecular sieves for gas separation. Their microporosity provides them with high surface area, and they can be converted into solid acids with superb catalytic activity.

The Al(III) atoms in zeolites replace Si(IV) atoms in a SiO_2 superstructure. To maintain a neutral charge, every aluminum atom must be accompanied by a counter-ion such as Na^+, K^+, H^+, NH_4^+, and so forth. Counter-ions can be swapped via ion exchange. When Na-Y zeolite is exchanged with an ammonium salt, the Na^+ ion is replaced by NH_4^+. When NH_4-Y is heated to the right temperature, the ammonium ion decomposes, releasing NH_3 (gas) and leaving behind the highly acidic H-Y zeolite.

The synthetic zeolites used in catalysts for hydrocracking include X, Y, mordenite, and ZSM-5. The latter is made by including a soluble organic template, such as a quaternary ammonium salt, in the mix of raw materials. ZSM-5 is used for catalytic dewaxing. Due to its unique pore structure, it selectively cracks waxy n-paraffins into lighter molecules. It is also used in FCC catalysts to increase propylene yields.

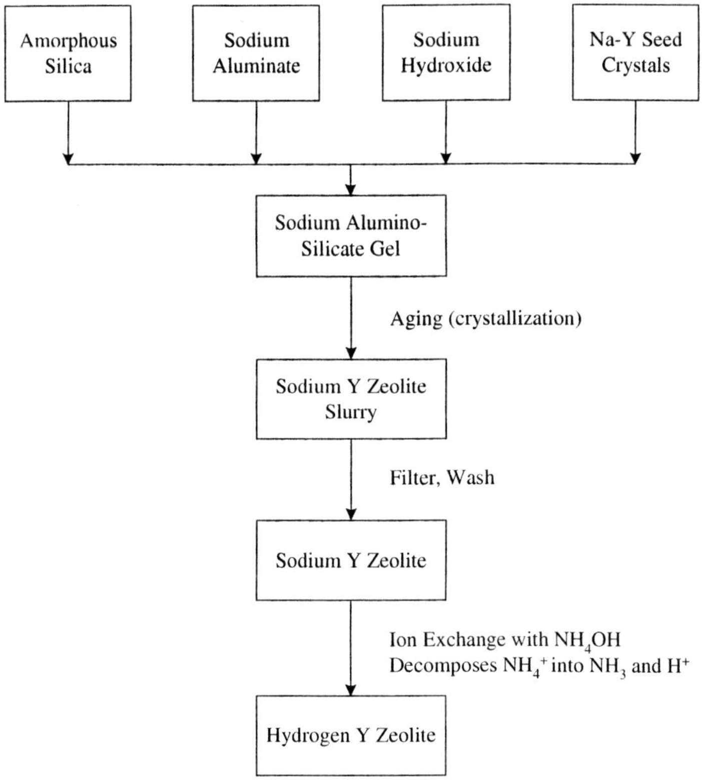

FIGURE 10.2 Synthesis procedure for H-Y zeolite.

10.5.1.2 Filtration, Washing, and Drying

Filtration, washing, and drying remove undesired impurities. In our $Al(OH)_3$ example, sodium nitrate is washed away with water. Sometimes ammonium hydroxide is added to expedite sodium removal. Subsequent air- and oven-drying removes most of the excess water and initiates the transformation of $Al(OH)_3$ into alumina (Al_2O_3).

10.5.1.3 Forming

Catalyst and support precursors can be formed into extrudates, spheres, or pellets. Extrudates are generated by forcing a paste (for example, formed by mixing powdered-alumina with water) through a die. Adding peptizing agents such as nitric acid increases the average pore size of the product. Raising the extrusion pressure tends to decrease the average pore size.

The resulting spaghetti-like strands are dried and broken into short pieces with a length/diameter ratio of 3 to 4. The particles are dried and then calcined. In our alumina example, calcination decomposes residual ammonium nitrate. It also hardens the particles and completes the conversion of $Al(OH)_3$ into Al_2O_3. The preferred alumina for catalyst supports is γ-alumina, also known as bohemite. This material has a high surface area, great physical strength, and a well-defined network of pores. If the calcination temperature gets too high, γ-alumina transforms into α-alumina or β-alumina, whose physical properties are far less desirable.

An extrudate cross section can be circular or shaped like a three- or four-leaf clover without the stem (Gruia, 2006). Compared to cylindrical extrudates, clover-leaf (multilobe) catalysts have a higher surface-to-volume ratio. In trickle-bed hydroprocessing reactors, they have less resistance to diffusion and a lower pressure drop. Spherical catalysts are made by (1) spray-drying slurries of catalyst precursors, (2) spraying liquid onto powders in a tilted rotating pan, or (c) dripping a silica-alumina slurry into hot oil (Magee and Dolbear, 1998). Pellets are made by compressing powders in a dye.

10.5.1.4 Impregnation

Impregnation is a common technique for distributing active metals within the pores of a catalyst support. Calcined supports are especially porous. Like sponges, they use capillary action to suck up aqueous solutions containing active metals. For some catalysts, the support is soaked in excess metal-containing solution, which saturates the pores fully.

In the incipient wetness method, precise amounts of solution are added — just enough to leave the support dry to the touch. After a drying step, additional solution may be added to increase loading of the same or different active metal.

10.5.1.5 Activation

Prior to use, most nonnoble-metal catalysts are activated (sulfided) by circulating hydrogen and a light, sulfur-containing start-up oil through the catalyst. Often, the start-up oil is spiked with dimethyl sulfide (CH_3–S–CH_3) or dimethyl disulfide (CH_3–S–S–CH_3). The temperature is raised slowly to the decomposition temperature of the sulfiding agent. The process continues until breakthrough, that is, the point at which significant amounts of H_2S appear in the recycle gas.

During dry sulfiding, a mixture containing 2 to 5 vol% H_2S in hydrogen is circulated through the catalyst. The temperature is increased slowly to the temperature at which the unit is expected to operate. The process continues until the exit gas contains the same amount of H_2S as the inlet gas.

Most manufacturers offer presulfided catalysts, which allow a refiner to shorten the start-up of a unit by 2 or 3 days. That may not seem like much, but for a 40,000 barrel/day FCC feed pretreater, it can generate up to U.S.$500,000 in extra income.

10.5.2 Noble-Metal Catalysts

Some hydrocracking catalysts contain small amounts of highly dispersed platinum or palladium. These noble metals are expensive, but their loading is low — 0.6

to 1.0 wt% — and their high hydrogenation activity justifies the cost. They are added to hydrocracking catalysts by impregnation with tetraammine complexes — $Pt(NH_3)_4^{2+}$ or $Pd(NH_3)_4^{2+}$. When the catalysts are heated in air to about 840°F (450°C), the complexes decompose, giving off ammonia and leaving behind divalent metal oxides.

In commercial hydrocrackers, catalysts containing noble-metal oxides are activated by direct reduction with high-pressure hydrogen at 700°F (350°C).

10.5.3 Hydrotreating Catalysts

Hydrotreating catalysts comprise oxides of either Mo or W and either Co or Ni on a support comprised of γ-alumina. Usually, CoMo catalysts are better for HDS while NiMo catalysts are better for HDN. NiW catalysts are especially active for the saturation of aromatics. Typical physical properties are shown in Table 10.6.

Hydrotreating catalyst particles are surprisingly small, with diameters of 1.5 to 3.0 mm and length/diameter ratios of 3 to 4. In many units, ceramic balls or successively larger catalyst particles are loaded on top of the first catalyst bed. This graded bed protects the bulk of the catalyst by filtering particulate matter out of the feed.

10.5.4 Hydrocracking Catalysts

Commercial hydrocracking catalysts comprise active metals on solid, highly acidic supports. The active metals are Pd, NiMo, and NiW, all of which catalyze both hydrogenation and dehydrogenation reactions. The most common supports are synthetic crystalline zeolites and amorphous silica-aluminas.

Hydrocracking catalyst shapes can be spherical or cylindrical, with gross dimensions similar to those for hydrotreating catalysts.

As already mentioned, in most hydrocrackers, the first few catalyst beds contain a high-activity HDN catalyst, which also is active for HDS, saturation of olefins, and saturation of aromatics. Other hydrocrackers use a bifunction catalyst — one that is active for both hydrotreating and hydrocracking — in all catalyst beds.

TABLE 10.6
Physical Properties for Hydrotreating Catalysts

Property	Low	High
Surface area, m^2/g	150	250
Pore volume, ml/g	0.5	1.0
Pore diameter (average), Å	75	250
Bulk density, lb/ft^3	30	60
Bulk density, kg/m^3	490	980
Co or Ni (as CoO or NiO), wt%	3	8
Mo or W (as MoO_3 or WO_3), wt%	10	30

10.5.5 Catalyst Cycle Life

Catalyst cycle life has a major impact on the economics of fixed-bed refinery units, including hydrotreaters and hydrocrackers. Cycles can be as short as 12 months and as long as 60 months. Two-year cycles are typical. At the start of a cycle, average reactor temperatures are low — 620 to 660°F (327 to 349°C). As the cycle proceeds, the catalyst deactivates and refiners must raise temperatures to maintain conversion. A catalyst cycle is terminated for one of the following reasons, whichever occurs first. Note that only one of the listed events relates directly to catalyst activity.

1. *The temperature required to achieve the unit's main process objective hits a metallurgical limit.* Or alternatively, the main process objective can be met only at reduced feed rate. To ensure safe operation, the maximum average reactor temperature is about 760°F (404°C) and the maximum peak temperature is about 800°F (427°C).
2. *Side reactions are starting to cause process or economic problems.* If the production of light gases exceeds the capacity of one or more towers in the downstream gas plant, operators must decrease feed rate or reduce conversion. Both options are expensive. Excess gas production consumes expensive hydrogen and converts it into low-value liquefied petroleum gas (LPG), which also is expensive. Running at high temperature decreases selectivity to middle distillates and increases aromatics in middle distillates. At some point, due to one or more of these factors, refinery-wide economics show that it is better to shut down for a catalyst change vs. trying to keep limping along — even though metallurgical limits have not yet been reached.
3. *The recycle compressor cannot overcome pressure drop across the unit.* The overall pressure drop is the difference in pressure between the recycle compressor suction and the recycle compressor discharge. At start-of-run, the pressure drop across the catalyst is low — 3 to 10 psi (0.2 to 0.7 bar) for each bed — but it increases as the run proceeds. Usually, the increase is largest in the first catalyst bed, which is most susceptible to fouling. Attempts to continue running a unit despite very high pressure drop can deform the quench-deck support beams inside a reactor.
4. *A related unit has to shut down for more than a few weeks.* Related units might include an upstream vacuum distillation unit, an upstream hydrogen source, or a downstream FCC unit. In refineries with enough intermediate tankage, hydroprocessing units can continue to run for a few days despite an interruption in the supply of liquid feed, but a loss of hydrogen supply can cause an immediate shutdown. At best, if the unit gets hydrogen from multiple sources, the feed rate must be reduced.

5. *Major process upsets.* Most process upsets are caused by sudden changes in feed quality. For a fixed-bed VGO hydrotreater, a slug of residue can poison part of the catalyst with trace metals such as Fe, Ni, V, and Si, or foul it with particulates, asphaltenes, or refractory carbon. In fixed-bed units, poisoning and fouling usually are confined to the top few feet of the first catalyst bed. If so, the ruined catalyst can be skimmed off the top and replaced during a brief, scheduled shutdown. A brief, scheduled shutdown does not require a cycle-ending catalyst change-out.
6. *Equipment failure.* Hardware problems occur most frequently in rotating equipment — pumps and compressors. Fortunately, many problems can be detected in advance, allowing operators to schedule a brief shutdown for preventive maintenance.

Process variables that increase or decrease the rate of catalyst deactivation are shown in Table 10.7.

Hydrogen keeps the catalyst clean by inhibiting coke formation. This explains why increasing the hydrogen partial pressure decreases the rate of catalyst deactivation.

Raising the temperature increases the rates of most hydrocracking reactions, including coke formation. Raising the hydrogen/oil ratio increases heat removal, which limits temperature rise.

If the feed rate goes up and targets for HDS, HDN, or conversion remain the same, the temperature must go up. If the feed rate goes up and the temperature does not, then HDS, HDN, or conversion will decrease.

TABLE 10.7
Factors Affecting Catalyst Cycle Life

	Effect on Cycle Life	Comment
Higher-H_2 partial pressure	+	
Higher recycle gas rate	+	Increases H_2 partial pressure
Higher makeup gas purity	+	Increases H_2 partial pressure
Increased purge of recycle gas	+	Increases H_2 partial pressure
Higher fresh feed rate	–	
Higher conversion	–	
Higher fresh feed endpoint	–	Increases rate of catalyst coking; can increase pressure-drop buildup rate
Higher fresh feed impurities[a]	–	Related to feed type and feed endpoint
Process upsets[b]	–	

[a] Deleterious feed impurities include sulfur, nitrogen, refractory carbon, asphaltenes, metals (nickel, vanadium, iron, silicon), and particulate matter (coke fines, FCC catalyst fines).

[b] Process upsets include "burps" in upstream units that feed the hydrocracker, equipment failures (typically loss of a feed pump or compressor), or temperature excursions requiring depressuring.

Increasing the feed endpoint or density tends to increase the amount of coke precursors in the feed. The precursors include asphaltenes, refractory carbon, and polynuclear aromatic hydrocarbons (PAH).

10.5.5.1 Catalyst Regeneration and Rejuvenation

After working 24/7 for a year or two (or in some cases five) in a fixed-bed hydroprocessing unit, the catalyst is spent. The entire unit is shut down and catalyst is removed. During the shutdown, which typically lasts 3 to 4 weeks, refiners inspect and repair equipment. Meanwhile, the catalyst is shipped to an off-site facility, where it is regenerated by controlled combustion in air, air plus oxygen, or air plus steam. During combustion, accumulated coke is converted to CO_2 and CO plus small amounts of SO_2 and NOx, which are formed from the sulfur and nitrogen in the coke. Typically, the temperature used for regeneration in air is 750 to 930°F (400 to 500°C).

The regenerated catalyst may also undergo rejuvenation, a wet process in which the active metals are chemically redispersed. A combination of regeneration and rejuvenation can restore a catalyst to more than 95% of its original activity.

Inevitably, some particles break apart during the unloading, transportation, regeneration, and rejuvenation of spent catalysts. If part of the catalyst is contaminated with Fe, Ni, V, or Si, that part cannot be regenerated. Typically, losses due to fragmentation and fouling amount to 10 to 15%.

In the bad old days, regeneration meant burning coke off the catalyst while it was still inside the reactor. Today, *in situ* regeneration is rare because it is hard to control and often gives poor results. A poor regeneration is costly, because afterwards the unit's performance will be poor. With a crippled catalyst, the unit may have to limp along for several months at lower feed rates and lower severity. Worst of all, the catalyst will not last long, which means that it will have to be regenerated or replaced sooner rather than later.

10.5.5.2 Catalyst Reclamation

Even though noble-metal hydrocracking catalysts contain only small amounts of Pd or Pt, these metals are so expensive that recovering the metals is more cost-effective than throwing them away. Other hydroprocessing catalysts contain Mo or W, Ni, or Co. Spent hydrotreating catalysts — especially those used to hydrotreat residue — can be very rich in vanadium, richer than many ores.

Reclamation companies convert these materials into salable products using different combinations of oxidation, pyrolysis, dissolution in acid or alkali, precipitation, extraction, or ion exchange. Depending on the process used, the salable products may include several of the materials shown in Table 10.8.

The book by Scherzer and Gruia (1996) provides a well-written description of catalyst reclamation processes used by four major companies — CRI-MET, Eurecat, Gulf Chemical, and TNO/Metrex.

TABLE 10.8
Some of the Materials Sold by Catalyst Reclamation Companies

Material	Formula
Palladium metal or chloride salt	Pd or Na_2PdCl_4
Platinum metal or chloride salt	Pt of Na_2PtCl_4
Molybdenum trisulfide	MoS_3
Molybdenum oxide	MoO_3
Ammonium molybdate	$(NH_4)_2Mo_4O_{13}\cdot 2H_2O$
Sodium molybdate	$Na_2MoO_4\cdot 2H_2O$
Tungsten trioxide	WO_3
Ammonium para-tungstate	$(NH_4)_{10}W_{12}O_{41}\cdot 5\ H_2O$
Sodium tungstate	$Na_2WO_4\cdot 2H_2O$
Vanadium pentoxide	V_2O_5
Sodium (meta) vanadate	$NaVO_3$
Nickel metal or chloride	Ni or $NiCl_2$
Cobalt metal or chloride	Co or $CoCl_2$
Nickel-cobalt concentrate	Ni_xCo_y
Iron-molybdenum concentrate	Fe_xMo_y
Alumina hydrate	$Al_2O_3\cdot 3H_2O$

10.6 PROCESS FLOW

10.6.1 TRICKLE-BED UNITS

Most hydrotreaters and hydrocrackers are trickle-bed units. A classic article by Satterfield describes the fundamental behavior of such units, in which mixtures of liquid and gaseous reactants pass down over fixed beds of catalyst. In hydroprocessing units, the liquid reactants are petroleum fractions, and the gaseous reactant is hydrogen.

Figure 10.3 shows a flow scheme for a once-through unit designed to process heavy gas oil feeds. Designs offered by major process licensors can differ in several areas, which correspond to the bold numbers on the diagram.

1. *Heaters.* Units with gas-only heaters mix hot gas with preheated liquid feed just before the reactants enter the first reactor. Other designs use a gas-plus-oil heater to bring the mixed fluids up to reaction temperature.
2. *Reactors, catalyst beds, and quench zones.* Addition of hydrogen typically occurs with heat release, and most hydroprocessing reactions are exothermic. The heat released in naphtha and kerosene hydrotreaters is relatively low, so units designed for these feeds may use just one reactor that contains a single catalyst bed. However, for heavier feeds or feeds that contain large amounts of sulfur, aromatics, or olefins, the

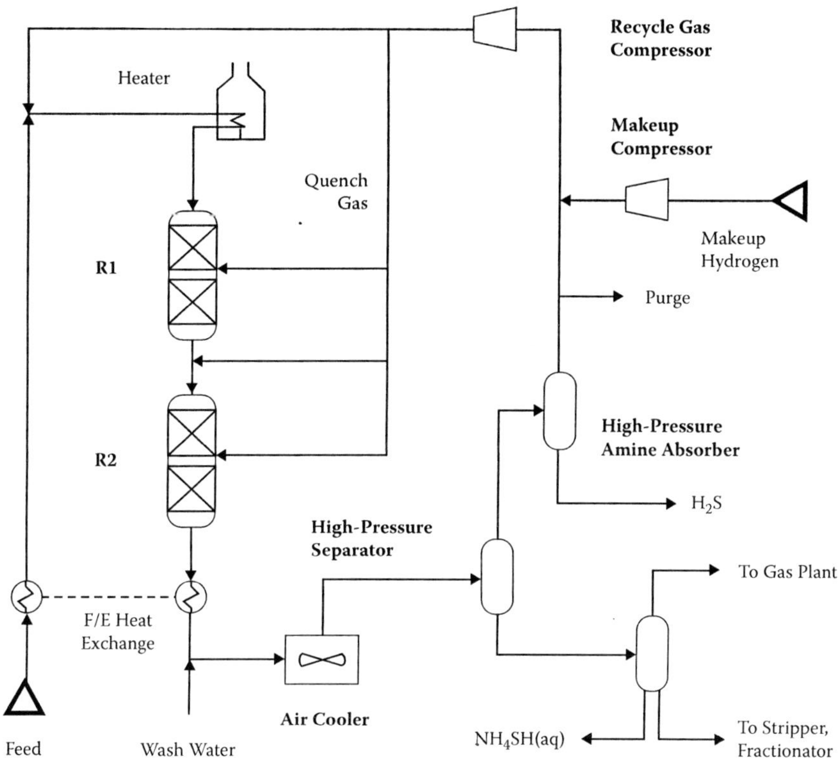

FIGURE 10.3 Once-through hydroprocessing unit: two separators and recycle gas scrubber.

total increase in temperature can exceed 180°F (100°C). It is unsafe to allow that much temperature rise in a single bed of catalyst. To divide the heat release into smaller, safer portions, commercial units use multiple catalyst beds with cooling in between. A unit can have one bed per reactor, or multiple beds in each reactor with quench zones in between. For simplicity, Figure 10.4 shows only 4 catalyst beds, but most hydrocrackers have more; some have as many as 30.

In a quench section (Figure 10.4), hot process fluids from the preceding bed are combined with relatively cold hydrogen-rich quench gas before the mixture passes into the next bed. We can think of a catalyst bed as a stack of thin, horizontal discs. Ideally, the top disc is the coolest, the bottom disc is the hottest, and at every point in each given disc, temperatures are identical. But in real units, the downward flow of reactants is never perfectly uniform, so the temperatures within the discs are different, especially near the bottom.

The difference between the highest and lowest temperature at the bottom of a catalyst bed is called the radial temperature difference (RTD).

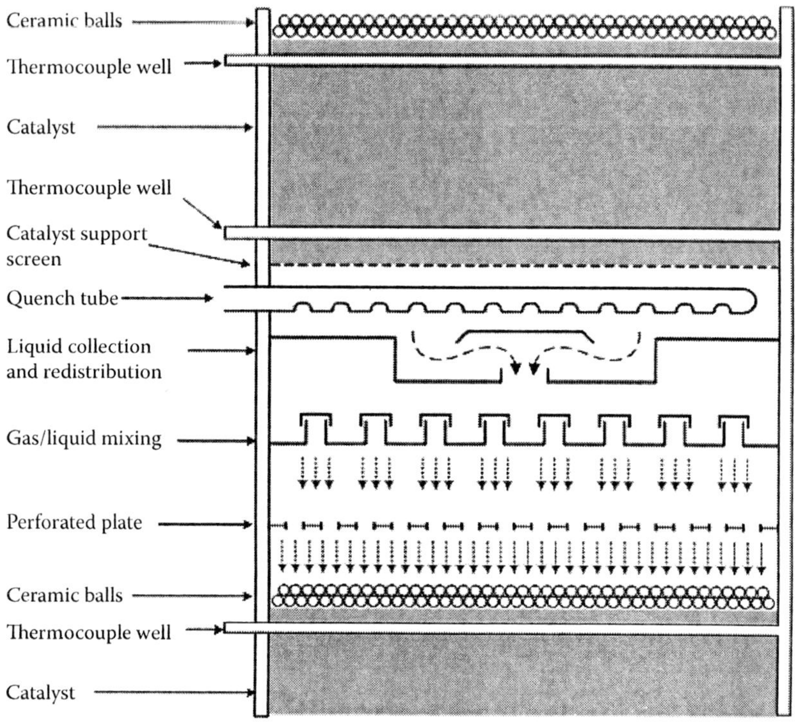

FIGURE 10.4 Hydroprocessing reactor: quench zone.

The truth is, we never know the actual highest and lowest temperatures, because we cannot place thermocouples everywhere. But if the measured RTD is small — less than 5°F (3°C) — we can assume that the actual RTD is also small, and that flow through the bed is nearly uniform. If the measured RTD is large, the actual RTD is almost certainly larger, and we have to be concerned about hot spots, flow blockages, and other potentially dangerous symptoms of maldistribution.

Modern quench sections are designed to do three things: (1) lower the overall temperature of the reacting fluids, (2) reduce radial maldistribution with radial mixing, and (3) redistribute the reactants and deliver them to the next bed. The major parts of a quench deck are the quench tube, the liquid collector and redistributor, the gas/liquid mixing zone, and the final distributor.

Quench tubes bring quench gas into the reactor. Some are very simple — just a tube with a series of holes in it. Others, such as the ExxonMobil spider vortex design, are more complex, distributing gas horizontally through several spokes to different parts of the quench deck.

In the liquid collector and redistributor, liquids are forced to flow down two angled slides into a raceway. The slides give the liquids some angular momentum, and the raceway gives them time to mix. More than anything else, this part of the quench deck reduces RTD.

In the gas/liquid mixing zone, a bubble-cap tray or similar device provides intimate contact between gases and liquids from the redistribution zone. The final distributor sends a fine spray of fluids down to the catalyst bed below.

In residue hydroprocessing units, heat release is high, but some licensors avoid using intrareactor quench because residue feeds often form lumps of coke-bonded catalyst in fixed-bed units. In reactors with complex internals, such lumps are very hard to remove during a catalyst change-out. Therefore, fixed-bed residue units often comprise three or more one-bed reactors in series with quench in between. In many cases, the first reactor is a guard bed filled with one or more catalysts designed to remove metals.

3. *Catalysts.* Hydrotreaters are loaded with either a CoMo HDS or NiMo HDN catalyst, or both. NiMo catalysts are better for the saturation of aromatics, which is required for the removal of hindered sulfur compounds during deep desulfurization. Therefore, some refiners load a layer of NiMo catalyst on top of a CoMo catalyst in diesel desulfurization units. Recently, catalyst manufacturers have been offering trimetallic (CoNiMo) hydrotreating catalysts.

   Most of the cracking in hydrocracking units is driven by catalysts with high acidity. The acidic sites are inhibited by organic nitrogen, so the first several catalyst beds in a hydrocracking unit typically contain a high-activity HDN catalyst. In a few units, all beds in a hydrocracker are filled with an amorphous dual-function catalyst, which catalyzes both HDN and cracking. This type of catalyst has a high selectivity for producing middle distillates from VGO.

   The last bed in a hydrocracker often contains a final layer of posttreat catalyst to remove mercaptans.

4. *Makeup and recycle hydrogen.* Compressors for makeup hydrogen are reciprocating machines, most of which are driven by electric motors. Recycle gas compressors can be reciprocating or centrifugal; the latter are often driven by steam. In naphtha hydrotreaters, the high-pressure off-gas can be purer than the makeup gas, because (a) conversion is nil and (b) liquids in the makeup gas are absorbed by the naphtha. In most other units, the makeup gas is purer than the recycle gas.

   Makeup hydrogen can enter the unit at the cold high-pressure separator (CHPS), at the suction of the recycle gas compressor, or at the discharge of the recycle gas compressor. If the makeup comes in at the CHPS, the makeup compressor discharge pressure is lower, which can reduce electricity costs. However, if part of the recycle gas is purged after leaving the CHPS, part of the incoming makeup gas goes right back out again. If the makeup comes in at the discharge of the recycle gas compressor, the discharge pressure of the makeup compressor is higher, but none of the high-purity makeup is lost with purge gas.

5. *High-pressure amine absorption.* Prior to the advent of ultra-low-sulfur fuels, it was rare to find hydroprocessing units with a high-pressure amine absorber to remove H_2S from the recycle gas. H_2S inhibits HDS reactions and lowers the purity of the recycle gas. For both of these reasons, high-pressure amine absorbers are now included in most new and revamped diesel hydrotreaters and mild hydrocrackers.
6. *Product cooling and separation.* Commercial units comprise a number of different product cooling and flash drum configurations. The simplest comprises a feed/effluent heat exchanger train, a large air- or water-cooled heat exchanger, and one or two flash drums.

   Heavy-feed units have at least a CHPS and a low-pressure separator (LPS). The CHPS overhead stream can go directly to the recycle gas system or through a high-pressure amine absorber for removal of H_2S. The CHPS bottoms go to the CLPS. Sometimes the pressure differential between the CHPS and CLPS is used to drive a power recovery turbine. As shown in Figure 10.5, some units include a hot high-pressure separator (HHPS) upstream from the CHPS. The HHPS overhead goes through a cooler to the CHPS, and HHPS bottoms go through a cooler to the LPS. This arrangement provides better heat recovery. In single-stage hydrocrackers with recycle of unconverted oil, hot separation minimizes fouling caused by the accumulation of PAH in the recycle oil.
7. *Wash water addition.* As mentioned above, HDS and HDN reactions produce H_2S and NH_3, respectively. Wash water is injected into the effluent from the last reactor to convert almost all of the NH_3 and some of the H_2S into aqueous ammonium bisulfide, $NH_4HS(aq)$. The $NH_4HS(aq)$ is rejected from the unit as sour water in the low-pressure flash drum.
8. *Fractionation.* For product fractionation, HDS units that treat naphtha or light gas oil may use a simple steam stripper to remove H_2S and traces of light hydrocarbons from the liquid product (CLPS bottoms). An absorber may be used to recover C_3+ compounds from the CLPS overhead.

   Conversion units may employ a full-fledged fractionation train, with a preflash tower to remove light ends; an atmospheric fractionator to separate light naphtha, heavy naphtha, middle distillates, and unconverted oil; and a vacuum tower to maximize the recovery of diesel. Some hydrocrackers use the atmospheric tower to produce full-range naphtha, which is then separated into light and heavy fractions in a naphtha splitter.
9. *Recycle of fractionator bottoms.* In full-conversion hydrocrackers, unconverted oil from the fractionator is recycled. Single-stage units

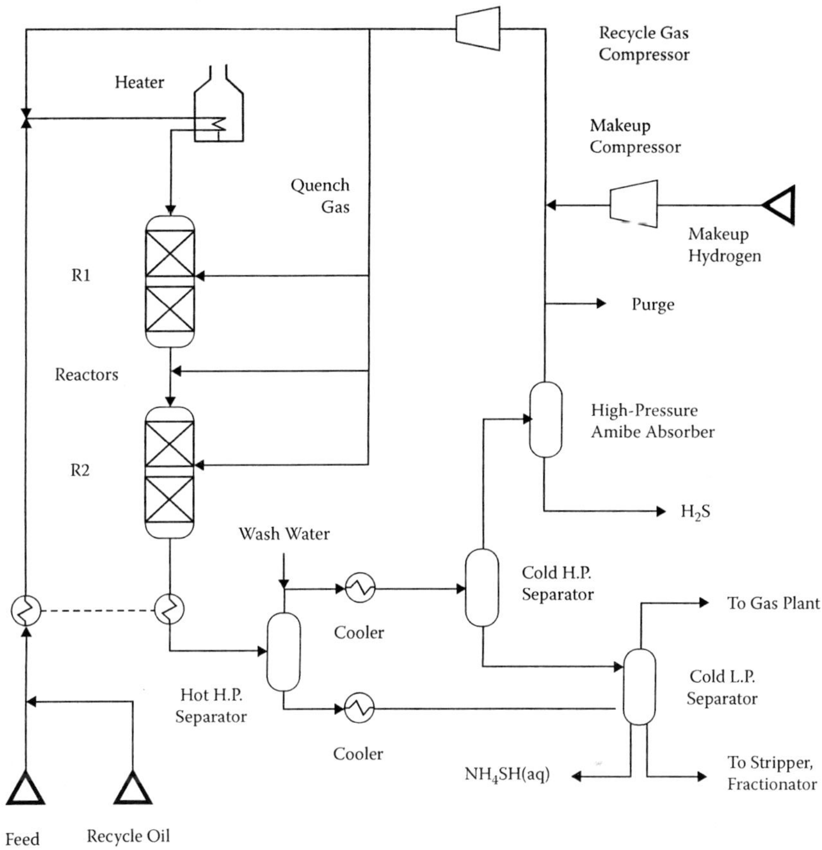

FIGURE 10.5 Single-stage hydrocracker: HHPS, recycle to R1.

with multiple reactors (Figure 10.5) send the recycled oil either to the hydrotreating reactor (R1) via the feed surge drum or to the hydrocracking reactor (R2). Recycle to R1 means that R1 must be larger, but recycle to R2 eliminates an expensive and troublesome high-pressure pump.

Figure 10.6 shows a two-stage hydrocracker. In these units, unconverted oil goes to a separate cracking reactor (R3) with its own high-pressure separator.

The unit shown in Figure 10.6 uses a single makeup and recycle gas system to supply all reactors. In other units, the second stage has a separate gas system. Units with a common recycle gas system need only one recycle compressor, but in units with two gas systems, the second stage can operate at lower pressure, which can reduce both investment and operating costs. Also, the second stage can use sweet gas (no H_2S) rather than sour, allowing the refiner to employ a wider range of catalysts.

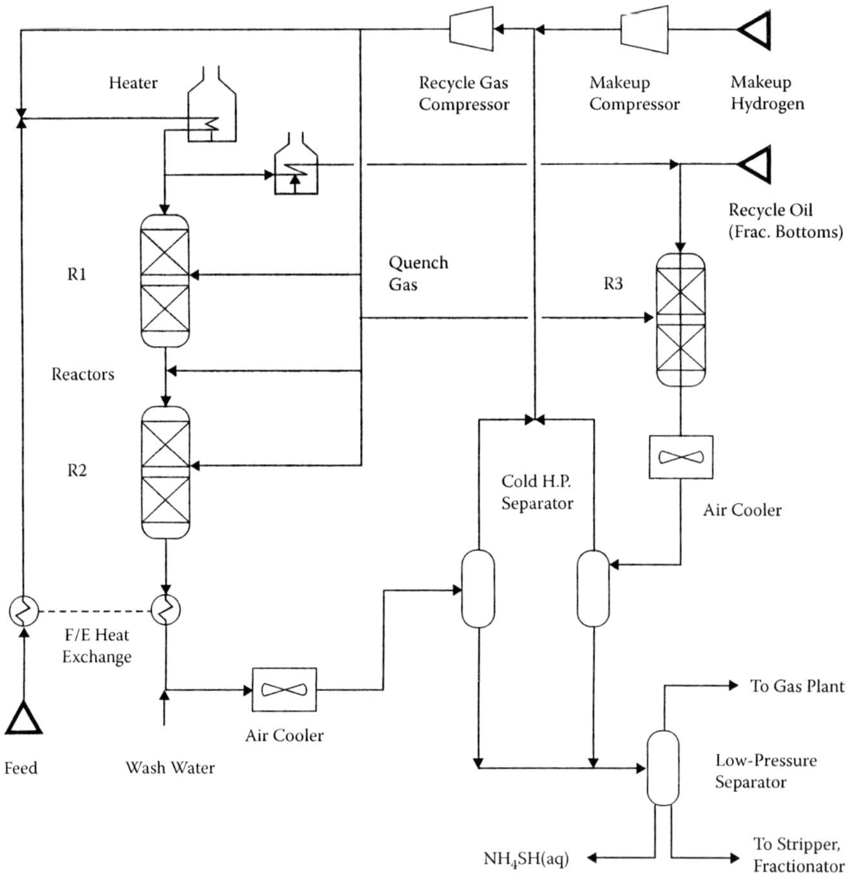

FIGURE 10.6 Two-stage hydrocracker: common recycle gas system.

Early fixed-bed hydrocrackers used a separate hydrotreat flow scheme, which resembles a two-stage design with nothing but hydrotreating catalyst in the first stage. This flow scheme is discussed in further detail by Gruia (2006).

10.6.2 Slurry-Phase Hydrocracking

Slurry-phase hydrocracking converts residue in the presence of hydrogen under severe process conditions — more than 840°F (450°C) and 2000 to 3000 psig (13,891 to 20,786 kPa). To prevent excessive coking, finely powdered additives made from carbon or iron salts are added to the liquid feed. Inside the reactor, the liquid/powder mixture behaves as a single phase due to the small size of the additive particles. Residue conversion can exceed 90%, and the quality of converted products is fairly good.

Unfortunately, the quality of the unconverted pitch is poor, so poor that it cannot be used as a fuel unless it is blended with something else — coal or heavy fuel oil. Even then, its high metals and sulfur content can create problems.

At the 5000 barrels/day CANMET demonstration plant in Canada, the pitch is sent to a cement kiln for use as a clinker. Other slurry-phase processes include COMBIcracking (developed by Veba Oel), Aurabon (UOP), and HDH Cracking (Intevep). Although several slurry-phase demonstration plants have been built, the pitch-disposal problem has kept it from gaining industry-wide acceptance.

10.6.3 Ebullating-Bed Units

In contrast to fixed-bed VGO hydrocrackers, ebullating-bed units can (and do) process residual oils. In ebullating-bed units (Figure 10.7), hydrogen-rich recycle

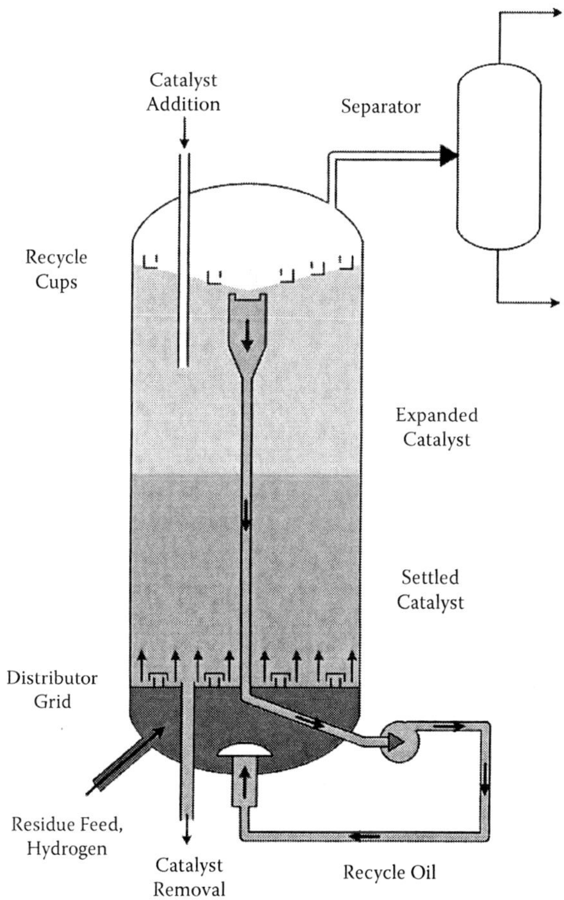

FIGURE 10.7 Ebullating-bed hydrocracking reactor.

gas is bubbled up through a mixture of oil and catalyst particles. This provides three-phase turbulent mixing, which is needed to ensure a uniform temperature distribution. At the top of the reactor, catalyst is disengaged from the process fluids, which are separated in downstream flash drums. Most of the catalyst is returned to the reactor. Some is withdrawn and replaced with fresh catalyst. The two major ebullating-bed processes are H-Oil, which is offered for license by Axens (IFP), and LC-Fining, which is offered by Chevron Lummus Global. Their main advantages are:

- High conversion of atmospheric residue, up to 90 vol%.
- Better product quality than many other residue conversion processes, especially delayed coking.
- Long run length. Catalyst life does not limit these units. Fresh catalyst is added and spent catalyst is removed continuously. Therefore, barring any mechanical problems, the units can run for a much longer time than fixed-bed residue units.

10.7 PROCESS CONDITIONS

For fixed-bed hydroprocessing units, the process conditions — pressure, temperature, space velocity, and catalyst — are determined by feed quality and process objectives. Table 10.9 shows typical process conditions for the hydrotreating of different feeds in fixed-bed hydrotreating units. The values shown are approximate.

TABLE 10.9
Typical Process Conditions for Hydrotreating Different Petroleum Fractions

	Naphtha	Kerosene	Diesel	VGO	Residue
WART[a]					
°F	530	550	575–600	680–700	700–725
°C	277	288	300–315	360–370	370–385
H_2 pressure[b]					
psig	250–450	250–600	600–800	800–2000	>2000
kPa	1825–3204	1825–4238	4238–5617	5617–13,891	>13,891
LHSV	5	4	2–3	0.8–1.5	0.5
H_2/oil ratio[c]					
scf/bbl	350	450	800	1200	>3000
M^3/m^3	60	80	140	210	>525

[a] Approximate weighted average reactor temperature at start of run.
[b] Approximate hydrogen partial pressure at the high-pressure separator.
[c] Approximate hydrogen-to-oil ratio at the first reactor inlet.

The H_2/oil ratios are for units in which off-gas from the high-pressure separator is recycled. For once-through naphtha hydrotreaters associated with catalytic reformers, the H_2/oil ratio can be much higher than 350 scf/bbl (60 m^3/m^3). For units that treat olefinic cracked stocks from FCC or coking units, H_2/oil ratios are higher to control the extra heat released by olefin saturation.

10.8 YIELDS AND PRODUCT PROPERTIES

Table 10.10 illustrates the yield flexibility of recycle hydrocracking. The ability to swing in just a day or two from 90 vol% full-range naphtha to >75 vol% full-range diesel provides unparalleled capability to respond to short-term changes in market conditions — if the refinery has sufficient blending, storage, and distribution capacity. To shift the product slate, operators adjust reactor temperatures and change cut points in the fractionation section.

For all process units, product specifications are set to meet refinery-wide objectives. For example, if a refinery wants to produce diesel fuel containing <15 wppm sulfur, and if its hydrocracker makes 40,000 barrels/day of sulfur-free middle distillate, the product sulfur specification for its 20,000 barrels/day distillate hydrotreater (DHT) could be as high as 45 wppm — if a blend of the two streams satisfies the requirements of ASTM D975, which is the standard specification for heavy-duty diesel fuel in the United States. In practice, the DHT sulfur target would be lower than 45 wppm to cushion the refinery against upsets and measurement error. For a diesel fuel containing 10 wppm sulfur, the analytical reproducibility for ASTM D5453 is ±1.8 wppm. For a diesel containing 50 wppm sulfur, the reproducibility is ±8.1 wppm. ASTM D5453 is an x-ray fluorescence method for measuring sulfur in distillate fuels, including ultra-low-sulfur diesel.

10.9 OVERVIEW OF ECONOMICS

10.9.1 Costs

Throughput, operating pressure, and process configuration — once-through or recycle of unconverted oil — are the major factors affecting construction costs for hydroprocessing units, which range from $1000 to $4000 per daily barrel. On this basis, a fully installed 25,000 barrels/day hydrocracker can cost between U.S.$40 million and U.S.$100 million. These estimates do not include costs for a hydrogen plant and off-site utilities.

For hydrotreaters, operating costs are roughly U.S.$1.7 per barrel. The cost of producing and compressing hydrogen accounts for 60 to 70% of this. For high-conversion hydrocrackers, operating costs are roughly U.S.$4.0 to U.S.$4.5 per barrel, of which 75 to 80% is due to hydrogen.

10.9.2 Benefits

Many hydrotreaters are stay-in-business investments, so it is difficult to quantify their upgrade value, which is the value of products minus costs — labor, materials

TABLE 10.10
Feed and Product Properties for a Flexible Single-Stage Hydrocracker

Feedstock Type		**Straight-Run Vacuum Gas Oil**	
Boiling range, °C		340–550	
Boiling range, °F		644–1022	
API gravity		22.0	
Specific gravity		0.9218	
Nitrogen, wppm		950	
Sulfur, wt%		2.5	
Product Objective	Naphtha	Jet	Diesel
Weighted average reactor temperature, °C	Base	–6	–12
Weighted average reactor temperature, °F	Base	–11	–22
Yields, vol% Fresh Feed			
C_4	11	8	7
C_5, 82°C (C_5, 180°F)	25	18	16
82°C+ (180°F+) Naphtha	90	29	21
Jet A-1 or diesel	—	69	77
Total C_4+	126	124	121
Chemical H_2 Consumption			
Nm^3/m^3	345	315	292
Scf/bbl	2050	1870	1730
Product Qualities			
C_5, 82°C			
RONC	79	79	80
Heavy Naphtha			
P/N/A	45/50/5	44/52/4	—
RONC	41	63	67
Endpoint, °C (°F)	216 (421)	121 (250)	118 (244)
Jet A-1			
Flash point, °C (°F)	—	38 (100)	—
Freeze point, °C (°F)	—	–48 (–54)	—
Smoke point, mm	—	34	—
FIA aromatics, vol%	—	7	—
Endpoint, °C (°F)	—	282 (540)	—
Diesel			
Cloud point, °C (°F)	—	—	–15 (5)
API gravity	—	—	44
Cetane number	—	—	55
Flash point, °C (°F)	—	—	52 (126)
Endpoint, °C (°F)	—	—	349 (660)

(liquid feed, hydrogen, catalysts, and chemicals), utilities, maintenance, and investment amortization. In some plants, the refinery planning linear program (LP) assigns equal value to treated and untreated naphtha, and even to treated and untreated distillates. This reflects the underlying assumption that the increase in value across a hydrotreater is equal to the cost of running the unit, that is, the upgrade value is zero. In other LPs, the naphtha hydrotreater (NHT) that pretreats catalytic reformer feed is lumped in with the reformer. Certainly, if a key naphtha or distillate hydrotreater shuts down, the refinery may have to run at a reduced rate, but that can be said of most units.

For an FCC feed pretreater, the upgrade value can be more than U.S.$3 per barrel if the calculation includes its positive impact on FCC yields. Usually, benefits to the FCC are greater than the value of conversion and volume swell in the hydrotreater itself. Typically, the upgrade value for a high-conversion VGO hydrocracker is U.S.$3 to U.S.$4 per barrel.

With hydroprocessing units, most refiners try to maximize feed rate while (1) meeting other process objectives and (2) maintaining a high on-stream factor. Some try to maximize conversion, while others just want to hit a key process target at minimum cost.

10.9.3 Catalyst Cycle Life

For fixed-bed units, catalyst cycle life dominates economics. Catalysts can not be changed if the units are operating, so shorter catalyst cycles mean decreased production. For a typical 25,000 barrels/day unit, 1 day of lost production can cost U.S.$100,000.

Here are some of the many economic trade-offs that must be considered when setting hydrocracker process targets:

- Higher feed rates and higher conversion are desirable economically, but they increase consumption of hydrogen and decrease catalyst cycle life.
- In units that can recycle fractionator bottoms, higher recycle oil rates can increase selectivity, but they may impose limits on fresh feed rate.
- For many recycle units, switching to once-through (zero recycle) operation is attractive economically if the unconverted oil (that is, the fractionator bottoms) goes to an FCC, olefins plant, or lube plant for further upgrading. Conversion goes down in the hydrocracker, but it may be possible to increase fresh feed rates without decreasing catalyst cycle life, and operating costs may go down due to decreased hydrogen consumption.

10.10 HYDROCRACKER–FLUID CATALYTIC CRACKER (FCC) COMPARISON

In a petroleum refinery, heavy molecules with low hydrogen-to-carbon ratios (H/C) are converted into light molecules with higher H/C ratios. The FCC process

TABLE 10.11
Comparison of Hydrocracking with FCC

	FCC	Hydrocracking
Operating pressure	Low	High, 1500–2800 psi
Operating temperature	High, 900–1000°F	Moderate, 600–780°F
Construction costs	Moderate	High
Volume swell	112–118 vol%	115–140 vol%
	Includes fuel gas FOEB	Fresh feed basis
Product olefins	High	Nil
Light naphtha octane (RONC)	>100	78–81
Heavy naphtha octane (RONC)	95–100	40–64
Distillate cetane index	Low	56–60
Distillate sulfur content	Moderate to high	Very low
Bottoms' sulfur content	Moderate to high	Very low

Note: FOEB = fuel oil equivalent barrels; RONC = research octane number clear (without tetraethyl lead).

increases H/C by rejecting carbon, while hydrocracking increases H/C by adding hydrogen. Consequently, FCC and hydrocracking have marked differences in operating conditions, volume swell, product yields, and product properties. Table 10.11 summarizes some of these differences.

10.11 OPERATIONAL ISSUES

Hydroprocessing — especially hydrocracking — is exothermic. Effective control of produced heat is the primary concern of designers, owners, and operators of hydrocracking units. In modern units, a high flux of recycle gas provides a sink for process heat. It also promotes plug flow and the transport of heat through the reactors. Most licensors recommend that the ratio of recycle gas to makeup gas should exceed 4:1.

During design, limits on temperature rise ($T_{rise} = T_{out} - T_{in}$) set the size of catalyst beds and determine the number and location of quench zones. During operation, when feeds (and maybe catalysts) are different, the T_{rise} is also different — sometimes dangerously so. A sudden spike in T_{rise} can lead to a temperature runaway or temperature excursion. These are dangerous. The rates of cracking reactions increase exponentially with temperature — the hotter they get, the faster they get hot. In a few cases, temperature runaways have melted holes in the stainless steel walls of hydrocracking reactors. This is remarkable, because the walls were more than 8 in. (20 cm) thick.

The best way to stop a temperature excursion is to depressure the unit by venting recycle gas through a special valve at the CHPS. This decelerates all hydrocracking reactions by rapidly reducing H_2 partial pressure in the reactors. Depressuring can also lead to catalyst maldistribution, decreased catalyst activity, or increased pressure drop. For these reasons, operators are extremely careful when restarting a unit after a temperature excursion.

Due to the presence of hydrogen, leaks in hydroprocessing units often cause fires. Such fires can be devastating, if not deadly. The replacement of a reactor and the reconstruction of other equipment damaged by the accident can take 12 months. The cost of lost production can exceed U.S.$50 million.

Safety concerns are responsible for several operating constraints, such as:

- An upper limit on temperature in the reactors. This and other temperature constraints prevent damage to the reactor.
- Upper limits on the T_{rise} in each bed and each reactor, and upper limits on the rate at which T_{rise} changes. These are designed to decrease the likelihood of temperature excursions.
- An upper limit on the velocity of fluid flow through elbows in high-pressure piping. This constraint emerged after erosion–corrosion cut a hole in a high-pressure pipe in a hydrocracker, causing a major accident.
- A lower limit on reserve quench gas — usually 15% of the total flow of recycle gas. Reserve quench provides a way to react quickly to nonemergency changes in T_{rise}.
- A lower limit on wash water injection. This ensures the near-total removal of ammonia from the system.

10.12 LICENSORS

Leading licensors of hydroprocessing technology are listed in Table 10.12.

Many engineering contractors will gladly build unlicensed hydrotreaters. However, for hydrocrackers and special-application hydrotreaters, especially those designed to meet clean-fuel specifications, refiners almost always select licensed technology from an experienced vendor willing to offer guarantees.

10.13 CONCLUSION

Advances in hydroprocessing are driven by competitive forces and clean-fuel regulations. These advances include improved catalysts, better reactor design, advanced process control, and online optimization. As clean-fuel regulations migrate from North America and the EU into the rest of the world, and as globalization of the oil industry continues apace, the need will continue for new (and better) hydroprocessing units. Hopefully, within a few years, this chapter will be obsolete and we will have to write an update.

TABLE 10.12
Leading Licensors of Hydroprocessing Technology

Company	Process Name	Description
Axens (IFP)	Prime-G	Gasoline desulfurization
	IFP hydrotreating	Naphtha, distillate, VGO hydrotreating
	IFP Hydrocracking	High-conversion fixed-bed hydrocracking
	T-Star	Ebullating-bed hydrotreating
	H-Oil	Ebullating-bed hydrocracking
CDTECH	CDHydro	Hydrotreating with catalytic distillation
	CDHDS	
Chevron Lummus	Isocracking	High-conversion hydrocracking
	RDS	Atmospheric residue hydrotreating
	VRDS	Vacuum residue hydrotreating
	OCR	On-stream catalyst replacement
	Isodewaxing	Catalytic dewaxing
	LC-Fining	Ebullating-bed hydrocracking
Criterion/ABB/Shell Global	SynSat	Distillate hydrotreating; aromatics saturation
	Deep gasoil HDS	Hydrotreating to make ultra-low-sulfur diesel
ExxonMobil	SCANfining	Hydrotreating to make low-sulfur gasoline
	OCTGAIN	Hydrotreating to make low-sulfur gasoline
	ULSD-Fining	Hydrotreating to make ultra-low-sulfur diesel
	MAXSAT	Saturation of aromatics in distillate streams
	LCO-fining	LCO hydrotreating
	GO-fining	FCC feed pretreating
	RESIDfining	Residue hydrotreating
	MIDW	Lube isomerization/dewaxing
Haldor Topsøe		Naphtha, distillate, VGO hydrotreating
KBR	MAK hydrotreating	Distillate and VGO hydrotreating
	MAK hydrocracking	Mild hydrocracking; FCC feed pretreatment
UOP	ISAL	Gasoline desulfurization
	Unifining	Naphtha hydrotreating
	Unionfining	Distillate, VGO, residue hydrotreating
	Unicracking	High-conversion VGO hydrocracking

REFERENCES

1. Gruia, A. 2006. Recent Advances in Hydrocracking. In *Practical Advances in Petroleum Processing*, Hsu, C.S. and Robinson, P.R. (Eds.). Springer, New York, chap. 8.
2. Klein, M.T. and Hou, G. 2006. Mechanistic Kinetic Modeling of Heavy Paraffin Hydrocracking. In *Practical Advances in Petroleum Processing*, Hsu, C.S. and Robinson, P.R. (Eds.). Springer, New York, chap. 20.
3. Lapinas, A.T., Klein, M.T., Gates, B.C., Macris, A., and Lyons, J.E. 1991 Catalytic hydrogenation and cracking of fluorene: Reaction pathways, kinetics, and mechanisms. *Ind. Eng. Chem. Res.* 30(42).
4. Magee, J.S. and Dolbear, G.E. 1998. *Petroleum Catalysis in Nontechnical Language*. PennWell, Tulsa, OK.
5. Mudt, D.R., Pederson, C.C., Jett, M.D., Marur, S., McIntyre, B., and Robinson, P.R. 2006. Refinery-Wide Optimization with Rigorous Models. In *Practical Advances in Petroleum Processing*, Hsu, C.S., and Robinson, P.R. (Eds.). Springer, New York, chap. 23.
6. Pedersen, C.C., Mudt, D.R., Bailey, J.K., and Ayala, J.S. 1995. Closed Loop Real Time Optimization of a Hydrocracker Complex, 1995 NPRA Computer Conference, CC-95-121, November 6-8.
7. Quann, R.J. and Jaffe, S.B. 1996. Building useful models of complex reaction systems in petroleum refining, *Chem. Eng. Sci.* 51:1615.
8. Quann, R.J. and Jaffe, S.B. 1992. Structure oriented lumping: Describing the chemistry of complex hydrocarbon mixtures, *I & EC Res.* 31:2483.
9. Satterfield, C.N. 1975. Trickle-bed reactors. *AIChE J.* 21(2):20.
10. Scherzer, J. and Gruia, A.J. 1996. *Hydrocracking Science and Technology,* Marcel Dekker, New York.
11. Stell, J. 2003. Worldwide refineries, capacities as of January 1, 2004. *Oil & Gas J.* 101(49), December 22.

11 Hydrogen Production

Miguel A. Valenzuela and Beatriz Zapata

CONTENTS

11.1 INTRODUCTION

Approximately 80% of the present world energy demand comes from fossil fuels (Das and Veziroglu, 2001). Unlike fossil fuels, hydrogen gas burns cleanly, without emitting any environmental pollutants (Fields, 2003). In addition, hydrogen is abundantly available in the universe and possesses the highest energy content per unit of weight (that is, 120.7 kJ/g) compared to any of the known fuels and could have an important role in reducing environmental emissions (Haryanto et al., 2005).

Although hydrogen is usually considered a fuel of the future, it has been used in large quantities as a feedstock in the petroleum refining, chemical, petrochemical, and synthetic fuel industries for the past 50 years. Examples include making ammonia and methanol and removing sulfur in petroleum refining for such products as reformulated fuels (Czuppon et al., 1996). Hydrogen gas is also used in the food processing, semiconductor, glass, and steel industries, as well as by electric utilities to cool the rotor and stator coils in large turbine generators (Ramachandran and Menon, 1998). In addition, one of the most important applications of liquid hydrogen has been as a fuel in space programs of many nations (Ohi, 2005).

In 2003, the total consumption of hydrogen was about 448×10^9 m^3 worldwide and about 84×10^9 m^3 in the U.S. Almost all of the hydrogen used is captive, that is, consumed at the refinery, chemical plant, or other industrial facility where it is produced. Nevertheless, hydrogen can be delivered as liquid (trucks) or gas (tube trailers and pipelines) (Ohi, 2005).

The shifting of fuels used all over the world from solid to liquid to gas, and the decarbonization trend that has accompanied it, implies that the transition to H_2 energy seems inevitable (Ogden, 1999; Dunn, 2002; Barreto et al., 2003; Sherif et al., 2005, Rand and Dell, 2005). Consequently, there has been a surge in funding devoted for research on the production, distribution, storage, and use of H_2 worldwide (Conte et al., 2001; Turner, 2004).

However, H_2 has its own problems, and there is a strong debate on the subject. Some critics doubt that H_2 is the right solution for the energy-related environmental, security, and sustainability issues. They argue that the current technology for producing H_2 as an energy carrier is too costly and wasteful of energy (Haryanto et al., 2005). There are two important pillars on which the H_2 economy rests: pollution-free sources for the H_2 generation and fuel cells for converting H_2 to useful energy efficiently (Romm, 2005).

In nature, hydrogen is always present in bound form, in organic compounds and water. Hydrogen can be produced from different sources, for example, coal, natural gas, liquefied petroleum gas (LPG), propane, methane, gasoline, light diesel, heavy residue, dry biomass, biomass-derived liquid fuel (such as methanol, ethanol, biodiesel), and nuclear power, as well as from water (Figure 11.1). Presently, hydrogen is mostly produced from fossil fuels (natural gas, oil, and coal). Steam reforming is the most widely used thermochemical process to produce hydrogen from raw materials, such as natural gas, coal, methanol, ethanol, or gasoline. Gasification and pyrolysis processes are used when the feedstocks are solids such as heavy and residual oils (Simbeck, 2003).

Indeed, refineries are large hydrogen consumers, and this consumption is increasing for enhancing the residue conversion capacity and for producing clean-burning gasoline and diesel fuels containing very low sulfur amounts and with a reduced aromatic and olefin content. Some hydrogen is sourced inside the refinery fences from catalytic reforming of naphtha, which produces most of the aromatics included in the gasoline pool. Other hydrogen is recovered from minor off-gas streams.

However, the growth of the refineries hydroprocessing capacity, which has increased more rapidly than crude capacity, has largely outpaced the by-product and off-gas hydrogen sources, and the hydrogen balance has been achieved through on-purpose production (Figure 11.2).

Certainly, the worldwide on-purpose hydrogen capacity has increased 70% between 1995 and 2003, and 96% of this on-purpose hydrogen is produced by steam reforming (76% from natural gas, 20% from light naphtha), while partial oxidation (gasification) of residues (petroleum coke, deasphalter pitch, residual oil) produces the remaining 4% (Basini, 2005).

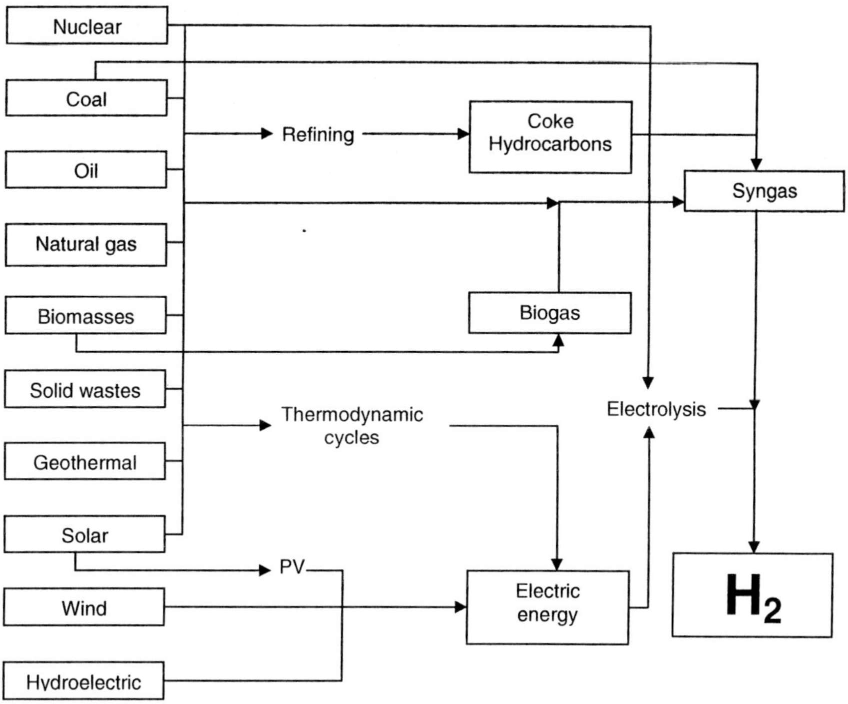

FIGURE 11.1 Different hydrogen production methods. (Adapted from Conte, M. et al., *J. Power Sources*, 100: 171, 2001. With permission.)

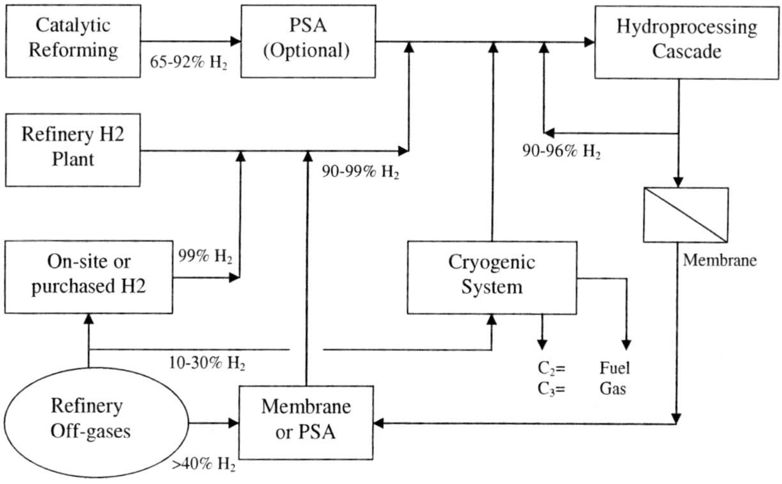

FIGURE 11.2 Refinery hydrogen network. (Adapted from Garland, R., Biasca, F.E., Chang, E., Bailey, R.T., Dickenson, R.L., and Simbeck, D.R., *Upgrading Heavy Crude Oils and Residues to Transportation Fuels: Technology, Economics and Outlook*, Phase 7, Mtn. View, CA, SFA Pacific, Inc., 2003. With permission.)

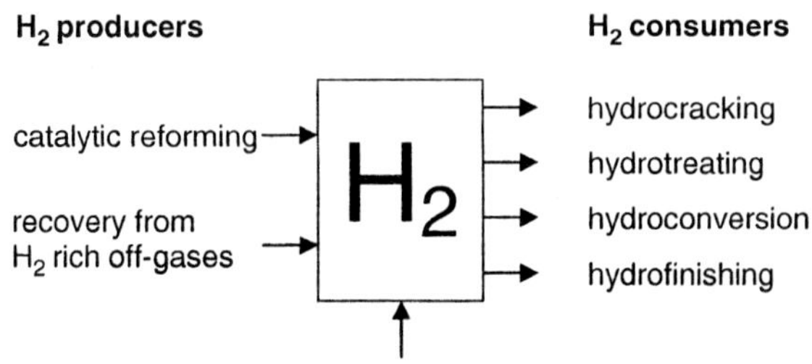

FIGURE 11.3 Schematic of refinery hydrogen balance. (Adapted from Rostrup-Nielsen, J.R. and Rostrup-Nielsen, T., *Cattech*, 106: 150, 2002. With permission.)

11.1.1 Refinery Hydrogen Balance

The hydrogen balance of refineries is complex, as illustrated in Figure 11.3 (Rostrup-Nielsen and Rostrup-Nielsen, 2002). The environmental objectives for providing better transportation fuels may lead to significant changes in the refinery industry. Specifications for reformulated gasoline have meant less aromatics and olefins and constraints on light hydrocarbons and sulfur. New legislation for diesel requires deep desulfurization to 10 to 50 ppm S. This is done by reacting the sulfur compounds into hydrogen sulfide, which is removed from the hydrocarbon stream. This has created a large requirement for more hydrotreating — hydrodesulfurization (HDS), hydrodenitrogenation (HDN), and hydrodemetallization (HDM) — and hydocracking. Traditionally, a major part of the hydrogen consumption in refineries was covered by hydrogen produced as a by-product from other refinery processes (for instance, the total hydrogen market in 1998 was 390×10^9 Nm^3/year + 110×10^9 Nm^3/year as coproduct), coming mainly from catalytic reforming. In catalytic reforming, the main reaction is the conversion of paraffins into aromatics and hydrogen. Because aromatics are not wanted in reformulated fuels, less hydrogen will become available from catalytic reforming. Similarly, the gasoline and diesel fractions from catalytic crackers are highly unsaturated.

In summary, there is a fast-growing need for increased hydrogen production capacity in refineries. Steam reforming of natural gas is the most common route to fill the gap, but gasification of heavy oil fractions and petcoke may play an increasing role, often combined with power generation (Damen et al., 2006). This need is being met mainly by the installation of steam reforming-based hydrogen plants (Garland et al., 2003).

11.1.2 Fuel Cells

Fuel cells are a viable alternative for clean energy generation. Over the past few years, automotive companies have announced new technologies or prototype vehicles adopting fuel cells in an effort to reduce atmosphere pollution (Wang and Zhang, 2005). A variety of fuel cells for different applications are under development (Zegers, 2006). The ideal fuel for the proton exchange membrane fuel cells (PEMFCs), considered the more advantageous device for mobile vehicles and for small stationary power units, is pure hydrogen, with less than 50 ppm carbon monoxide content, as dictated by the poisoning limit of the Pt fuel cell catalyst. Figure 11.4 shows a proposed scheme of different processes to reform gas streams to be used in fuel cells.

Therefore, the paramount issue facing fuel cells, which provides power for the mobile vehicles, right now is how to get the hydrogen to the vehicles (Farrauto et al., 2003). Partial oxidation is presently considered an alternative to steam reforming for the generation of hydrogen from fossil fuels in decentralized applications (Wang and Zhang, 2005). An important example is the generation of hydrogen for stationary or mobile fuel cells; other applications have also been proposed, such as hydrogen injection into gas turbine combustors for flame stabilization or on-site hydrogen production for metallurgical treatments. While methane is the fuel of choice for stationary applications, liquid hydrocarbons (LPG, gasoline, and diesel) are preferred for mobile applications (Ferreira-Aparicio et al., 2005).

From an economical point of view, coal, natural gas, and oil will continue to be used as a cheap feedstock to produce hydrogen (Rostrup-Nielsen, 2004); however, a requirement of hydrogen produced from fossil fuels will have to include CO_2 sequestration (Rostrup-Nielsen, 2005).

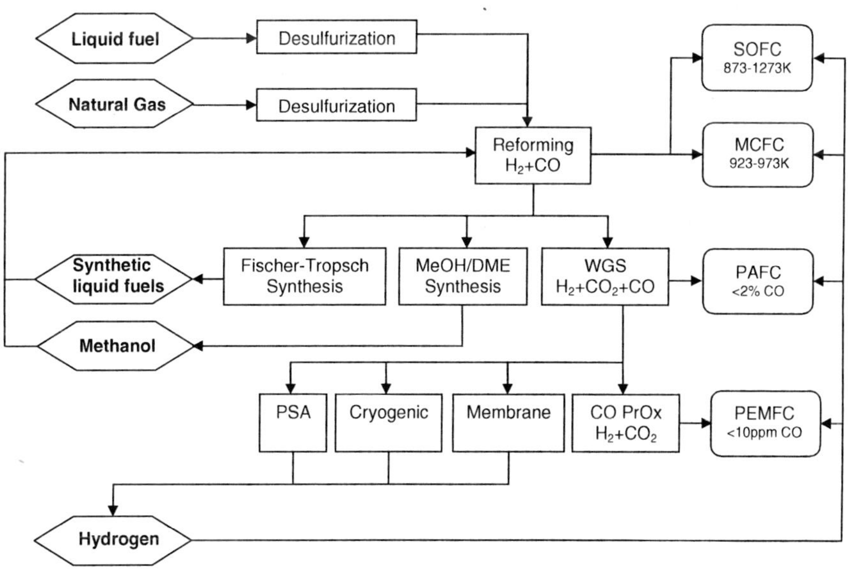

FIGURE 11.4 Scheme of the application of different processes to reform gas streams before its use in fuel cells. (Adapted from Ferreira-Aparicio, P. and Benito, J.M., *Catal. Rev.*, 47: 491, 2005. With permission.)

This chapter discusses process innovations and factors that enhance hydrogen production from conventional and alternative routes, as well as the miscellaneous options of them using fossil fuels as feedstock.

11.2 HYDROGEN PRODUCTION

11.2.1 Hydrocarbon Gasification

Economical and environmental considerations have led petroleum refiners to intensify efforts to maximize production of high-quality products from crude oil. To meet this goal, a number of refinery processes have been developed to upgrade heavy residues. Some of these conversion processes (for example, hydrotreating or hydrocracking) require considerable quantities of hydrogen. Today most of the hydrogen is produced by steam reforming of natural gas (83%) or naphtha (14%). Only 3% is obtained by gasification of heavy residues (Simbeck, 2003). However, though the capital investment for such a plant is considerably higher than for a unit based on steam reforming of natural or refinery gas, gasification of residues can still be advantageous, particularly where natural gas or other light hydrocarbons are not readily available or are expensive.

The gasification process is an alternative to steam reforming (SR). However, as already mentioned, it still has a minor utilization for refinery on-purpose hydrogen production. The process is based on very exothermic reactions produced inside a combustion chamber. It has a unique flexibility with respect to the possibility of utilizing various feedstocks, ranging from natural gas to deasphalter pitch to petroleum coke. Its relatively low diffusion is related to the high capital costs (which can double those of an SR analogous capacity) and the oxygen consumption features, making its economics competitive with those of SR only for large-scale applications (>250,000 Nm^3/h hydrogen). However, the diffusion of gasification is expected to increase due to the falling demand of heavy residues and the possibility of realizing very large plants producing hydrogen for the refinery and synthesis gas for integrated gasification combined cycle (IGCC) (Holopainen, 1993).

A gasification process is a way to convert fossil fuels, biomass, and wastes into either a combustible gas or a synthesis gas for subsequent utilization. It offers the potential for both clean power and chemical production (Figure 11.5). A gasification process can be fired with coal, coal with biomass and wastes, and refinery residues, as well as natural gas (Furimsky, 1998).

11.2.1.1 Technology Status

There are some 160 modern gasification plants in operation and 35 at the planning stage around the world. Electricity, ammonia, oxy-chemicals, syngas, methanol, and hydrogen are the primary products obtained by gasification technology, as summarized in Table 11.1. The feedstocks include coal, natural gas, refinery residues, and biomass/wastes in combination with coal (Minchener, 2005).

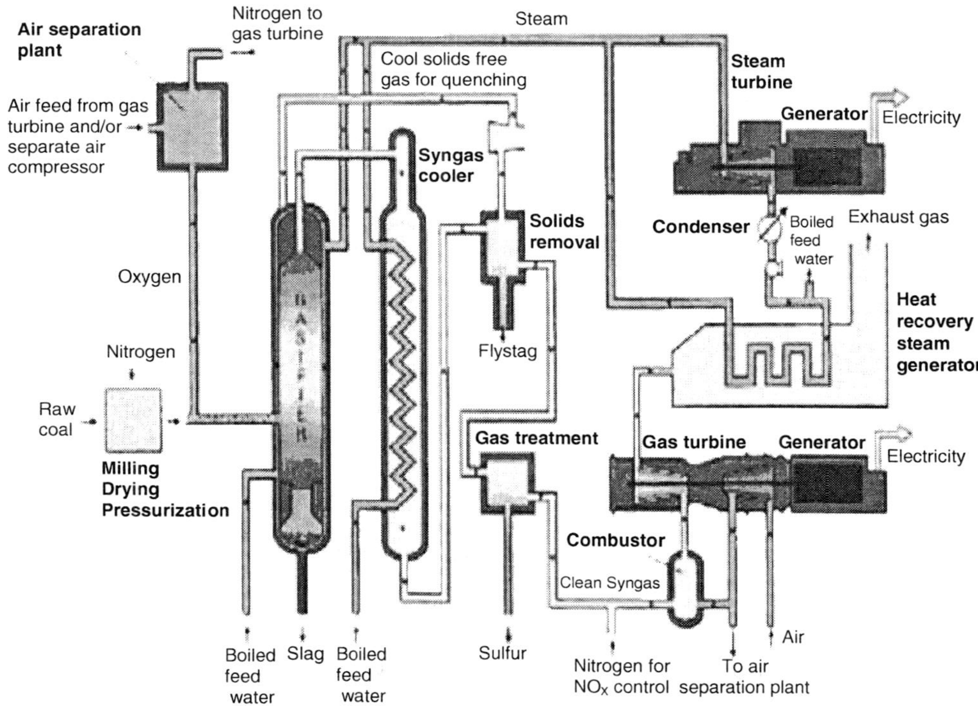

FIGURE 11.5 Scheme of a IGCC plant. (Taken from Minchener, A.J., *Fuel*, 84, 2222, 2005.)

TABLE 11.1
Primary Products Produced through Fossil Fuel Gasification

Product	Primary Product: Operating Plant	Primary Product: Planned Plant	Secondary Product
Electricity	35	25	6
Hydrogen	11	1	11
Ammonia	34	3	1
Syngas	14	1	2
Methanol	12	1	11
Oxy-chemicals	22	0	1
Carbon dioxide	7	0	5
Others (FT liquids, fuel gas)	25	4	0
Total	**160**	**35**	**37**

Source: Adapted from Minchener, A.J., *Fuel*, 84: 2222–2235, 2005. (With permission.)

Table 11.2 shows the feedstocks used in gasification plants. As can be seen, the large majority of operational plants to date are based in the use of fuels, and the planned plants will use mainly coal and coal/petcoke. Concerning refinery residues, these can take several forms, depending on the design of the refineries and their specific products (Speight, 1986).

The primary bottoms that comprise most of the fuels of interest for energy applications include atmospheric distillation residue, vacuum distillation residue, residual tar from the solvent deasphalting/visbreaking process, and petroleum coke. These fuels are used extensively to produce chemicals and gases, although power production has been integrated with the more recent units (Figure 11.5).

TABLE 11.2
Feedstocks Used in Gasification Plants

Feedstock	Operational Plant	Planned Plant
Coal	27	14
Coal/petcoke	3	1
Petcoke	5	7
Natural gas	22	0
Biomass	12	3
Fuel oil/heavy petroleum residues	29	2
Municipal waste	5	0
Naphtha	5	0
Vacuum residue	12	2
Unknown	40	6
Total	**160**	**35**

Source: Adapted from Minchener, A.J., *Fuel*, 84: 2222–2235, 2005. (With permission.)

There are three technology variants, classified by gasifier configurations according to their flow geometry: entrained flow, fluidized bed, and moving bed. The first one, most commonly used for coal gasification, pulverized coal particles, and gases, flows concurrently at high speed. In the case of the fluidized-bed gasifier, the feedstocks are suspended in the gas flow and mixed with those undergoing gasification. In the moving-bed gasifier, also called the fixed-bed gasifier, gases flow relatively slowly up the bed of the feedstock. Both concurrent and countercurrent technologies are available, but the former is more common. Each has advantages and disadvantages together with differing commercial track records. In general terms, with regard to suppliers, Shell and Texaco entrained-flow gasifiers are used in nearly 75% of the 160 operational plants. Of the rest, Lurgi moving-bed gasification technologies are also used to a significant extent. For the planned gasification plants, it is understood that approximately 75% of these will also use either Texaco or Shell designs. The suppliers of the major gasification installations are listed in Table 11.3.

11.2.1.2 Texaco and Shell Designs

Texaco and Shell began gasification processes (or noncatalytic partial oxidation) for syngas production from heavy fuel oil or sour crude (Keller, 1990). In both schemes, fuel feed is partially burned in noncatalytic reactors to supply sufficient heat for the endothermic feed stream reaction with the balance of the fuel. Both produce gas composed primarily of H_2 and CO, plus some CO_2, H_2S, small amounts of residual CH_4, and soot (Figure 11.6).

The chemical reaction is quite complex, especially when using crude petroleum, which contains straight-chain, branched-chain, cyclic, and complex cyclic compounds. An ideal case and the process designer goal is shown in the general equation (Peña et al., 1996)

$$C_xH_y + (x/2)\,O_2 \rightarrow xCO + (y/2)H_2 \tag{11.1}$$

Other reactions, such as carbon deposition, combustion, steam reforming, water–gas shift, and Boudouard equilibrium, are also implicated. Although the chemistry of the two processes is essentially similar, there are differences in operating pressure, reactor gas cooling system, soot removal, and feedstock versatility. Apparently, both technologies minimize these differences. Both Texaco and Shell processes are in use by many ammonia producers, but certain problems occur before steady-state operation. The Texaco high-pressure process has been found to contain small amounts of formic acid, which corrodes certain parts in the quench and carbon separation equipment and piping, requiring stainless steel to solve the problem. Similarly, erosion has been found in Shell waste heat boiler inlets, depending upon the ash of the feedstock.

In the Shell process, the feed enters the homogenizer, where carbon agglomerates from soot recovery are dispersed; the discharge then flows to the reactor. In the reactor, the feed, after mixing with the preheated steam, is cracked to carbon, methane, and hydrocarbon radicals during a short residence time. Then the

TABLE 11.3
Technology Suppliers for Gasification Projects Worldwide

Technology Supplier	Gasifier Type	Solid Fuel Feed Type	Oxidant	Major Installations
Chevron Texaco, U.S.	Entrained flow	Water slurry	O_2	Tampa Electric IGCC Plant, Cool Water IGCC Plant, Chevron Texaco Eldorado IGCC Plant, Eastman Chemical, Ube Industries, Motiva Enterprises, Deer Park
Global Energy E-Gas, U.S.	Entrained flow	Water slurry	O_2	Wabash River IGCC Plant, Louisiana Gasification Technology IGCC Plant
Shell, United States/Netherlands	Entrained flow	N2 carrier/dry	O_2	Demkolec IGCC Plant (Buggenum, Netherlands) Shell Pernis IGCC Plant (Netherlands, Harburg)
Lurgi, Germany	Moving bed	Dry	Air	Sasol Chemical Industries, Great Plains Plants
British Gas/Lurgi, Germany, U.K.	Moving bed	Dry	O_2	Global energy power/methanol plant, Germany
Prenflo/Uhde, Germany	Entrained flow	Dry	O_2	Elcogas, Puertollano IGCC Plant (Spain), Furstenhausen in Saarland
Noell/GSP, Germany	Entrained flow	Dry	O_2	Schwarze Pumpe, Germany
HT Winkler (HTW) RWE, Rheinbraun/Uhde Germany	Fluidized bed	Dry	Air or O_2	None
KRW, U.S.	Fluidized bed	Dry	Air or O_2	Sierra Pacific (Nevada, United States)

Source: Adapted from Minchener, A.J., *Fuel*, 84: 2222–2235, 2005. (With permission)

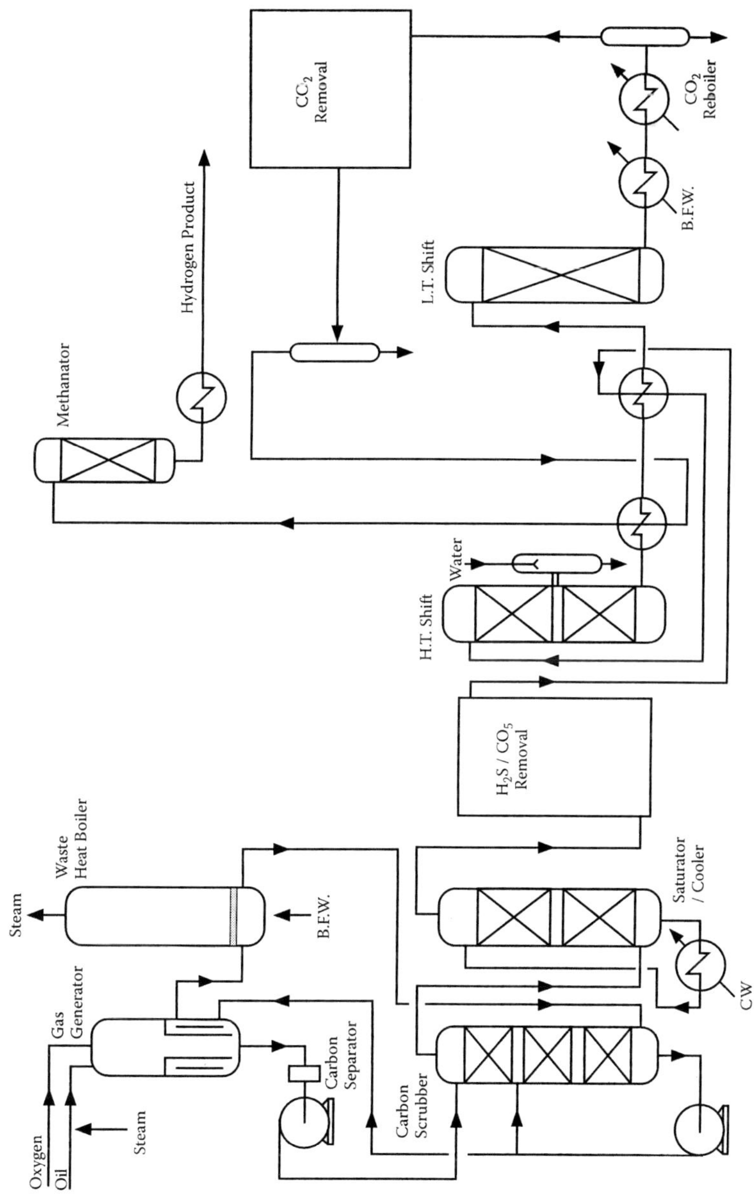

FIGURE 11.6 Typical flow scheme for hydrogen production from fuel oil by noncatalytic partial oxidation. (Adapted from Balthasar, W., *Int. J. Hydrogen Energy*, 9: 649, 1984. With permission.)

hydrocarbons react with oxygen to form CO_2 and H_2O and exothermic heat, which in turn are used for the hydrocarbon steam endothermic reaction. Finally, the product gases reach an equilibrium composition for the water–gas shift reaction. The synthesis gas thus produced contains H_2S and COS, which are removed prior to high-temperature shift processing. These reactions are carried out in a Shell gasifier by injecting preheated heavy oil, preheated oxygen, and steam through a specially designed burner into a closed combustion vessel. This refractory-lined steel vessel serves as a pressure reactor. The reactor is operated at 25 to 60 bar at temperatures of 1280 to 1380°C. Gasification of heavy fuel oil produces gases with about 46.7% H_2, 48.1% CO, and 3.8% CO_2, in addition to small amounts of N_2, CH_4, H_2S, and COS (Balthasar, 1984). The Shell and Texaco processes incorporate waste heat recovery from very high-temperature effluent gases, carbon removal from product gases, and recycle or recovered carbon. Purification and separation hydrogen in both processes are similar in principle to that of steam reforming, but modified to remove sulfur from synthesis gas and using a sulfur-tolerant catalyst ($CoMo/Al_2O_3$) in the high-temperature shift.

The Shell and Texaco processes have similar feedstocks, chemistry, and product gases, but they do differ in their technology and processing approach. In the Shell process, the hydrocarbon feed, steam, and oxygen are all preheated and fed to the reactor burner as separate streams that mix at the burner exit, while in the Texaco process, the hydrocarbon feed is mixed with steam and the preheated compounds. Depending upon the processing requirement, the operation pressure for the Texaco process is 80 bar at 1450°C, which is somewhat higher than that of the Shell process, which uses up to 60 bar. The Shell process uses a waste heat boiler, while the Texaco process generally quenches gases with water to recover heat. The two processes differ in their carbon or soot recovery in naphtha and in their recycling procedure for carbon. The Shell process is used to remove CO_2 and H_2S from synthesis gases before HTS for CO conversion, while the Texaco process employs a sulfur-tolerant HTS catalyst. Chevron-Texaco technology has lately focused on designing residuum oil supercritical extraction plants, while Shell is focusing on the design of large-scale (three-train) plants for combined hydrogen and power production. However, in the future, several processes different from those of Texaco and Shell may evolve. For instance, the Lurgi MPG and Babcock Borsing Power Noell gasification processes were both developed in East Germany and are quite similar to those of Texaco.

The comparison of gasification technologies with the conventional SR process reveals the following economic implications. There are some regions where the economics of the hydrogen manufacture by gasification are more attractive than hydrogen manufacture by SR. The main economic factors in the region that would dictate the choice of gasification are: (1) regional shortages of natural gas will result in higher SR manufacturing costs from either increased prices of gas or the need for emergency-feed storage facilities, and (2) increasingly restrictive air pollution regulations on SO_2 emissions will generally reduce the value placed on high-S fuel oil, and thus decrease the manufacturing costs for gasification that can readily use this fuel (Peña, 1996).

11.2.2 Steam Reforming

The most economical route to produce hydrogen is by steam reforming of hydrocarbons, which covers a wide capacity range (Rostrup-Nielse, 2002). Typical capacities range from 10,000 to 100,000 Nm^3/h or larger. Although refineries are the main users of hydrogen and the major contributors to new hydrogen plants at large capacities, an emerging market for fuel cell applications with typical capacities from 5 to 1000 Nm^3/h hydrogen is now appearing.

The principal reactions for converting hydrocarbons into hydrogen by steam reforming are:

$$CH_4 + H_2O \rightarrow CO + 3H_2 \quad (H^\circ_{298} = 206 \text{ kJ/mol}) \tag{11.2}$$

$$CO + H_2O \rightarrow CO_2 + H_2 \quad (H^\circ_{298} = -41 \text{ kJ/mol}) \tag{11.3}$$

$$C_nH_m + nH_2O \rightarrow nCO + [(m + 2n)/2]\ H_2 \tag{11.4}$$

The steam reforming reaction (Equation 11.2) is endothermic and takes place in a high-alloy tube reformer loaded with a nickel-based catalyst, which is placed inside a furnace equipped with burners in the side or at the top of the furnace. The process is typically operated with excess steam-to-carbon ratios (above 2.5) at temperatures at about 750 to 900°C, depending on the use of the gas. The exit raw synthesis gas composition is a mixture of H_2, CO, CO_2, and unreacted methane, which is near the equilibrium of the steam reforming reaction and water–gas shift (WGS) reaction (Equation 11.3). A typical outlet gas is shown in Table 11.4 (Song and Guo, 2006). Synthesis gas obtained from steam reforming is usually hydrogen rich. However, an increase in temperature, decrease in pressure, and reduction of the steam-to-carbon ratio will lower the H_2/CO ratio (Rostrup-Nielsen et al., 2002). Changing the operation conditions, H_2/CH_4 ratio,

TABLE 11.4
Typical Reformer Furnace Outlet Gas Compositions

Furnace outlet temperature, °C	890.0
Furnace outlet pressure, bar	24.0
S/C ratio	3.0
Component, vol%	
Hydrogen	51.0
Carbon monoxide	10.4
Carbon dioxide	5.0
Methane	2.0
Water vapor	31.6

Source: Adapted from Song, X. and Guo, Z., *Energy Conv. Manag.*, 47: 560, 2006. With permission.

and CO_2/CH_4 ratio favors the carbon formation reactions (methane decomposition and Bouduard reaction, Equations 11.5 and 11.6, respectively):

$$CH_4 \rightarrow C + 2H_2 \quad (H°_{298} = 75 \text{ kJ/mol}) \tag{11.5}$$

$$2CO \rightarrow C + CO_2 \quad (H°_{298} = -172 \text{ kJ/mol}) \tag{11.6}$$

Steam reforming is an energy-efficient and reliable technology; its efficiency corresponds to 80% of the ideal thermodynamically achievable efficiency, defined as the ratio between the lower heating values (LHVs) of the product synthesis gas and the feed gas. The larger SR furnaces can contain more than 600 tubes (each with a diameter of 100 to 150 mm and a length of 10 to 13 m) and can produce a synthesis gas stream sufficient for recovering more than 250,000 Nm^3/h hydrogen (Basini, 2005). In Figure 11.7 is presented a complete scheme of reactions, including catalysts and operation conditions, that are carried out in a conventional steam reforming plant using natural gas as a feedstock (Farrauto et al., 2003).

The steam reformer operates as an adiabatic reactor. Thus, nonuniform temperature along the reactor impacts the chemistry of the whole process, represented by means of Equation 11.1 through Equation 11.6. Since the reaction

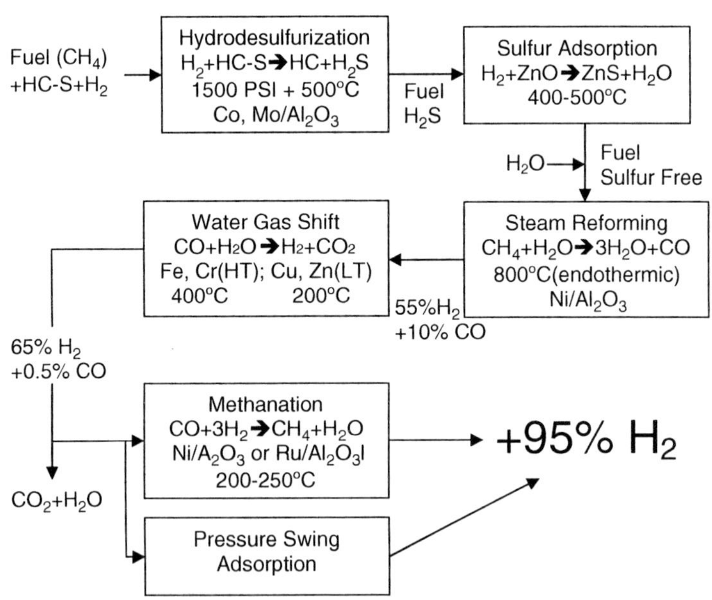

FIGURE 11.7 Scheme of reactions, catalysts, and operation conditions carried out in a conventional steam reforming process. (Adapted from Farrauto, R. et al., *Annu. Rev. Mater. Res.*, 33: 1, 2003. With permission.)

proceeds with an increase in the net number of moles of product, it is favored at low pressures. However, SR reactors are usually operated at pressures above 20 atm in order to avoid additional compression steps because many customers of modern H_2 plants require the product at high pressure. Steam reforming is a well-established technology, although continuous improvement in material for reformer tubes, better control of carbon limits, better catalysts regarding sulfur tolerance and carbon deposition, as well as new process schemes have improved significantly, lowering plant costs (Figure 11.8) (Ferreira-Aparicio et al., 2005; Rostrup-Nielsen, 2005).

Refinery naphtha streams can also be used as feedstock for steam reforming plants. Naphtha consists of a wide range of hydrocarbons with an initial boiling point of 44 to 56°C and a final boiling point of 103 to 154°C. Naphtha is used in many steam reformers throughout the world; there are 58 large steam naphtha reforming (SNR) plants in operation and 3 under development (Simbeck, 2003). Naphtha is preferentially used as a feedstock in SR in the Asia–Pacific region, where natural gas is not readily available. Obviously, the SNR process must use elevated steam-to-carbon ratios in comparison with the steam methane reforming (SMR) process, and special catalysts must be used to prevent cracking and carbon formation (Czuppon et al., 1996).

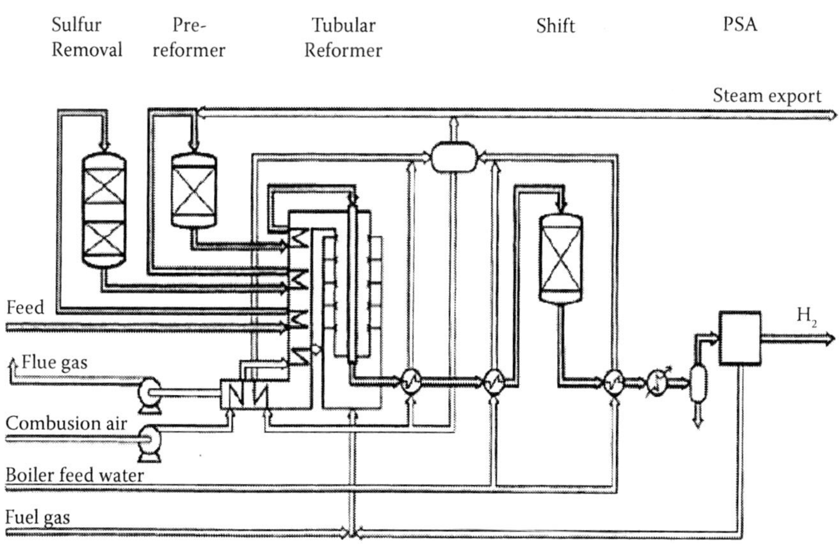

FIGURE 11.8 Typical process layout for a hydrogen plant based on advanced tubular steam reforming technology, including shift conversion followed by pressure swing adsorption (PSA) to delivery pressure. (Adapted from Ferreira-Aparicio, P. and Benito, J.M., *Catal. Rev.*, 47: 491, 2005. With permission.)

11.2.2.1 Autothermal Reforming

Autothermal reforming is a combination of steam reforming and partial oxidation processes in a single reactor. By this route, the energy for reforming is provided by the partial oxidation of the hydrocarbon feedstock. In a first step, natural gas and steam are mixed with oxygen under substoichiometric conditions,

$$CH_4 + 3/2\ O_2 \rightarrow CO + 2H_2O \quad (H°_{298} = 519\ kJ/mol) \tag{11.7}$$

within the burner, where partial oxidation reactions take place, producing the required heat for the subsequent endothermic reactions. After that, in the same reactor is placed downstream the reforming catalyst (Ni supported on Mg-Al spinel), where steam reacts with the remaining fuel to produce synthesis gas. The oxygen/fuel ratio is adjusted to determine the operation temperature, and the composition of the gaseous effluent will be determined by the thermodynamic equilibrium at the exit pressure and temperature. The heart of the autothermal reforming process is the reactor (Figure 11.9). It is a refractory-lined cylindrical vessel swaged to a smaller diameter at the top to provide a combustion zone. A specially designed oxygen burner is installed in this section: the catalyst is contained at the bottom of the reactor, which has a larger diameter than the top (Song and Guo, 2006; Ferreira-Aparicio et al., 2005).

11.2.2.2 New Trend in Reforming Processes

The main problems to be solved in the conventional fixed-bed steam reformers are thermodynamic equilibrium limitation, diffusion limitation, carbon formation/catalyst

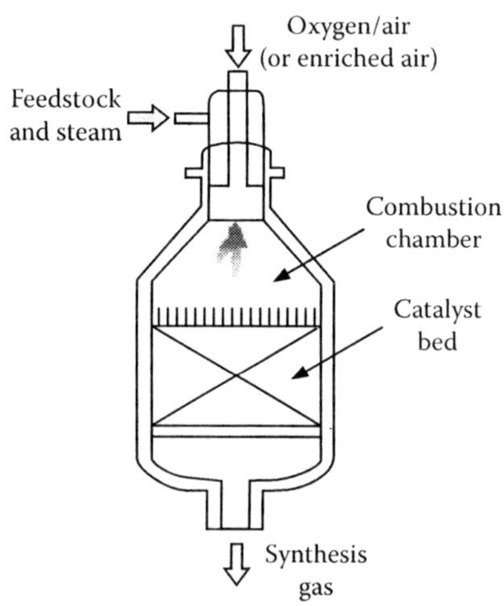

FIGURE 11.9 Diagram of an ATR reactor. (Adapted from Song, X. and Guo, Z., *Energy Conv. Manag.*, 47: 560, 2006. With permission.)

deactivation, heat transfer limitation, and environmental pollution/CO_2 emission (Chen and Elnashaie, 2005).

Although SR is a well-established process, improvements concerning new catalysts, reactor engineering, and process modeling are being applied in new plants around the world (Dybkjær, 2005). In the case of steam reforming of higher hydrocarbons, several configuration processes, including circulating fluidized-bed membrane reformer (CFBMR), have been reported (Chen and Elnashaie, 2005). The CFBMR may be referred to as the third-generation reformer; the other ones were the bubbling fluidized-bed membrane reformer (BFBMR) and the fixed-bed steam reformer (FBR), as the second- and first-generation reformers, respectively. According to the investigation results of Chen and Elnashaie, having analyzed autothermal and nonautothermal configurations for the efficient production of hydrogen using the CFBMR process, the autothermal operation with direct contact between cold feeds (water and hydrocarbon) and hot circulating catalyst may be the best configuration (Figure 11.10) to optimize hydrogen production and energy consumption. Recent developments for on-site hydrogen production, including companies, development status, and characteristics, are found in Ferreira-Aparicio et al. (2005).

11.2.3 Catalytic Partial Oxidation

Catalytic partial oxidation (CPO) is presently considered an alternative to steam reforming for the generation of hydrogen from fossil fuels in decentralized applications (Armor, 2005). This process is thought to be used in the generation of hydrogen for stationary or mobile fuel cells; other applications have been proposed in gas turbine combustors and metallurgical treatments (Beretta and Forzatti, 2004). While methane is the fuel of choice for stationary applications, liquid hydrocarbons (LPG, gasoline, and diesel) are preferred for mobile applications (Trimm and Onsan, 2001). The successful production of syngas from the partial oxidation of methane, ethane, n-butano, and higher hydrocarbons over Rh catalysts has been studied in detail by the extensive work of Schmidt and coworkers on short-contact-time reactors (Bharadwaj and Schmidt, 1995; Hohn and Schmidt, 2001; Schmidt et al., 2003).

The overall reaction of a general fuel, C_xH_y, and air is the formation of syngas:

$$C_xH_y + (x/2)\ O_2 \rightarrow xCO + (y/2)\ H_2 \tag{11.8}$$

A competitive reaction is the highly exothermic total oxidation of the fuel:

$$C_xH_y + [x + (y/4)]\ O_2 \rightarrow xCO_2 + (y/2)\ H_2O \tag{11.9}$$

Other possible reactions include olefin formation, steam reforming, and water–gas shift (O'Connor et al., 2000). In the case of methane used as a feedstock in CPO, the use of rhodium catalysts leads to a high conversion of the mixture CH_4/O_2 to CO and H_2 at contact times of a few milliseconds under adiabatic conditions (that is, at temperatures higher than 800°C) (Beretta and Forzatti, 2004).

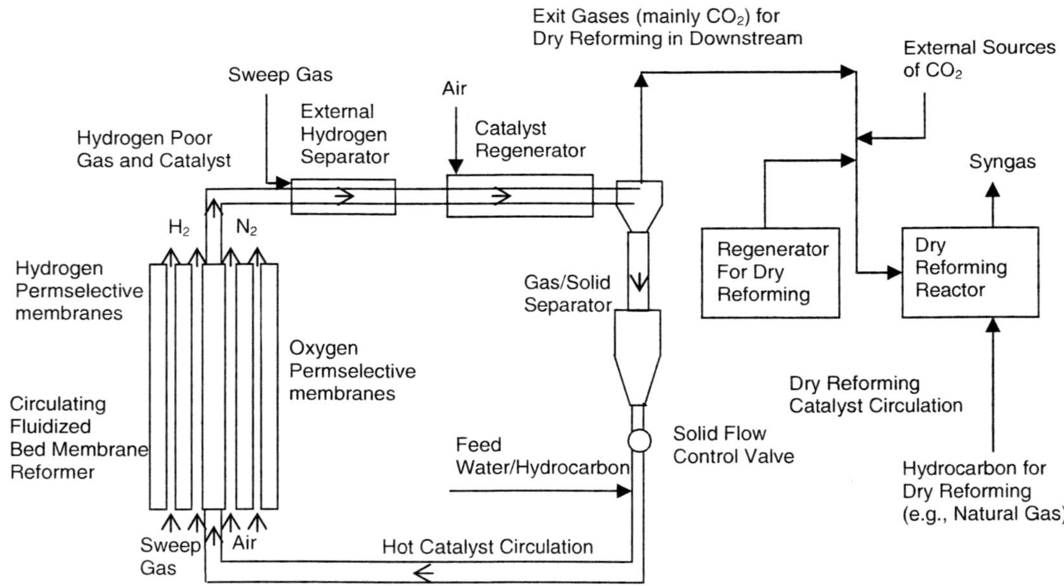

FIGURE 11.10 Novel process for efficient hydrogen production by steam reforming of hydrocarbons. (Adapted from Chen, Z. and Elnashaie., S.S.E.H., *AICHE J.*, 51: 1467, 2005. With permission.)

11.2.4 Hydrocarbon Decomposition

If methane of other hydrocarbons are heated in the absence of air, they will decompose to produce hydrogen and carbon:

$$CH_4 \rightarrow C + H_2 \quad \Delta H = 75 \text{ kJ/mol} \tag{11.10}$$

$$C_nH_m \rightarrow nC + (m/2)H_2 \tag{11.11}$$

Universal Oil Products (UOP) developed the Hypro process for the thermal cracking of methane of natural gas (Cox and Williamson, 1977). A fluidized-bed reactor is used in which methane is decomposed at about 850°C and carbon is deposited onto a Ni/Al_2O_3 catalyst. The catalyst is continuously removed and fed to a regenerator, where the carbon is burnt by injecting air and supplemental natural gas. In contrast to the Hypro process, an improved process described by UTC uses an expendable nickel catalyst or metal–glass fiber material on which to collect the produced carbon (Dicks, 1996).

Catalytic methane decomposition (CMD) over supported metal catalysts produces very pure hydrogen without the formation of carbon oxides, which eliminates the operations of separating the gaseous mixtures (Muradov, 1993; Steinberg, 1999). CMD is therefore a useful application in some specific cases, such as hydrogen fuel cells, where CO-free hydrogen is required to avoid deactivation of the platinum electrode (Couttenye et al., 2005). Because this reaction is mildly endothermic, the temperature must be at about 600°C for it to proceed at a reasonable rate. CMD has been achieved using a plasma, solar radiation, molten metal bath, and thermal reactor with no catalyst, metal catalysts, and carbon catalysts (Dunker et al., 2006). In addition to avoiding CO_2 production, other advantages of the CMD reaction are that the process is less endothermic than that of SR (Equation 11.2) and the solid carbon may have value as a replacement for carbon black.

Metal catalysts used in CMD present the highest activity; however, the activity is lost quickly as the active sites become covered with carbon (Valenzuela et al., 2004; Otsuka and Takenaka, 2004). Regeneration of the catalyst requires oxidation of the carbon, and thus all the carbon is converted to CO_2. Although carbon catalysts are less active than metal catalysts, previous work using certain carbon catalysts has shown promising results (Muradov, 2001).

11.2.5 Synthesis Gas Generation

Hydrogen and synthesis gas have been extensively utilized for more than 70 years in chemical and refinery industries. Their uses are becoming more complex, being influenced by strategic, political, economical, and sustainability considerations. Synthesis gas is a mixture of hydrogen and carbon monoxide, usually containing carbon dioxide, used as an intermediate product for further synthesis (Figure 11.11). The present use of SG is primarily for the manufacture of ammonia and methanol,

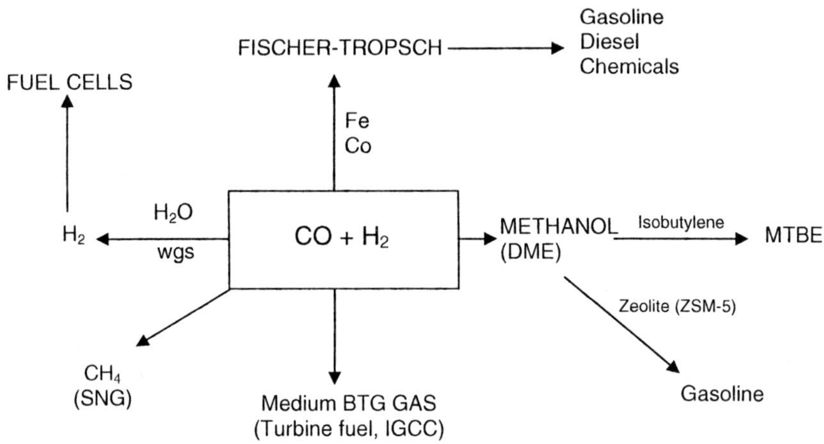

FIGURE 11.11 The use of synthesis gas as a chemical feedstock. (Adapted from *Kirk-Othmer Encyclopedia of Chemical Technology*, 4th ed., John Wiley & Sons, New York, 1996. With permission.)

followed by pure hydrogen for hydrotreating in refineries (Rostrup-Nielsen et al., 2002). Depending on the reaction conditions and catalyst used, different chemicals may be produced on a large industrial scale (Wender, 1996). The growing demand for downstream products is the driving force for synthesis gas manufacture; its total daily capacity is at about 120 millions of Nm^3, an energy equivalent of more than 200,000 barrels of oil per day (Song and Guo, 2005).

Synthesis gas made from coke was distributed in large cities as "town gas" for cooking, heating, and lighting, but because it is dangerously CO rich, it has been replaced with natural gas (NG). However, recently, the syngas produced from heavy hydrocarbons and coal has been used for feeding large IGCC plants producing electric energy (Basini, 2005). The generation of synthesis gas may, from a point of view of feedstock used, be categorized in two different types of processes: one based on natural gas or high hydrocarbons (naphtha, residual oil, petroleum coke), and the other on coal. In Table 11.5 reaction schemes of the different processes to obtain syngas are reported. However, natural gas and high hydrocarbons will remain the major feedstock for the manufacture of synthesis gas due to its lower investment compared with a coal-based plant.

With the decline of the oil resource, natural gas seems to be the most suitable for synthesis gas production. Nevertheless, economically viable processes based on natural gas for manufacturing synthesis gas typically produce a mixture gas that is too rich in hydrogen to meet the stoichiometric ratios required by major synthesis gas-based petrochemicals. Thus, H_2/CO adjustment units such as the pressure swing adsorption (PSA) unit or the reverse shift of hydrogen and CO_2 unit are necessary prior to downstream synthesis.

In Figure 11.12, the variation of H_2/CO ratios as a function of the feedstock to different processes is presented. Synthesis gas manufacture may be responsible

TABLE 11.5
Comparison of Synthesis Gas Production from Different Routes

Process Name	Reaction	H°298K (kJ/mol)	Industrial Application	Advantages	Associated Problems
Steam reforming	CH4 + H_2O = CO + $3H_2$	206	H_2 production; synthesis gas production	Low carbon deposition; suitable for high-pressure processes; easy separation of the products	High H_2/CO ratio; need separation for follow-up F-T or methanol synthesis; energy-intensive process
CO_2 reforming (dry reforming)	CO_2 + CH_4 = 2CO + $2H_2$	247	Synthesis gas or H2 production	Use of two greenhouse gases, i.e., CO_2 as the feedstock; high availability in some gas fields	Energy-intensive process; low H_2/CO ratio; more H_2 is needed for follow-up F-T or methanol process; easy carbon deposition
Partial oxidation (oxy-reforming)	CH_4 + $1/2O_2$ = CO + $2H_2$	−36	Synthesis gas or H_2 production	Mild exothermic reaction; energy saving; H_2-to-CO ratio = ~2 (suitable for methanol or F-T synthesis)	Hot spot may occur in the catalyst bed
Methanol steam reforming	CH_3OH + H_2O = CO_2 + $3H_2$	49	H_2 production	High yield of H_2	Energy-intensive process
Methanol oxy-reforming	CH_3OH + $1/2H_2$ = CO_2 + $2H_2$	−192	H_2 production	Exothermic reaction; save energy	Hot spot may occur in the catalyst bed

Source: Adapted from York, A.P.E. et al., *Top. Catal.*, 22: 345–358, 2003. (With permission.)

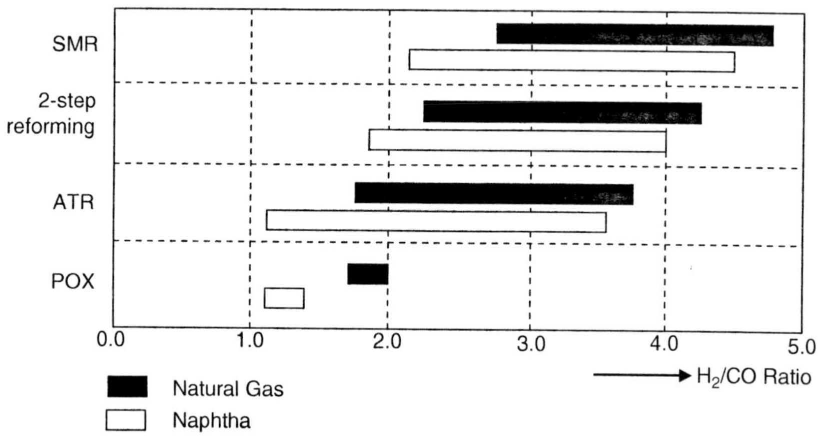

FIGURE 11.12 H_2/CO ratios from various syngas processes. (Adapted from Rostrup-Nielsen, J.R. et al., *Adv. Catal.*, 47: 65, 2002. With permission.)

for ca. 60% of the investments of large-scale gas conversion plants based on natural gas. Therefore, research in syngas generation is now focused on reducing the operating costs and saving energy. Direct production of flexible H_2/CO synthesis gas can avoid the H_2/CO adjustment units (Figure 11.13). Hence, there is a strong incentive to find solutions for providing flexible H_2/CO ratios.

In more recent years the synthesis gas produced from coal, petroleum coke, and deasphalter bottoms has been used for feeding large IGCC units producing electric energy (Basini, 2005). Relevant initiatives are also ongoing for realizing syngas-to-liquid (GTL) processes converting NG into high-quality energy vectors (gas oil, naphtha, methanol, and dimethylether) (Basini, 2005).

Consequently, hydrogen and synthesis gas mixtures are assuming the role of secondary energy vectors. Figure 11.14 shows how the pool of the primary energy vectors (hydrocarbons, coke, fissile fuels, and renewable fuels) can be transformed into thermal, mechanical, and electric energy through the intermediate formation of hydrogen and synthesis.

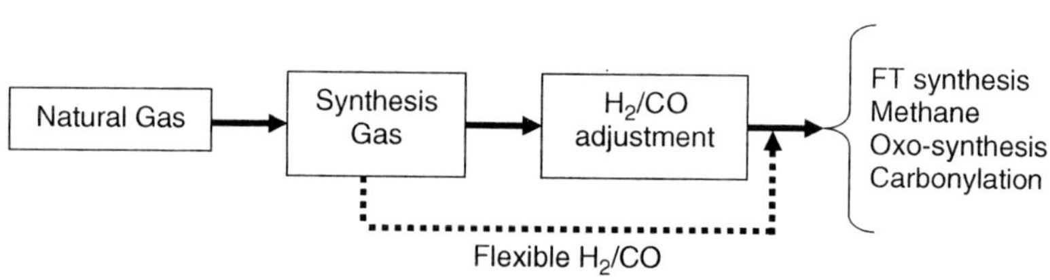

FIGURE 11.13 Synthesis gas route from natural gas to chemicals. (Adapted from Song, X. and Guo, Z., *Energy Conv. Manag.*, 47: 560, 2006. With permission.)

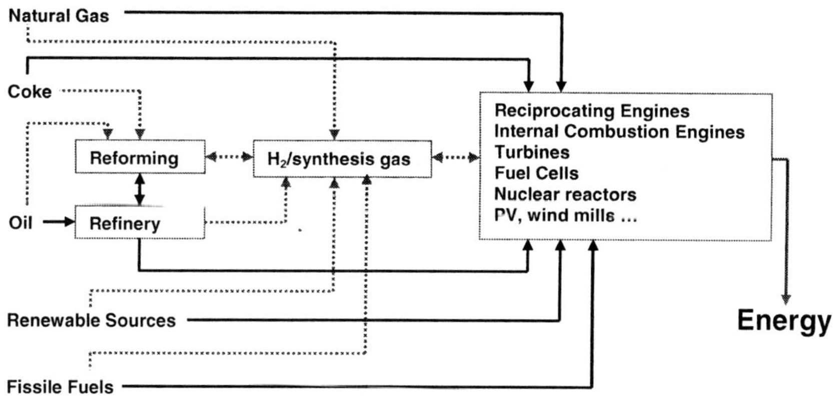

FIGURE 11.14 Parallel energy production chains in which the primary sources are either directly transformed (solid lines) or via the intermediate formation of H_2 and synthesis gas (dotted lines). (Adapted from Basini, L., *Catal. Today*, 106: 34, 2005. With permission.)

11.2.5.1 GTL Applications

The economics of GTL are different from those of oil refining. In a refinery, the cost of the raw materials dominates while the capital costs are responsible for a relatively small part of the production costs. Instead, the capital costs have a major impact with respect to raw materials costs in determining the economics of the GTL processes. These comprise three major sections, represented in Figure 11.15. The first section includes the production of synthesis gas. The second is constituted by the synthesis of the liquid fuels (gas oil, gasoline, naphtha, waxes, MeOH/DME). The third concerns a final product upgrading, separation, and purification.

A detailed study comparing syngas generation technologies from natural gas feedstock can be found in Wilhelm et al. (2001). According to this work, two-step reforming and ATR should be the technologies of choice for large-scale GTL plants.

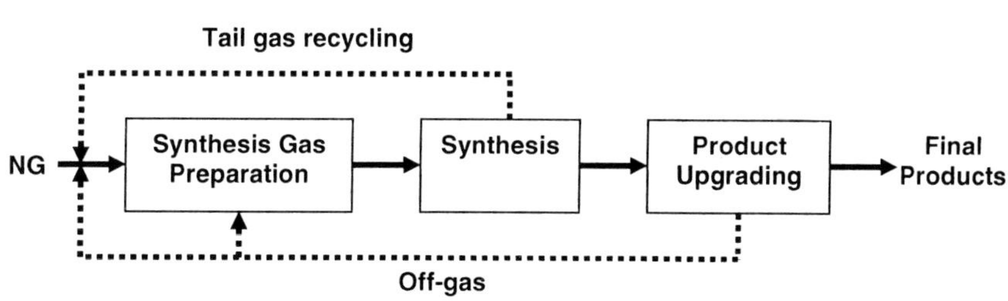

FIGURE 11.15 Simplified GTL scheme including the three main block processes and showing main recycle options to the synthesis gas generation. (Adapted from Basini, L., *Catal. Today*, 106: 34, 2005. With permission.)

REFERENCES

Armor, J.N. (2005). Catalysis and the hydrogen economy. *Catal. Lett.*, 101:131.

Balthasar, W. (1984). Hydrogen production and technology: today, tomorrow and beyond. *Int. J. Hydrogen Energy*, 9:649.

Barreto, L., Makihira, A., and Riahi, K. (2003). The hydrogen economy in the 21st century: a sustainable development scenario. *Int. J. Hydrogen Energy*, 28:267.

Basini, L. (2005). Issues in H_2 and synthesis gas technologies for refinery, GTL and distributed industrial needs. *Catal. Today*, 106:34 and references therein.

Beretta, A. and Forzatti, P. (2004). Partial oxidation of light paraffins to synthesis gas in short contact-time reactors. *Chem. Eng. J.*, 99:219.

Bharadwaj, S.S. and Schmidt, L.D. (1995). Catalytic partial oxidation of natural gas to syngas. *Fuel Process. Technol.*, 42:109.

Chen, Z. and Elnashaie, S.S.E.H. (2005). Optimization of reforming parameter and configuration for hydrogen production. *AICHE J.*, 51:1467.

Conte, M., Iacobazzi, A., Ronchetti, M., and Vellone, R. (2001). Hydrogen economy for a sustainable development: state-of-the-art and technological perspectives. *J. Power Sources*, 100:171.

Couttenye, R.A., Hoz de Vila, M., and Suib, S.L. (2005). Decomposition of methane with an autocatalytically reduced nickel catalyst. *J. Catal.*, 233:317.

Cox, K. and Williamson, K. (1977). *Hydrogen: Its Technology and Implications*, Vol. 1. CRC Press, Boca Raton, FL.

Czuppon, T.A., Knez, S.A., and Newsome, D.S. (1996). *Kirk-Othmer Encyclopedia of Chemical Technology* 4th ed., John Wiley & Sons, New York, 13, p. 884.

Damen, K., Troost, M., Faaij, A., and Turkenburg, W. (2006). A comparison of electricity and hydrogen production systems with CO_2 capture and storage. Part A. Review and selection of promising conversion and capture technologies. *Prog. Energy Comb. Sci.*, 32:215.

Das, D. and Veziroglu, T.N. (2001). Hydrogen production by biological processes: a survey of literature. *Int. J. Hydrogen Energy*, 26:13.

Dicks, A.L. (1996). Hydrogen generation from natural gas for the fuel cells systems of tomorrow. *J. Power Sources*, 61:113.

Dunker, A.M., Kumar, S., and Mulawa, P.A. (2006). Production of hydrogen by thermal decomposition of methane in a fluidized-bed reactor: effects of catalyst, temperature, and residence time. *Int. J. Hydrogen Energy*, 31:473 and references therein.

Dunn, S. (2002). Hydrogen futures: toward a sustainable energy system. *Int'l. J. Hydrog. Energy*, 27:235–264.

Dybkjær, I. (2005). What are the options for hydrogen plant revamps? *Hydrocarbon Process.*, 84:63.

Farrauto, R., Hwang, S., Shore, L., Ruettinger, W., Lampert, J., Giroux, T., Liu, Y., and Ilinich, O. (2003). New materials needs for hydrocarbon fuel processing: generating hydrogen for the PEM fuel cell. *Annu. Rev. Mater. Res.*, 33:1.

Ferreira-Aparicio, P., Benito, J.M., and Sanz, J.L. (2005). New trends in reforming technologies: from hydrogen industrial plants to multifuel microreformers. *Catal. Rev.*, 47:491.

Fields, S. (2003). Making the best of biomass: hydrogen for fuel cells. *Environ. Health Perspect.*, 111:A38.

Furimsky, E. (1998). Gasification of oil sand coke: review. *Fuel Process. Technol.*, 56:263.

Garland, R., Biasca, F.E., Chang, E. Bailey, R.T., Dickenson, R.L., Johnson, H.E., and Simbeck, D.R. (2003). Upgrading heavy crude oils and residues to transportation fuels: technology, economics and outlook, Phase 7, SFA Pacific, Inc., Mtn. View, CA, www.sfapacific.com.

Haryanto, A., Fernando, S., Murali, N., and Adhikari, S. (2005). Current status of hydrogen production techniques by steam reforming of ethanol. *Energy Fuels*, 19:2098 and references therein.

Hohn, K.L. and Schmidt, L.D. (2001). Partial oxidation of methane to syngas at high space velocities over Rh-coated spheres. *Appl. Catal. A Gen.*, 211:53.

Holopainen, O. (1993). IGCC plant employing heavy-petroleum residues. *Bioresource Technol.,* 46:125.

Keller, J. (1990). Diversification of feedstocks and products: recent trends in the development of solid fuel gasification using the Texaco and the HTW process. *Fuel Process. Technol.*, 24:247.

Minchener, A.J. (2005). Coal gasification for advanced power generation. *Fuel*, 84:2222.

Muradov, N.Z. (1993). How to produce hydrogen from fossil fuels without CO_2 emission. *Int. J. Hydrogen Energy*, 18:211.

Muradov, N.Z. (2001). Hydrogen via methane decomposition: an application for decarbonisation of fossil fuels. *Int. J. Hydrogen Energy*, 26:1165.

O'Connor, R.P., Klein, E.J., and Schmidt, L.D. (2000). High yields of synthesis gas by millisecond partial oxidation of higher hydrocarbons. *Catal. Lett.*, 70:99.

Ogden, J.M. (1999). Prospects for building a hydrogen energy infrastructure. *Annu. Rev. Energy Environ.*, 24:227.

Ohi, J. (2005). Hydrogen energy cycle: an overview. *J. Mater. Res.*, 20:3180.

Otsuka, K. and Takenaka, S. (2004). Production of hydrogen from methane by a CO_2 emission-suppressed process: methane decomposition and gasification of carbon nanofibers. *Catal. Surv. Asia*, 8:77.

Peña, M.A., Gomez, J.P., and Fierro, J.L.G. (1996). New catalytic routes for syngas and hydrogen production. *Appl. Catal. A Gen.*, 144:7.

Ramachandran, R. and Menon, R.K. (1998). An overview of industrial uses of hydrogen. *Int. J. Hydrogen Energy*, 23:593.

Rand, D.A.J. and Dell, R.M. (2005). The hydrogen economy: a threat or an opportunity for lead-acid batteries? *J. Power Sourc.*, 144:568.

Romm, J.J. (2005). *The Hype about Hydrogen: Fact and Fiction in the Race to Save the Climate.* Island Press, Washington DC.

Rostrup-Nielsen, J.R. (2004). Fuels and energy for the future. *Catal. Rev.*, 46:247.

Rostrup-Nielsen, T. (2005). Manufacture of hydrogen. *Catal. Today*, 106:293.

Rostrup-Nielsen, J.R., and Rostrup-Nielsen, T. (2002). Large-scale hydrogen production. *Cattech*, 106:150

Rostrup-Nielsen, J.R., and Sehested, J., and Norskov, J.K. (2002). Hydrogen and synthesis gas by steam and CO_2 reforming. *Adv. Catal.*, 47:65.

Schmidt, L.D., Klein, E.J., Leclerc, C.A., Krummenakcher, J.J., and West, K.N. (2003). Syngas in millisecond reactors: higher alkanes and fast lightoff. *Chem. Eng. Sci.*, 58:1037.

Sherif, S.A., Barbir, F., and Veziroglu, T.N. (2005). Wind energy and the hydrogen economy: review of the technology. *Solar Energy*, 78:647.

Simbeck, D.R. (2003). *Upgrading Heavy Crude Oils and Residues to Transportation Fuels: Technology, Economics and Outlook*, Phase 7. SFA Pacific, Inc. Mtn. View, CA.

Song, X. and Guo, Z. (2005). A new process for synthesis gas by co-gasyfing coal and natural gas. *Fuel*, 84:525.

Song, X. and Guo, Z. (2006). Technologies for direct production of flexible H_2/CO synthesis gas. *Energy Conv. Manag.*, 47:560.

Speight, J.G. (1986). Upgrading heavy feedstocks. *Annu. Rev. Energy*, 11:253.

Steinberg, M. (1999). Fossil fuels decarbonization technology for mitigating global warming. *Int. J. Hydrogen Energy*, 24:771.

Turner, J.A. (2004). Sustainable hydrogen production. *Science*, 305:972.

Trimm, D.L. and Onsan, Z.I. (2001). Onboard fuel conversion for hydrogen-fuel-driven vehicles. *Catal Rev.*, 43:31.

Valenzuela, M.A., González, O., Córdova, I., Flores, S., and Wang, J.A. (2004). Hydrogen production by methane decomposition on nickel/zinc aluminate catalysts. *Chem. Eng. Trans.*, 4:61.

Wang, Y.H. and Zhang, J.C. (2005). Hydrogen production on Ni-Pd-Ce/γ-Al_2O_3 by partial oxidation and steam reforming of hydrocarbons for potential applications in fuel cells. *Fuel*, 84:1926.

Wender, I. (1996). Reactions of synthesis gas. *Fuel Process. Technol.*, 48:189.

Wilhelm, D.J., Simbeck, D.R., Karp, A.D., and Dickenson, R.L. (2001). Syngas production for gas-to-liquids applications: technologies, issues and outlook. *Fuel Process. Technol.*, 71:139.

Zegers, P. (2006). Fuel cell commercialization: the key to a hydrogen economy. *J. Power Sources*, 154: 497.

Index

D

E

F

G

H

I

J

K

L

M

S

T

U

V

Z